Smelting and Utilization Technology of Ironsands

Zhenyang Wang · Lei Zhang · Jianliang Zhang

Smelting and Utilization Technology of Ironsands

Science Press
Beijing

Springer

Zhenyang Wang
School of Metallurgical and Ecological Engineering
University of Science and Technology Beijing
Beijing, China

Jianliang Zhang
School of Metallurgical and Ecological Engineering
University of Science and Technology Beijing
Beijing, China

Lei Zhang
Research Institute of Macro-Safety Science
University of Science and Technology Beijing
Beijing, China

ISBN 978-981-95-2867-7 ISBN 978-981-95-2868-4 (eBook)
https://doi.org/10.1007/978-981-95-2868-4

Jointly published with Science Press
The print edition is not for sale in China (Mainland). Customers from China (Mainland) please order the print book from: Science Press.
ISBN of the Co-Publisher's edition: 978-7-03-072324-6

This Springer imprint is published by the registered company Springer Nature Singapore Pte Ltd.
The registered company address is: 152 Beach Road, #21-01/04 Gateway East, Singapore 189721, Singapore

Preface I

Ironsands are a type of abundant, easily mined, and cost-effective coastal sand ore. This book explores various aspects of ironsands, including their mineral properties, milling, magnetic separation, and pre-oxidation treatments. It also examines the impact of ironsands on the sintering and pelletizing processes, as well as their influence on the quality of sinter and pelletized ore. Additionally, the book addresses the application of ironsands in gas-based direct reduction processes and high-temperature melting and splitting operations. The process of extracting titanium ferro-alloys from titanium-containing slag via the alumino-thermal method is also discussed.

The book presents a comprehensive analysis of the fundamental mechanisms underlying the extraction of valuable elements from ironsands, examines the impact of these processes on product quality, and provides recommendations for optimizing smelting techniques. Finally, it includes a specific case study on the short-flow industrial testing of ironsands.

This book is intended for production and engineering technicians in the metallurgical industry, as well as personnel in national and local government industrial management and science and technology departments. It is also suitable as a reference for universities and colleges offering related courses in metallurgy.

Beijing, China
April 2025

Zhenyang Wang

Preface II

Ironsands, a type of coastal sand ore, is characterized by high contents of iron and titanium, abundant reserves, ease of mining, low extraction cost, and widespread distribution, thus exhibiting significant potential for industrial utilization. In view of these advantages, the author has conducted systematic research over many years on the smelting technologies and application strategies for ironsands. The present book compiles these research findings, structured according to distinct metallurgical processes, with the aim of providing a technical foundation and process reference for the utilization of ironsands in both iron and titanium metallurgy.

The primary objective of this book is to explore viable technological routes and process parameters for integrating ironsands into both the conventional long ironmaking process based on blast furnaces and emerging non-blast furnace short processes. Accordingly, this work aims to establish a comprehensive technical framework for the smelting and utilization of ironsands. Chapter 2 reviews the historical development of ironsands research since the 18th century and outlines the current understanding of key scientific and technical issues. Chapter 3 investigates the fundamental mineralogical properties of ironsands to provide baseline data for subsequent process design. Chapters 4 and 7 analyze the relationship between particle size and magnetic separation behavior, as well as the enhancement of reduction kinetics through pre-oxidation treatments. Chapters 5 and 6 examine the impacts of ironsands on sintering and pelletizing behavior, optimizing operational parameters to ensure product quality. Chapters 8–10 shift focus to non-blast furnace technologies, including gas-based direct reduction and high-temperature smelting after pre-reduction, aiming to recover both metallic iron and titanium-rich slag. The titanium-rich slag is then utilized as a feedstock for titanium-iron alloy production via aluminothermic reduction. Chapter 11 presents industrial case studies. Chapter 1 offers a concise summary of each chapter's content, while Chap. 12 consolidates the key conclusions of the entire book for the reader's reference.

Due to the author's limited expertise, there may be deficiencies or oversights in this volume. Constructive criticism and corrections from readers are sincerely welcomed and deeply appreciated.

Beijing, China

Zhenyang Wang
Lei Zhang
Jianliang Zhang

Competing Interests The authors have no competing interests to declare that are relevant to the content of this manuscript.

Contents

Chapter 1
Introduction

In 2013, China's total iron ore consumption reached 1.086 billion tons, with imported ore accounting for 819 million tons, representing approximately 75% of total demand. Among the imports, more than 85% originated from the world's three major producers—BHP Billiton and Rio Tinto of Australia, and Vale of Brazil. That year, the average import price of iron ore stood at USD 134 per ton, significantly higher than the USD 26.6 per ton in 2000. However, 2014 saw a sharp downturn in the iron ore market, with prices falling by 25% to an average of USD 100.42 per ton, and plummeting to USD 75.61 by December. Despite the price decline, China's dependence on iron ore continued to rise. Annual imports increased to 933 million tons (up 13.9%), including 556 million tons from Australia and 176 million tons from Brazil, resulting in a dependency rate of 78.5%. This trend continued in 2015 and 2016, with imports reaching 1 billion tons and 80%–89% dependency. Figure 1.1 illustrates the trend in iron ore prices over recent years.

In order to expand the sources of iron and titanium ore resources, this book focuses on ironsands—a coastal sand ore known for its large reserves, ease of mining, and low cost—as the subject of research. The goal is to explore efficient methods for extracting iron and titanium from ironsands through studies of mineral properties, magnetic separation, pre-oxidation treatment, sintering, pelletizing, and metallurgical properties. This research aims to enable the efficient extraction of iron and titanium resources, utilizing processes such as gas-based direct reduction and high-temperature melting and splitting. The ultimate objective is to produce titanium and iron end-products in the form of ferrihydrite, high-titanium slag, and titanium-iron alloys. This would facilitate the low-cost utilization of ironsands, broaden the sources of iron and titanium resources, and contribute to the development of relevant application technologies.

From a historical perspective, research on ironsands dates back to 1806, when Mitchel [1] discovered and documented small, round, spherical particles of iron ore with a slight metallic luster, presumed to be ironsands. Since then, research on ironsands has continued uninterrupted for over 200 years, covering aspects such

Z. Wang et al., *Smelting and Utilization Technology of Ironsands*,
https://doi.org/10.1007/978-981-95-2868-4_1

Fig. 1.1 Vale iron ore price trend

as mining, beneficiation, mineral properties, and applications. In recent studies, including those by Jung [2] in *Metallurgical and Materials Transactions B*, the carbothermal reduction of ironsands using $CaO/CaCO_3$ has been explored.

In China, extensive research on vanadium-titanium magnetite from the Panzhihua region was conducted in the 1980s, leading to significant breakthroughs. This research resulted in the successful and stable smelting of vanadium-titanium magnetite in blast furnaces, with the TiO_2 content in the slag reaching up to 26 wt%. However, due to resource distribution and exploration challenges, China's research on ironsands began later. In 2010, Shen [3] at Chongqing University investigated the use of ironsands as iron-containing raw materials for carbon-containing pellet preparation, applying it to the direct reduction process with recommended process parameters. In 2014, Zhang [4] from Northeastern University and in 2015, Li [5] from the University of Science and Technology of Beijing conducted systematic research on the application of ironsands in gas-based reduction processes, providing detailed process parameter data.

Chapter 2 of this book reviews the historical development of ironsands research, from its discovery, exploration, and extraction to its application in the nineteenth and twentieth centuries, presenting a comprehensive summary of current research on ironsands both domestically and internationally in the twenty-first century. Chapter 3 aims to provide a thorough understanding of the basic characteristics of ironsands, offering data support for subsequent chapters. It examines the composition, particle size, physical phase, microscopic morphology, and high-temperature pyrolysis characteristics of ironsands, both at ambient and elevated temperatures. This chapter also evaluates the feasibility of using ironsands in industrial production based on composition and particle size, and it provides insights into the crystal structure, formation mechanisms, and phase transformation relationships within ironsands. Additionally, five key reactions occurring during the pyrolysis of ironsands are identified, along with the conditions under which they take place.

Compared to common iron ores like hematite and magnetite, the smelting of ironsands is more challenging, primarily due to their lower grade of valuable elements and poorer reduction kinetics. To address these issues, this book proposes preprocessing techniques for ironsands:

(1) Enhancing the grade of titanium and iron through magnetic separation;
(2) Improving reduction kinetics by applying pre-oxidation treatment.

Chapter 4 optimizes the magnetic separation of ironsands, considering both concentrate grade and recovery rate. Parameters such as grinding time, particle size, and magnetic separation intensity are optimized. Additionally, the book analyzes the micro-dissociation behavior of ironsands during grinding and the forces acting on mineral particles during magnetic separation. The appropriate grinding time and magnetic separation parameters are further clarified from both microscopic and mechanical perspectives.

References

1. Mitchel G. An account of a new semi-metallic substance, called menacane, and its ores. Trans R Irish Acad. 1806:11–30.
2. Jung SM. Effects of $CaO/CaCO_3$ on the carbothermic reduction of titanomagnetite ores. Metall Mater Trans B-Proc Metall Mater Proc Sci. 2015;46(3):1162–74.
3. Shen WH. Study on direct reduction of carbon-containing pellets using iron-bearing ironsand. Chongqing: Chongqing University; 2010.
4. Zhang P. Fundamental research on gas-based reduction of Indonesian vanadium titanomagnetite sand ore. Shenyang: Northeastern University; 2014.
5. Li YQ. Fundamental research on gas-based reduction of Indonesian ironsand pellets. b: University of Science and Technology Beijing; 2015.

Chapter 2
Research History and Key Areas in Ironsand Studies

This chapter is divided into three sections. Sections 2.1 and 2.2 provide a chronological review of marine sand deposit research prior to the year 2000. Sections 2.3–2.9 outline the post-2000 research progress, highlighting unresolved or insufficiently studied issues that are discussed in greater depth in subsequent chapters. Section 2.10 focuses on the global distribution of titanium resources and the development status of the titanium industry.

2.1 Research in the Nineteenth Century

In the nineteenth century, research on ironsands primarily focused on a preliminary understanding of the minerals themselves, including their physical properties such as color, density, hardness, morphology, and magnetic characteristics. The studies also explored the geographical distribution and unique features of ironsands in different regions, along with a preliminary analysis of their chemical composition and content. The recognition of the characteristics of ironsands during this period, coupled with the exploration of large-scale ironsand deposits, laid a solid foundation for the theoretical research and practical applications of ironsands in the twentieth century.

The earliest literature on ironsands that can be traced dates back to 1806, where Mitchel [1] described a metal-associated ore. He proposed that the phase composition of this ore was comprised of rutile, ferro-rutile, and titanium iron sand. The ore was primarily black or dark brown, with small spherical particles being the dominant form. It exhibited a slight metallic luster, was hard and opaque, and had a density of approximately 4500 kg/m^3. Mitchel also provided details on the distribution of this ore, which was mainly found in regions such as Roseno (Hungary), Salzburg (Austria), Burgess (Spain), and Siberia. Based on Mitchel's description, it can be inferred that the characteristics of this iron ore deposit align closely with those of the ironsands studied in this book. This piece of literature represents the earliest

Z. Wang et al., *Smelting and Utilization Technology of Ironsands*,
https://doi.org/10.1007/978-981-95-2868-4_2

documented research on ironsands available to the author. However, due to the limitations in detection technology at the time, Mitchel's research largely focused on macroscopic characteristics, basic physical and chemical properties, and geological distribution, without offering in-depth analysis.

In 1810, Thomas [2] published a paper describing a mineral found in the Doll River in Aberdeen, England. This mineral was distributed across several riverbed areas of the Doll River and was used by local residents to polish writing paper, thus earning the name "iron sand." This marked the first recorded use of the term "iron sand" in English literature. According to Thomas's description, the black mineral particles were mixed with some white, slightly red, or brown quartz, feldspar, and mica particles. When exposed to a magnet, these mineral particles could be attracted, facilitating the separation of useful minerals from gangue. Additionally, gangue separation could also be achieved by placing the minerals on a tilted plate and submerging them in water. The mineral itself had few edges, a rough surface, a slight metallic luster, and was opaque. After being ground into a powder, it turned a brownish-black color, and the powder was strongly attracted to magnets. Its density was approximately 4765 kg/m^3. Notably, in this work, Thomas became the first to employ acid–base titration to study the phase composition and elemental content of ironsands. Initially, hydrochloric acid was used for dissolution, followed by wet techniques such as precipitation, filtration, evaporation, and condensation to analyze the types and content of elements in the ironsands. The results of this analysis are presented in Table 2.1.

Between 1810 and 1866, researchers such as Thomson [3], Weaver [4, 5], Smith [6], Forbes [7], and others investigated the magnetic properties and geological distribution of ironsands. They also performed preliminary quantitative analyses of the chemical composition and content of ironsands from various regions, building upon the work of earlier scholars.

It was not until 1875 that Davidson [8] made a significant contribution by distinguishing the four oxides—Fe_2O_3, FeO, TiO_2, and MgO—present in ironsands through studies conducted in Berwick County, northern England. Using wet chemical analysis, Davidson quantified the content of these oxides, with the results summarized in Table 2.2.

Table 2.1 Chemical composition and corresponding contents of ironsands from the River Dore, United Kingdom

Black iron oxide	White titanium oxide	Arsenide	Silicon and aluminum	Total
98.70	12.65	1	1.5	113.85

Table 2.2 Chemical composition and corresponding contents of ironsands from Berwickshire, Northern England

Fe_2O_3	FeO	TiO_2	MgO
75.6	17.6	6.2	0.2

In 1878, Wright [9] conducted the first investigation into the reduction characteristics of ironsands through weight gain studies and presented a series of groundbreaking findings. These included the observation that FeO is more difficult to reduce than Fe_2O_3 and Fe_3O_4, making it a more stable iron oxide. Wright also discovered that the mass ratio of Fe to O in FeO does not follow the 1:1 atomic ratio, with the actual mass of iron being lower than the theoretical value. This anomaly was attributed to lattice defects in the iron oxide, which Wright suggested were caused by oxidation, leading to the formation of oxides in different valence states and thereby reducing the iron content. In his experiments, Wright noted that the resulting product had lower quality than expected, and speculated that it consisted of compounds such as $Fe_{16}O_{17}=14FeOFe_2O_3=13FeOFe_3O_4$.

Wright also investigated the beneficiation and reduction of New Zealand ironsands, concluding that when finely ground, ironsands could easily separate gangue from iron oxides. He observed that at the boiling point of mercury, neither CO nor H_2 affected the mass of the ironsands. However, at the boiling point of sulfur, both CO and H_2 were able to reduce the mass of the ironsands, indicating that the reduction conditions for ironsands are more stringent than for other ordinary iron ores. Furthermore, during the reduction process, Wright found that H_2 reduced ironsands more rapidly than CO due to its lower atomic mass and higher diffusion capacity. Based on this, he speculated that the starting temperature for H_2 reduction of ironsands should be lower than that for CO. Although Wright attempted to determine the exact starting temperature for the reduction reaction, temperature control limitations only allowed him to provide a temperature range. Despite these constraints on temperature control and reducing gas composition, Wright's study represented an early attempt to explore how reduction reaction parameters—such as temperature, gas composition, and time—affect the quality of reduction products.

2.2 Research in the Twentieth Century

Research on ironsands in the twentieth century can be characterized by the simultaneous advancement of theoretical research and practical applications. On the one hand, the development of advanced detection technologies, such as polarized mineral microscopy, X-ray diffraction, and electron probes, enabled clear characterization of the microstructure and composition of ironsands. On the other hand, advancements in mining technology led to a shift in the exploration and extraction of ironsands, moving from land-based to offshore sources. By the end of the twentieth century, several regions, including New Zealand, South Africa, Japan, and Norway, had developed mature ironmaking processes that utilized ironsands. Consequently, research on ironsands gradually shifted toward optimizing smelting processes and parameters, as well as exploring comprehensive utilization of the iron and titanium resources contained in ironsands.

Notable scholars such as John Hopkins, Moncton [10], Jukes Brown [11], Kepu [12], McKenzie [13], and Heskett [14] conducted detailed studies on the distribution of ironsands from the perspectives of mineralogy and geological exploration. They also emphasized the application potential of the iron elements found in ironsands.

In 1923, Scott [15, 16] and Thorpe [17] employed wet methods, including hot acid dissolution, filtration, and precipitation, to treat New Zealand ironsands. They observed the formation of a white oxide precipitate, which they identified as a new elemental oxide. Using one of the few X-ray facilities available at the time, they confirmed the existence of element 72—hafnium—and published their findings in *Nature*. This marked the beginning of research on the trace elements in ironsands.

In 1924, a study by Bishop [18] and Inouye [19] highlighted Japan's ability to utilize ironsands as a raw material for ironmaking, alongside other iron minerals such as magnetite, hematite, limonite, and ilmenite, to produce pig iron. This signaled the entry of ironsands into industrial applications.

In 1937, Wylie [20] investigated the mineralization and weathering evolution of ironsands along the coastal regions of the Northern Archipelago of New Zealand. The study concluded that New Zealand ironsands primarily originate from volcanic rocks, formed by volcanic eruptions and subsequently cooled by air or seawater. Erosion by air and seawater led to the formation of large reserves and multi-mineralized ironsands, with titanium magnetite and titanium iron ore being the dominant phases. In the same year, Gibbs [21] identified vanadium resources in ironsands and conducted chemical testing to determine the vanadium content.

From 1938 to 1949, research on ironsands primarily focused on geological exploration, mineralogical analysis, and the enhancement of ore dressing techniques to improve the grade of valuable elements. Studies also explored the application of vanadium and titanium resources in ironsands. Notable contributions during this period include: Kirkaldy [22], who studied the geological distribution of ironsands in Haslemere and Midhurst, and Hull [23], who conducted mineralogical analyses of ironsands on Taranaki Island. Additionally, Hutton [24–26] analyzed the vanadium and titanium resources in ironsands from Patea, Fitzroy, and Taranaki, while Junfu Suto [27] conducted investigations on the Tertiary marine sand deposits in Japan from geological and mineralogical perspectives. Modriniak [28] performed geophysical research on ironsands in the Whakatane region of New Zealand, and Mason [29] and Ongley [30] proposed methods for utilizing the iron, vanadium, and titanium resources in New Zealand ironsands. Fleming [31] and Finch [32] conducted detailed studies on the marine sand geology of Wanganui, and Beck [33] explored the geology of ironsands in Waitara.

Due to the low grade and magnetic properties of ironsands, Okasa [34] studied the grade and recovery rate of coastal ironsands in Aomori Prefecture, Japan, both before and after magnetic separation in 1950. The original ore grade was TFe: 19.69 wt% and TiO_2: 4.49 wt%; after fine grinding (100–150 mesh) and magnetic separation, the concentrate grades were TFe: 52.24–64.80 wt% and TiO_2: 40.44 wt%. The recovery rates were 77.7–84.7% for iron and 46.2% for titanium, indicating that the concentrate had industrial utilization value.

In 1955, Martin [35] introduced the origin, composition, and reserves of New Zealand's ironsands. He summarized the reasons behind past failures in industrial production and highlighted the challenges and opportunities presented by the comprehensive utilization of iron, titanium, and vanadium resources in ironsands. This marked a shift in the focus of ironsands research, expanding from iron to the broader, more comprehensive utilization of these resources.

In 1958, Nicholson [36, 37] estimated that approximately 600 million tons of ironsands with iron grades above 58 wt% were present in the Mokao River region of northern New Zealand. Additionally, he identified over 90 million tons of ilmenite resources with TiO_2 grades ranging from 44–45 wt% in Calamia and Jackson Bay in southern New Zealand, underscoring the abundance of New Zealand's ironsands mineral resources.

During the 1960s, research on ironsands continued to focus on mining, beneficiation, enrichment, and utilization. McDougall [38] measured the concentration of ironsands at depths of 5 to 50 fathoms along the west coast of northern New Zealand, finding the highest concentration (30 wt%) in shallow coastal areas. Beyond 50 fathoms, the concentration of ironsands significantly decreased, and its distribution was found to be closely related to wave characteristics post-formation. Figure 2.1 illustrates records from the mining process at the time. In 1962, Hirst [39] and Maung [40] introduced methods for the enrichment and beneficiation of ironsands from the Singatoka and Raglan regions, while Ross [41] focused on preliminary experiments on the surface magnetic permeability of ironsands in the Taranaki area.

Between 1964 and 1968, Wright published four key papers that used polarized mineralogical microscopy, electron probes, thermomagnetic curves, and X-ray diffraction techniques to systematically analyze the microstructure, thermomagnetic properties, and evolution of iron–titanium compounds in ironsands. In 1964, Wright [42] studied the homogeneous and heterogeneous mineral phase structures of ironsands, focusing on the types of oxides formed under various Titanium–iron solid solution ratios. Using a ternary phase diagram, he characterized the oxidation pathway

Fig. 2.1 Research process for ironsands mining

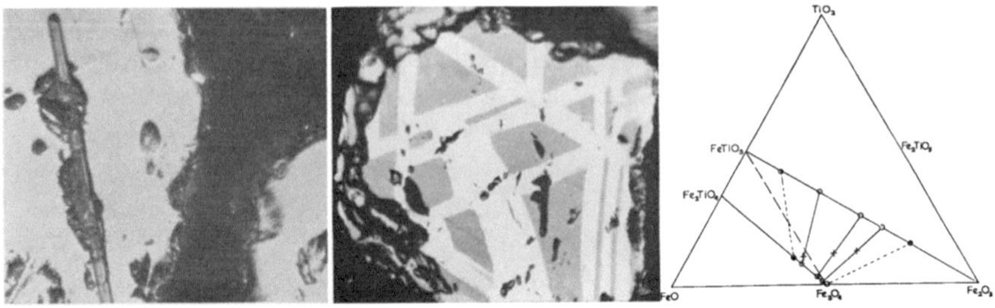

Fig. 2.2 Homogeneous and inhomogeneous structures of ironsands and ternary composition during oxidation

of ironsands, as shown in Fig. 2.2. Wright also described the dissolution and re-dissolution of aluminum spinel during oxidation. By comparing titanium magnetite from volcanic rocks with that from ironsands, he confirmed that the latter originated from volcanic rocks and was formed through erosion and oxidation by seawater and the atmosphere.

In 1965, Wright [43] employed the electron microprobe technique to accurately analyze the elemental composition (Fe, Ti, Mg, Al, Mn) of New Zealand ironsands. He proposed that in the homogeneous fraction, the elemental content exhibited minimal variation across different sites, with an average of 61 wt% Fe and 8 wt% Ti. Calcium, primarily present in apatite and silicate, was found not to be part of the titanium magnetite structure. Wright suggested that non-homogeneous streaks were the result of the phased oxidation of the homogeneous fraction. In 1967, Wright [44] heated ironsands in air, observing that homogeneous titanomagnetite particles oxidized into homogeneous titanohematite particles, while non-homogeneous titanomagnetite remained unchanged, with titanium only present in the titanohematite phases in the final product. He also noted that pseudobrookite was only temporarily present in natural ironsands and during oxidation, owing to the low TiO_2 content in these sands.

In 1968, Wright [45] discovered that titanium rapidly accumulated in rhombohedral laminates during the early stages of ironsand formation, when the iron content in the region was low. As the oxidation process progressed, with low Mg, Al, and Mn content in the ore grains, these elements began to enrich in the homogeneous cubic phase where titanium was almost absent. Meanwhile, the Fe–Ti ratio increased in the rhombic crystalline phase, eventually leading to the formation of a homogeneous titano-hematite structure. Under conditions of higher Mg, Al, and Mn content in the ore grains, Mg–Al–Mn spinel eventually dissolved into the titano-hematite. Furthermore, the homogeneous and non-homogeneous particles in the ironsands were largely similar in composition. With the subsequent publication of research by Gow [46], it was further confirmed that ironsands formed from volcanic rocks are the primary constituent of ironsand deposits.

During the 1970s, the focus of ironsands research shifted from purely geological exploration, mining, beneficiation, and mineral structure to practical industrial applications.

In 1971, Jessop [47] published a paper highlighting the growing interest in using ironsands as an iron-bearing mineral in iron and steel smelting. In 1976, a study by Pajalich [48] on ironsands resources in Indonesia revealed that, in addition to South Africa and New Zealand, Indonesia also possessed significant reserves of ironsands. In 1974, McAdam [49, 50] conducted an initial study on the structural changes in the physical phases of ironsands under reducing gas conditions. Also in 1974, McConnell [51] reported that New Zealand Steel Limited was mining two ilmenite ironsand deposits: one on the west coast of the northern peninsula (the Waikato North mine), which provided minerals for New Zealand's Direct Reduction and Electric Furnace Steelmaking processes, and the other at the Tehacho mine, where ilmenite ore was extracted from an offshore suspension processing site and exported directly to Japan. Two distinct utilization routes for ilmenite were observed during this period: one at Glenbrook, where it was used as an iron-bearing raw material for steel billet production through direct reduction and EAF (Electric Arc Furnace) melt separation processes; and the other in Japan, where it was used as a sintering raw material for blast furnace operations. This period thus marked the integration of ironsands into both long and short ironmaking processes.

In 1975, Christie [52], Foley-Fisher [53], and Ingram [54] conducted studies on New Zealand ironsands, focusing on grain size distribution, sedimentary lithology, mineralogy, beneficiation processes, and their applications in the steel industry.

Between 1979 and 1981, Watson [55–57] and Low [58] provided detailed descriptions of the newly developed Waipipi Ironsands Mine, located in the northwest of New Zealand. Meanwhile, Carter [59] reviewed and summarized the distribution and concentration of ironsands on the continental shelf off the western coast of New Zealand. Falconer [60] explored the utilization of the magnetic properties of ironsands for industrial mining, reflecting the increasing focus during this period on the practical applications of ironsands.

From 1982 to 1983, Watson [61, 62] published consecutive studies on re-election roughing and magnetic selection techniques for ironsands, which laid the groundwork for applying these materials in industrial production. Smith [63] assessed the feasibility of applying ironsands in the ironmaking process, concluding that their use in both blast furnace and non-blast furnace processes was entirely feasible.

In 1985, Hukkanen [64] proposed that titanomagnetite could be utilized as a composite raw material for the production of iron, steel, vanadium, and titanium. After appropriate milling, titanomagnetite can be upgraded through weak magnetic separation, followed by reduction and grading to produce iron, vanadium, and titanium products. Notably, the reduction of iron and vanadium had already been successfully implemented at the High Weald Iron and Vanadium Plant in South Africa, producing steel and vanadium slag, which was subsequently roasted and leached to yield high-grade V_2O_5. In New Zealand, the "New Zealand Method" was developed for the efficient use of ironsands, where iron is separated via direct reduction and furnace smelting after the mining and initial purification stages. This process also yields a high-grade TiO_2 product, synthetic rutile, a technique also realized in Australia. Another method for utilizing ironsands involves the production of high-grade foundry pig iron and titanium slag in electric furnaces, known as the Sorel process, successfully implemented in Canada and South Africa. In Norway, efforts focused on the high-recovery recycling of titanomagnetite's main valuable components.

After 1989, a new large ironsands deposit in New Zealand gained attention, prompting studies by Stokes [65] and Lawton [66] on the reserves and geology of this deposit. Figure 2.3 illustrates the distribution points of ironsand deposits in northwest New Zealand as of 1993. The application of ironsands in blast furnace steelmaking was further investigated by Bates [67], while MacKenzie [68] proposed their use in the preparation of ferrite. Additionally, Mangkusubroto [69] highlighted Indonesia's abundant ironsands resources once again.

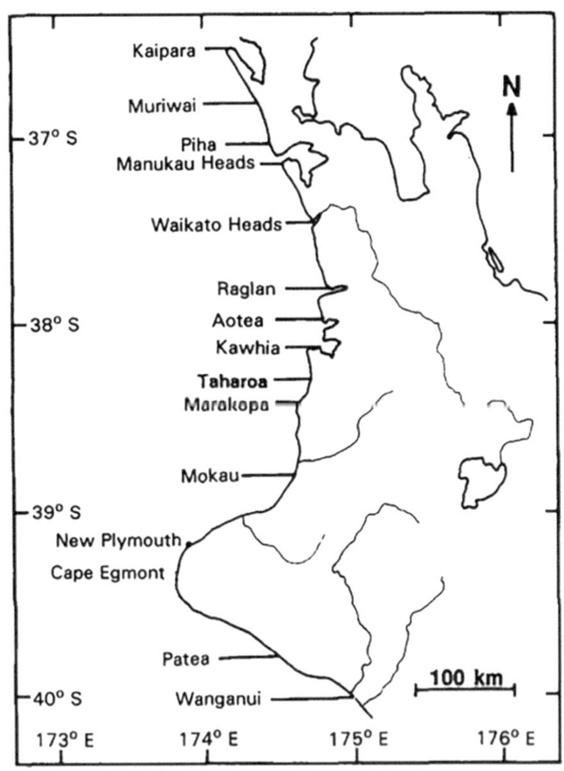

Fig. 2.3 Distribution of ironsands deposits in northwestern New Zealand (1993)

2.3 Current Studies on the Mineral Phase Structure of Ironsands

In 2010, Carolyn et al. [70] published in the American Mineralogist the occupancy of iron, titanium, and oxygen atoms in a magnetite-Titanium–iron spinel solid solution (the main physical phase of ironsands minerals).

$FeO–TiO_2–Fe_2O_3$ are in solid solution with each other in three main systems: one is the titanomagnetite system, i.e., the magnetite-titanomagnetite solid solution system ($Fe_3O_4–Fe_2TiO_4$), and in the ironsands, the homogeneous portion as well as the non-homogeneous portion of the substrate, i.e., titanomagnetite in this cubic spinel structure, abbreviated as TTM; and the second one is the titanomagnetite system, i.e., the αhematite-titanomagnetite solid solution system ($\alpha Fe_2O_3–FeTiO_3$). $\alpha Fe_2O_3–FeTiO_3$ solid solution system, in the ironsands, the non-homogeneous particles of the layered part (or called striated part), that is, this kind of rhombohedral crystal system of titanomagnetite, abbreviated as TTH. in the dry and below the temperature of the solid line of the cooling process, the titanomagnetite with the interaction of water, air, undergoes oxidation and the magneto-extreme mineralisation, and finally forms in the low-temperature, low-pressure conditions, very stable Titanomagnetite is very stable at low temperature and low pressure. The third is the Fe_2TiO_5-$FeTi_2O_5$ solid solution system. The first solid solution system, however, is the main mineral structure of ironsands and influences the magnetic properties of the rocks in which they are found, as well as of the Earth's crust as a whole.

Titanomagnetite has a spinel structure (space group Fd—3 m), which can be expressed in the usual formula as $A_8B_{16}O_{32}$, where A represents the + 2-valent cation, occupying the tetrahedral position and presenting $[AO_4]$ tetrahedra, and B represents the + 3-valent cation, occupying the octahedral position and presenting $[BO_6]$ octahedra, and the presence of 32 per cell of the oxygen atoms, presenting a cubic closest stacking, with the stacking direction perpendicular to the tertiary axis direction [111]. Along the tertiary axis direction, the $[AO_4]$ tetrahedra are connected to the $[BO_6]$ octahedra in a co-corner-top manner, with each corner-top of the unit cell shared by one $[AO_4]$ tetrahedron and three $[BO_6]$ octahedra. This arrangement is based on the principle of minimum lattice energy, and when cations in the lower valence states occupy the tetrahedral positions, the minimum electrostatic energy is reached, resulting in the formation of orthospinel structures. Whereas in anti-spinel structures, such as in Fe_3O_4, Fe^{3+} possesses a stronger manifestation of its ability to occupy the tetrahedra, resulting in Fe^{2+} and other Fe^{3+} occupying the octahedral vacancies.

Titanomagnetite is a solid solution formed by magnetite and titanomagnetite spinel, where Fe^{3+} in the magnetite is gradually replaced by Ti^{4+} in the titanomagnetite spinel, accompanied by the conversion of Fe^{3+} to Fe^{2+}, thus maintaining an electrovalent equilibrium in the final state:

$$Fe_A^{3+}\left(Fe^{2+}\ Fe^{2+}\right)_B O_4 - Fe_A^{2+}\left(Fe^{2+}\ Ti^{4+}\right)_B O_4$$

In the cubic spinel structure of titanomagnetite, the way and occupancy of Fe^{3+} being replaced by Fe^{2+} and Ti^{4+} in different degrees has been a hot topic of research, and the related researchers have used many means, such as X-rays, powder diffraction, Mössbauer spectroscopy, transmission electron microscopy, X-ray absorption spectroscopy and other detection means, trying to accurately characterize the problem of the ionic occupancy of ilmenite in titanomagnetite, but the problem has not yet been more clearly described, and different theories and models still exist, see Table 2.3.

As for the generation mechanism of layered titanium hematite phase, there are two theories, one is that the layered area is a miscible zone of magnetite and titanomagnetite spinel, and the other is that titanium enrichment occurs due to the migration of ironsands mine cations caused by the oxidation of seawater and air, which results in the dissolution of TTH from TTM due to the lower solubility, i.e., the TTH phase is dissolved from the TTM phase The TTH phase is produced by dissolving from the TTM phase. The above two theories suggest that the mechanism of TTH generation is still unclear.

2.4 Current Studies on the Magnetic Separation of Ironsands

Certain ores have magnetic properties, and the discovery of this ore is very early.

As early as 550 B.C., Socrates in ancient Greece described magnets as not only attracting iron coils, but also imparting the same ability to attract magnets themselves. In short, they attract each other.

The use of magnetism in ores, however, is a more recent development.

The idea of using magnetism to separate magnetic minerals from non-magnetic veins was proposed by Ball, Norton, Edison and others in the middle of the nineteenth century, and it was only at the end of the nineteenth century that the idea was initially applied. Later on, thanks to the improved knowledge of the basic principles of magnetism and the development of permanent magnet materials, magnetic separation has been involved from rough solid materials to fluids, and from strong magnetic materials to antimagnetic materials.

The earliest use of permanent magnets to generate a magnetic field is characterised by the fact that it does not require any energy or cooling liquid and the magnetic induction strength is always stable. The disadvantage is that the magnetic induction strength cannot be adjusted and varies with temperature. Later, the more economical ferrite materials appeared, including barium ferrite and strontium ferrite, characterised by low raw material prices, simple manufacturing process, very low cost, low density, light weight, and high coercivity, the main drawback is that the remanent magnetism is not high, and the ambient temperature has a greater impact on the magnetic properties. In addition, a magnetic field can also be generated using rare earth-iron alloys, which have high magnetic properties and high magnetic product energy, but the manufacturing cost is also relatively high.

Table 2.3 Site occupancy of Fe^{3+}, Fe^{2+} and Ti^{4+} ions in titanomagnetite

Model	Sample	Technique	Conclusions
A [71]	Ferromagnetic minerals separated from volcanic rocks	Chemical composition, cell parameters, magnetic saturation, Curie temperature	$Fe^{3+}_{1-x}Fe^{2+}_{x}(Fe^{2+}Fe^{3+}_{1-x}Ti^{4+}_{x})O^{2-}_{4}$
B [72]	Theoretical research	Theoretical and empirical speculation	Ti^{4+} always occupies the octahedral vacancies, keeping the octahedral vacancies filled with Fe^{2+}
C [73]	Fe_2O_3, TiO_2 and Fe powders were heated to 1150 °C under vacuum and held for 6h	Chemical analysis, cell parameters, magnetic saturation, Curie temperature	When $x \leq 0.5$ $Fe^{3+}_{1-a}Ti^{4+}_{a}\left(Fe^{2+}_{1+x}Fe^{3+}_{1-2x+a}Ti^{4+}_{x-a}\right)O^{2-}_{4}$; When $x \geq 0.5$ $Fe^{3+}_{2-2x}Fe^{2+}_{2x-1-a}Ti^{4+}_{a}(Fe^{2+}_{2-x+a}Ti^{4+}_{x-a})O^{2-}_{4}$;
D [74]	Mixture of sintered Fe_2O_3, TiO_2, Fe	Magnetic Saturation, Electrical Conductivity	When $0 < x < 0.2$, Fe^{3+} occupies preferentially in the tetrahedron; When $0.2 < x < 0.8$, $Fe^{2+}_{x-0.2}Fe^{3+}_{1.2-x}\left(Fe^{2+}_{1.2}Fe^{3+}_{0.8-x}Ti^{4+}_{x}\right)O^{2-}_{4}$; When $0.8 < x < 1$, Fe^{3+} preferentially occupies the octahedron
E [75]	Fe_2O_3, TiO_2 and sponge iron in silver foil heated to 930 °C and held for 4–7 days	Chemical composition, single cell parameters, thermogravimetric parameters, neutron diffraction, hysteresis curves, magnetic saturation	Fe^{2+}–Fe^{3+} occupancy shows a linear relationship with composition

(continued)

Table 2.3 (continued)

Model	Sample	Technique	Conclusions
F [76]	Single crystal growth in CO–CO_2 atmosphere	Chemical composition, magnetization curve, hysteresis curve	When $0 \leq x \leq 0.2$ $Fe^{3+}\left(Fe^{3+}_{1-2x}Fe^{2+}_{1+x}Ti^{4+}_{x}\right)O_4^{2-}$; When $0.2 \leq x \leq 1$ $Fe^{3+}_{1.25-1.25x}Fe^{2+}_{1.25x-0.25}\left(Fe^{3+}_{0.75-0.75x}Fe^{2+}_{1.25-0.25x}Ti^{4+}_{x}\right)O_4^{2-}$;
G [77]	Mix Fe_2O_3 and TiO_2 stoichiometrically and heat to 1000 °C	Mössbauer spectrum	$Fe^{3+}_{f(3-x)}Fe^{2+}_{1-f(3-x)}(\left(Fe^{3+}_{2(1-x)-f(3-x)}Fe^{2+}_{x+f(3-x)}Ti_x\right)O_4$; When $0 \leq x < 0.2$, $f = 0.333(1-0.25x)$; When $0.2 \leq x < 0.5$, $f = 0.35(1-0.57x)$; When $0.5 \leq x < 1$, $f = 0.50(1-x)$;
H [78]	Single crystal growth in an atmosphere of CO_2 and H_2 and slow cooling from 1200 to 900 °C	Cell parameters, chemical composition, Mössbauer spectrum	Fe^{3+} tetrahedron + Fe^{3+} octahedron = Fe^{2+} tetrahedron + Ti^{4-} octahedron
I [70]	Stoichiometrically mixed Fe_2O_3, TiO_2 and Fe in a silver foil heated to 897 °C and held for 7 days	Chemical composition, XMCD, XAS, Mössbauer spectrum, cell parameters	Ti^{4+} always occupies the octahedral vacancies, and for $0 < x < 0.45$, Fe^{2+} gradually increases in the octahedral vacancies, and Fe^{2+} preferentially occupies the octahedral vacancies until $x > 0.4$, after which Fe^{2+} gradually increases in the tetrahedral vacancies, accompanied by a decrease in Fe^{2+} in the octahedral vacancies

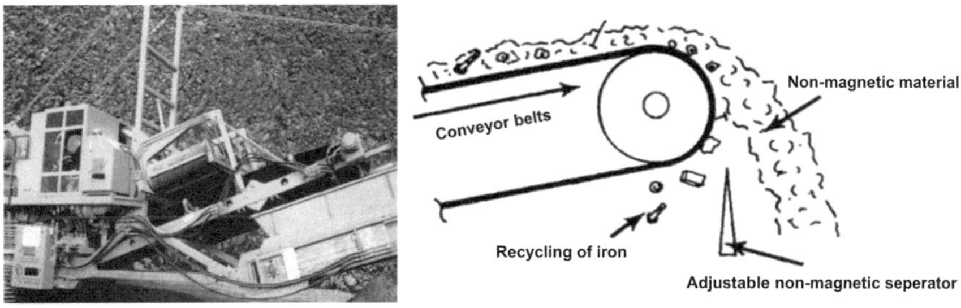

Fig. 2.4 Process flow for iron recovery from non-magnetic materials via magnetic separation

When higher magnetic induction strength and magnetic field gradient are required, permanent magnets cannot be realised, and then electromagnets should be used, including conventional iron-core coil-type magnets, solenoidal magnets, alternating-field magnets, pulsating-field magnets, pulsating-field magnets, and superconducting magnets.

Figures 2.4, 2.5, 2.6 and 2.7 show the current applications of several large magnetic separators.

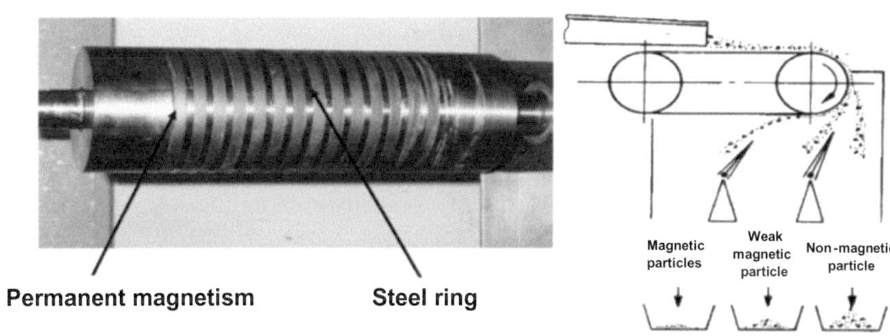

Fig. 2.5 Schematic diagram of a permanent magnetic rolling separator

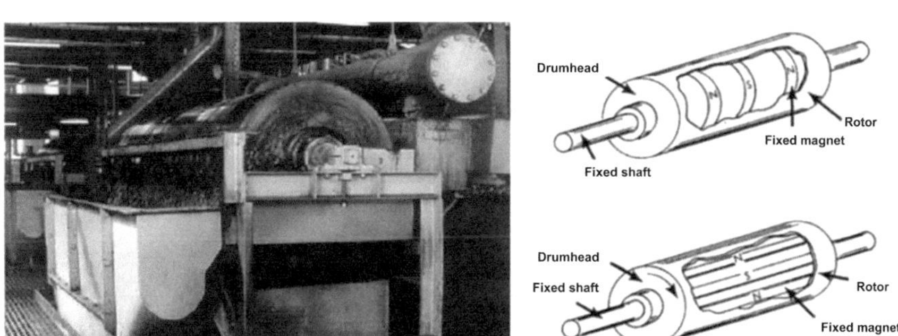

Fig. 2.6 Structural representation of a drum magnetic separator

Fig. 2.7 High-gradient magnetic separator (left) and superconducting magnetic separator (right)

Magnetic minerals and non-magnetic minerals, or strongly magnetic minerals and weakly magnetic minerals, are ultimately separated by differences in the forces applied to them. The forces on magnetic particles in a magnetic field can be quantitatively described by Eq. 2.1.

$$F_{\mathrm{m}} = \frac{k}{\mu_0} \cdot V \cdot B \cdot \nabla B \tag{2.1}$$

where k denotes the volume magnetisation of the particles; μ_0 denotes the vacuum permeability; V denotes the volume of the particles; B denotes the magnetic induction; $\bigtriangledown B$ denotes the magnetic field strength gradient

In the process of magnetic separation, in addition to the magnetic field force, it is also subject to other forces, including gravity, inertia force, fluid resistance, surface tension, etc., to which the particles are subjected. The calculation of the gravity force F_g is shown in Eq. 2.2.

$$F_g = \rho \cdot V \cdot g \tag{2.2}$$

where ρ denotes the density of the particles; g is the gravitational acceleration.

The resistance F_d exerted by the fluid on the mineral particles can be expressed by Eq. 2.3.

$$F_d = 6 \cdot \pi \cdot \eta \cdot b \cdot v_p \tag{2.3}$$

where η is the kinetic viscosity index; b is the particle radius; and v_p is the velocity of the particle relative to the fluid.

When the magnetic force on the magnetic particles is greater than the resistance, the magnetic particles can be separated and extracted. When the magnetic force on the magnetic particles is less than the resistance, the magnetic particles will not be extracted but will go into the tailings, so the particles can be separated according to the difference of the force on them. The conditions of magnetic separation are shown in Eqs. 2.4 and 2.5.

$$F_{\text{magnetic force}}^{\text{magnetic particle}} > \sum F_{\text{resistances}}^{\text{magnetic particle}} \tag{2.4}$$

$$F_{\text{magnetic force}}^{\text{non - magnetic particles}} < \sum F_{\text{resistances}}^{\text{non - magnetic particles}} \tag{2.5}$$

From Eqs. 2.1, 2.2 and 2.3, the particle radius b has a large influence on the force, such as F_{m} is proportional to b^3, F_{g} is proportional to b^3, and F_d is proportional to b. In dry magnetic separation, the particle radius has little influence in dry magnetic separation because F_d is negligible and the particle radius is comparable to the magnetic field force and gravity, but for wet magnetic separation, the influence of particle radius on the force is more significant.

Currently, Kasama [79] and Gehring [80] have conducted extensive investigations into the effect of laminated TTH material melted onto the TTM substrate on the magnetic properties of ironsands. However, studies addressing the fine grinding and dissociation of ironsands, the recovery of iron and titanium following wet magnetic separation, and the actual production processes involved in magnetic separation remain relatively scarce. Therefore, Chap. 4 focuses on theoretical analysis and experimental research related to these issues.

2.5 Current Studies on the Pre-oxidation of Ironsands

Pre-oxidation is a conventional method for treating minerals, involving heating to a specified temperature in air or oxygen-enriched conditions. This process promotes the oxidation of the minerals, altering their chemical composition or crystal structure, thereby making them more amenable to subsequent treatments and processing.

Currently, pre-oxidation is employed to enhance the reduction rate of ilmenite in the production of titanium dioxide or titanium sponge [81]. It has also been applied to selectively enrich titanium-containing blast furnace slag and recover titanium-rich phases through precipitation [82]. The discovery that pre-oxidation treatment increases the reduction rate of ironsands was first made by McAdam [83, 84].

Park [85], in contrast, provided a preliminary explanation of the mechanism behind pre-oxidation. By studying the reduction characteristics of hematite, magnetite, and ironsands, Park found that, under identical temperature and reducing atmosphere conditions, hematite was the most easily reduced, followed by magnetite, while ironsands proved more difficult to reduce. The reduction degree of these three ores under the same reaction conditions is illustrated in Fig. 2.8 [86].

When magnetite is pre-oxidized to hematite, its reducibility increases significantly. This enhancement is primarily due to the conversion of oxygen atoms from the face-centered cubic structure of magnetite to the densely packed hexagonal structure of hematite during the crystallographic transformation. This transformation results

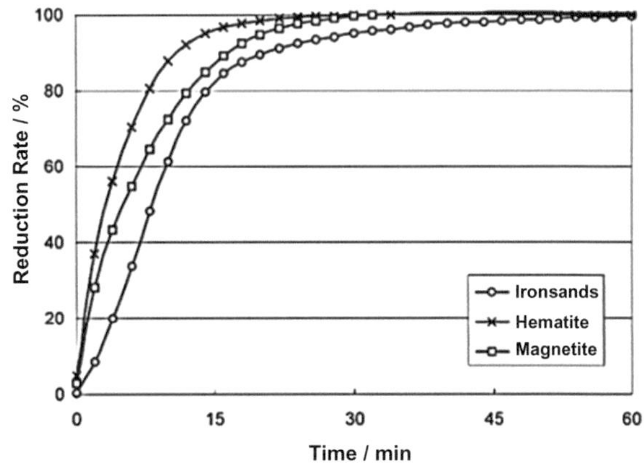

Fig. 2.8 Comparative analysis of the reduction behavior of ironsands, hematite, and magnetite under 75 vol% CO–25 vol% Ar at 1100 °C

in a 25% increase in volume, which opens up the crystal structure and facilitates reduction. In this context, Park E. investigated the morphological changes and the physical phase composition of ironsands after pre-oxidation. He also compared the reduction rate and reduction degree of raw and pre-oxidized ironsands under a CO reducing atmosphere, as illustrated in Figs. 2.9, 2.10 and 2.11. These figures demonstrate that pre-oxidation treatment enhances both the reduction rate and the reduction degree of ironsands, confirming that pre-oxidation improves the reduction process of ironsands.

Furthermore, Park investigated the physical phase changes in ironsands during the pre-oxidation process by heating the sample in a muffle furnace at a ramping rate of 200 °C/hour under atmospheric conditions, followed by X-ray diffraction (XRD) analysis. He observed that titanomagnetite (TTM) remained the primary physical phase in ironsands up to 800 °C. The oxidation process occurred in two stages

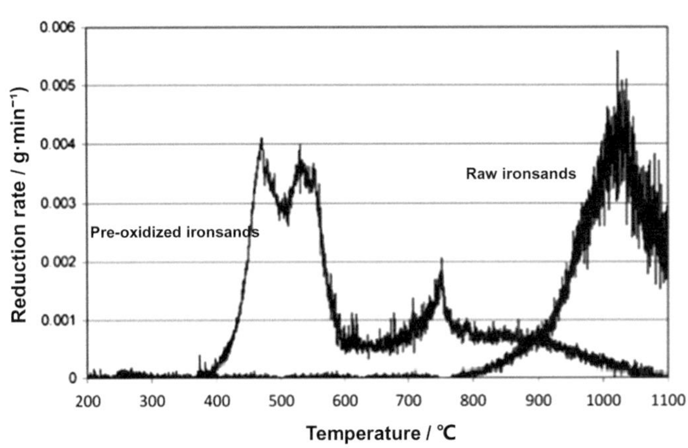

Fig. 2.9 Comparison of reduction rates between pre-oxidized and raw ironsands under non-isothermal conditions (75 vol% CO–25 vol% Ar)

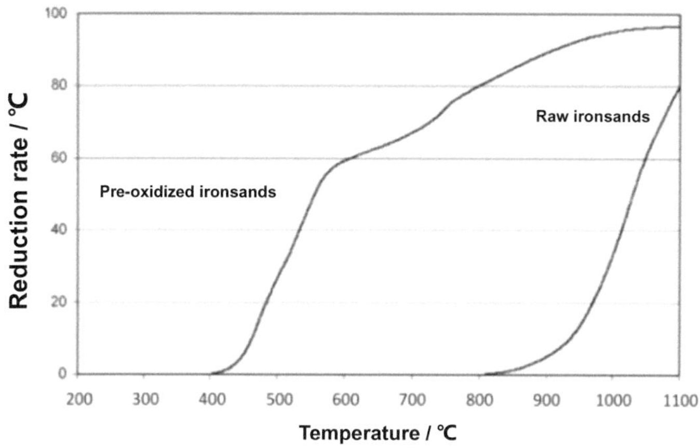

Fig. 2.10 Comparative reduction behavior of pre-oxidized and raw ironsands under non-isothermal conditions (75 vol% CO–25 vol% Ar)

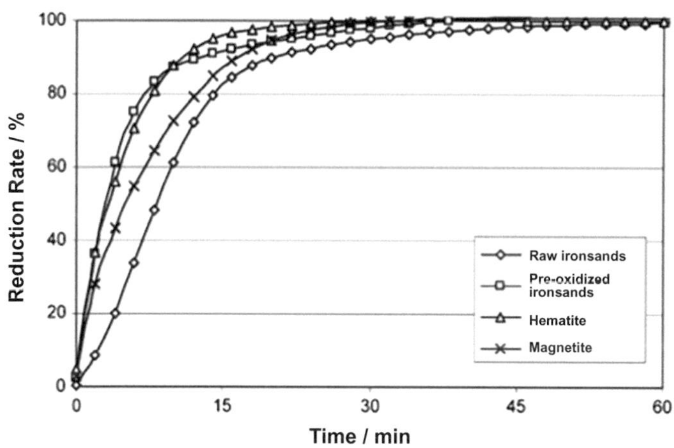

Fig. 2.11 Comparative reduction of raw ironsands, pre-oxidized ironsands, hematite, and magnetite under isothermal conditions (75 vol% CO–25 vol% Ar)

before reaching 800 °C. At 600 °C, Fe_3O_4 in TTM was oxidized to pseudohematite, with changes in composition but no alteration in structure. It was only between 600 and 800 °C that TTM began to oxidize to titanomagnetite-hematite (TTH). However, during isothermal oxidation above 1000 °C, pseudohematite was not observed, as it is unstable at high temperatures. The oxidation rate accelerated between 800 and 1000 °C, while the mineral phase of the ironsands remained largely unchanged above 1000–1100 °C. Based on these observations, Park derived equations to describe the changes in physical phase during the oxidation of ironsands, as shown in Eqs. 2.6 to 2.8.

$$Fe_{3-x}Ti_xO_4 + \frac{1}{4}O_2 = \frac{3}{2}Fe_{2-\frac{2}{3}x}Ti_{\frac{2}{3}x}O_3 \tag{2.6}$$

$$Fe_{2-\frac{2}{3}x}Ti_{\frac{2}{3}x}O_3 + \frac{x}{6}O_2 = (1-x)Fe_2O_3 + \frac{2}{3}xFe_2TiO_5 \tag{2.7}$$

$$\begin{aligned} Fe_{2-\frac{2}{3}x}Ti_{\frac{2}{3}x}O_3 + \frac{1}{6}\left(\frac{2x-3\delta}{2-3\delta}\right)O_2 = 2\left(\frac{1-x}{2-3\delta}\right)&Fe_{2-\delta}Ti_\delta O_3 \\ + \frac{2}{3}\left(\frac{2x-3\delta}{2-3\delta}\right)&Fe_2TiO_5 \end{aligned} \tag{2.8}$$

Park attributed the enhanced reducibility of ironsands due to pre-oxidation to the transformation of cubic spinel TTM to rhombic TTH. This lattice transformation is believed to contribute to the improved reductivity of the ironsands. However, the explanation remains somewhat general, as it lacks a detailed analysis from the perspective of microscopic morphology, cell size, and other relevant factors. Additionally, the equations describing the reactions and phase changes during pre-oxidation are speculative and have not been validated through thermogravimetric analysis.

2.6 Current Studies on the Gas-Based Reduction of Ironsands

Since 2000, numerous studies have been conducted on the reduction mechanisms of ironsands, covering a wide range of topics. These include the reduction reactions between graphitic carbon and ironsands [87–90], the differences in reduction behavior between ironsands in circulating fluidized beds and fixed beds [91], and the reduction of ironsands using a mixture of methane and hydrogen [92].

In 2003, Park published a study on the reduction of ironsands using carbon monoxide [86], followed by a study on hydrogen reduction of ironsands in 2004 [93]. These two papers provide a detailed examination of the changes in physical phase, micro-morphology, reduction rate, and reduction degree of ironsands under the influence of these two reducing gases.

2.6.1 *Reduction of Ironsands Using Carbon Monoxide*

Park selected New Zealand ironsands as the research subject and, by comparing the changes in physical phase composition and micro-morphology of hematite, magnetite, and ironsands under carbon monoxide reduction conditions, conducted a more detailed analysis of the reduction characteristics of ironsands in a carbon monoxide atmosphere. The composition of the New Zealand ironsands studied is presented in Table 2.4.

The experimental setup consisted of a vertical tubular resistance furnace, functioning as a fixed-bed reactor. The temperature range of 1000–1100 °C was selected, as carbon buildup reactions are negligible above 1000 °C, and low-melting-point compounds are less likely to form. The reduction process utilized a $CO–CO_2–Ar$ gas mixture, which was first passed through high-purity calcium sulfate and 4A molecular sieves to remove impurities such as moisture. The gas was then directed through a mass flow meter to maintain a controlled flow rate before entering the furnace. The composition of the gas products was analyzed and recorded in real-time using an online mass spectrometer. Additional analytical techniques employed included X-ray diffraction, electron microprobe analysis, scanning electron microscopy, and thermogravimetric analysis.

From a micro-morphological perspective, the reduction of homogeneous ironsands particles begins at the outer surface and gradually progresses toward the center. The reduction of these particles is generally completed within 30 min. Since these particles represent the primary physical phase of ironsands, this time frame is reflected in the reduction curves. Changes in the micro-morphology of the homogeneous ore particles are shown in Fig. 2.12.

The reduction of non-homogeneous particles begins with the layered TTH phase. The reduction of TTH starts after approximately 3 min, by which point the homogeneous particles have not yet begun to reduce. The reduction of the layered TTH phase occurs rapidly, and as a result, the non-homogeneous particles are completely reduced within 15 min, as shown in Fig. 2.13.

The reason why the TTH phase with a layered structure is easier to reduce than the homogeneous portion of TTM is that the reduction of TTH involves a transition from a rhombic to a cubic crystal system. This structural transformation opens up the particle framework, accelerating the nucleation and growth of metallic iron, thereby facilitating the reduction process.

Park E. also provided a schematic equation for the physical phase changes during the reduction of ironsands, as shown in Eq. 2.9. During the reduction, iron is separated by post-reduction migration from the TTM phase, forming a Ti–rich TTM phase. This is followed by the formation of titanite-iron spinel, which, after passing through ilmenite and pseudobrookite phases, eventually forms particles with a mosaic distribution of metallic iron and veinlets, covered by metallic Fe.

$$\begin{aligned} & Fe_{3-x}Ti_xO_4 \rightarrow FeO + Fe_2TiO_4 \rightarrow Fe + Fe_2TiO_4 \rightarrow Fe + FeTiO_3 \rightarrow Fe \\ & \quad + FeTi_2O_5 \rightarrow Fe + TiO_2 \end{aligned} \tag{2.9}$$

Table 2.4 Chemical composition of New Zealand ironsands/wt%

TFe	Fe^{2+}	Fe^{3+}	TiO_2	Al_2O_3	MgO	SiO_2	CaO	Mn
57.2	24.2	33.0	7.43	3.59	2.94	2.17	0.67	0.51

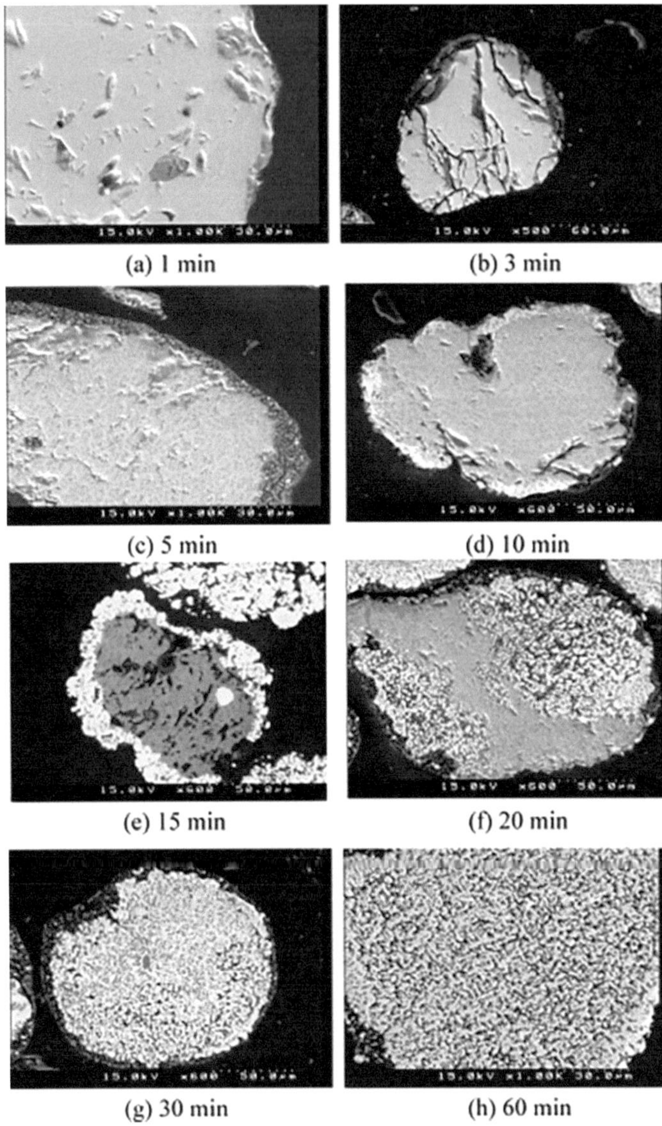

Fig. 2.12 Microstructural evolution of homogeneous ironsands particles during reduction in 75 vol% CO–Ar at 1100 °C

The higher thermodynamic stability of ironsands can be attributed to the following: In the TTM lattice, as shown in Eq. 2.10, one Ti^{4+} ion and one Fe^{2+} ion replace two Fe^{3+} ions, resulting in a lower concentration and activity of Fe^{3+}, while the concentration and activity of Fe^{2+} in TTM are higher compared to those in

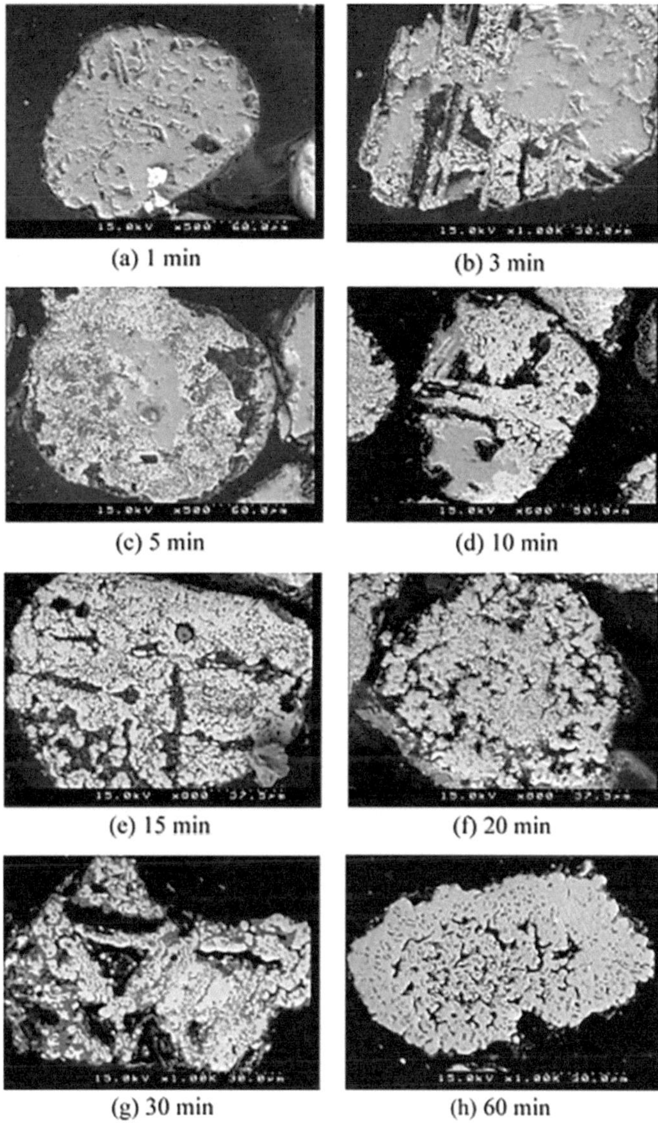

Fig. 2.13 Microstructural changes in non-homogeneous ironsands particles during reduction in 75 vol% CO–25 vol% Ar at 1100 °C

magnetite. This explains the high thermodynamic stability of Fe^{3+} in TTM. Moreover, the reduction process involves the removal of oxygen, which leads to the Fe^{3+} to Fe^{2+} transition. This transition induces an increase in Fe^{2+} concentration within the lattice, thereby accelerating the reduction of Fe^{2+} to metallic Fe. Furthermore, this process obstructs the transition from Fe^{3+} to Fe^{2+}. Ti^{4+} ions, which are coordinated by six oxygen atoms in the octahedral lattice, hinder the removal of oxygen atoms, thus contributing to the stabilization of the structure.

$$\underbrace{\left(\left[Fe^{3+}\right]_{tetrahedron}\left[Fe^{3+}, Fe^{2+}\right]_{octahedron}O_4\right)_{1-x}}_{\text{magnetite}} \cdot \underbrace{\left(\left[Fe^{2+}\right]_{tetrahedron}\left[Fe^{2+}, Ti^{4+}\right]_{octahedron}O_4\right)_{x}}_{\text{titanium - iron spinel}} \tag{2.10}$$

2.6.2 Reduction of Ironsands Using Hydrogen

The study of hydrogen reduction of ironsands is rarely reported; however, hydrogen, along with carbon monoxide, is a key component in the direct reduction process. Therefore, it is essential to investigate the mechanism of hydrogen reduction of ironsands.

In 2004, Park conducted a study on the hydrogen reduction of ironsands [93]. Unlike previous studies that focused on carbon monoxide, this study provided precise solubility data for the New Zealand ironsands used in the experiments. Specifically, the homogeneous particles were titanomagnetite (TTM), a solid solution system of magnetite (Fe_3O_4) and titanomagnetite spinel (Fe_2TiO_4) ($Fe_{3-x}Ti_xO_4$; $x = 0.27 \pm 0.02$), while the non-homogeneous particles consisted of layered titanohematite (TTH; $Fe_{2-y}T_yO_3$), which was formed with a matrix of titanomagnetite.

The experimental setup included a laboratory vertical tubular resistance furnace, which served as a fixed-bed reactor. The reduction gas was a mixture of hydrogen (H_2) and argon (Ar), with high-purity calcium sulfate and 4A molecular sieves used to remove impurities. The gas flow was regulated by a proton flow meter. For analysis, X-ray diffraction, scanning electron microscopy, and thermogravimetric analysis were employed. The raw material used in the experiments was again New Zealand ironsands, and its composition is presented in Table 2.4.

Park E. first analyzed the thermodynamics of ironsands reduction, as shown in Eqs. 2.11 and 2.12. The equilibrium phases under varying hydrogen partial pressures and temperatures are presented in Fig. 2.14. The results indicated that under low reduction potential and low temperature, the equilibrium phases consisted mainly of metallic iron and Titanium–iron spinel. As the reduction potential and temperature were increased, the equilibrium phases transitioned to metallic iron and titano-crystalline iron oxide. Upon further increases in reduction potential and temperature, the equilibrium phases shifted to metallic iron and ilmenite. Ultimately, with continued increase in reduction potential and temperature, the equilibrium phase reached metallic iron and titanium dioxide.

$$\begin{gathered} Fe_2TiO_4 + H_2 = Fe + FeTiO_3 + H_2O \\ \Delta G^{\theta}(J/mol) = 24116.4 - 7.49T \end{gathered} \tag{2.11}$$

$$FeTiO_3 + H_2 = Fe + TiO_2 + H_2O$$
$$\Delta G^{\theta}(J/mol) = 34916.4 - 7.07T \quad (2.12)$$

The variation in the reduction degree of ironsands with time at different temperatures and reducing atmospheres is illustrated in Fig. 2.15. From Fig. 2.15a, it is evident that ironsands are almost fully reduced at temperatures above 900 °C. In contrast, at 700–800 °C, after 2 h of reduction, the degree of reduction is only 70%, compared to 95% at higher temperatures. As the temperature increases from 700 to 1000 °C, the reduction rate increases significantly. However, when the temperature rises further to 1100 °C, the rate of reduction slows down. From Fig. 2.15b, it can be observed that increasing the H_2 content from 5 vol% to 10 vol% significantly enhances the reduction rate. However, increasing the H_2 content to 50% does not have a substantial effect on further increasing the reduction rate.

The microscopic morphology of the reduced ironsands after 60 min in a 25 vol% H_2-75 vol% Ar atmosphere at different temperature conditions is shown in Fig. 2.16. It can be concluded that at low temperatures, iron primarily grows in the form of iron whiskers. As the temperature increases, the rate of reduction rises due to enhanced sintering, and iron is predominantly present as aggregates of fine metallic iron particles, with a noticeable trend towards pore shrinkage.

The microscopic morphology of the homogeneous and non-homogeneous particles of ironsands at different reduction times, under a 25 vol% H_2-75 vol% Ar reduction atmosphere at 900 °C, is depicted in Figs. 2.17 and 2.18. The reduction of homogeneous particles begins at the outer surface and progresses inward, with a clear demarcation between the reduced metallic iron and the unreduced portion

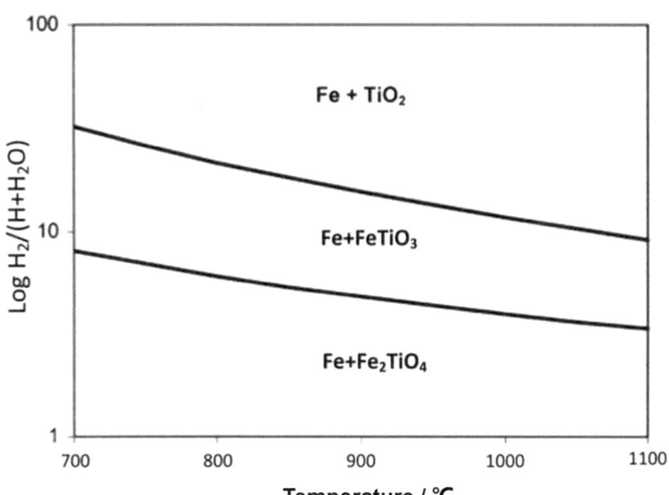

Fig. 2.14 Equilibrium phase composition of ironsands under varying temperatures and hydrogen partial pressures

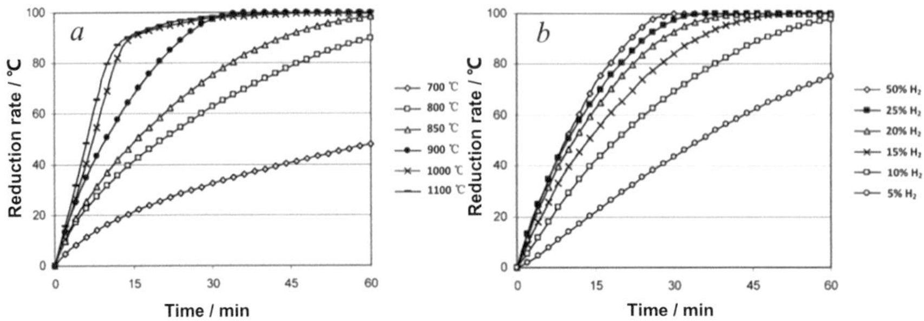

Fig. 2.15 Reduction degree of ironsands as a function of time under different temperatures and reducing atmospheres

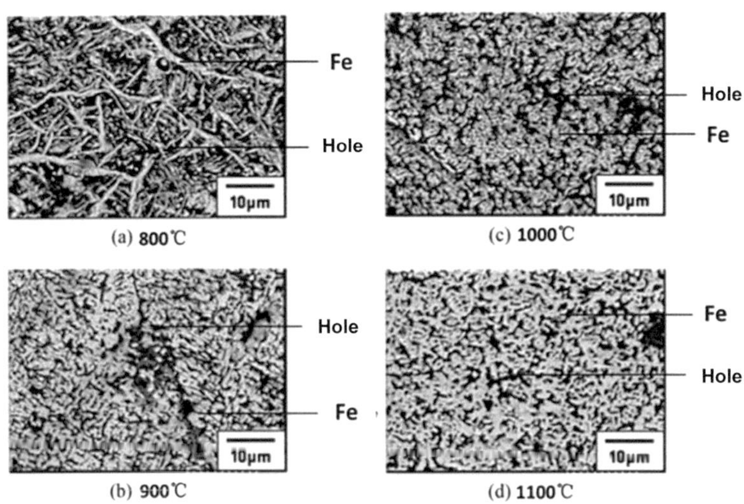

Fig. 2.16 Micromorphology of ironsands reduced in 25 vol% H_2–75 vol% Ar for 60 min at various temperatures

throughout the process. For non-homogeneous particles, metallic iron preferentially nucleates and grows from the TTH phase, resulting in a needle-like structure. These non-homogeneous particles are fully reduced within 15 min, although the boundary between the TTM and TTH phases remains distinct.

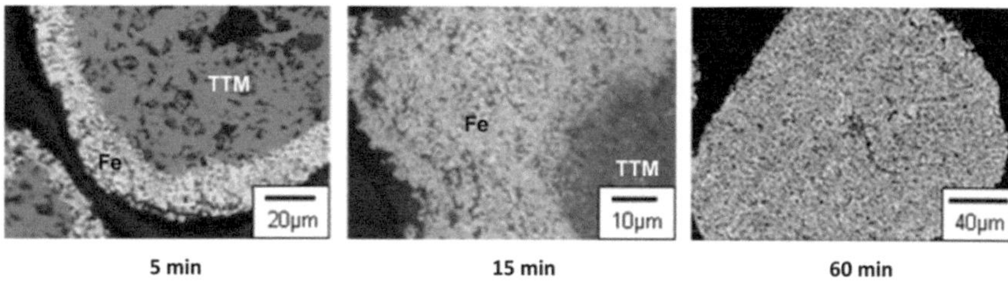

Fig. 2.17 Micromorphology of homogeneous ironsands particles reduced in 25 vol% H_2–75 vol% Ar at 900 °C for different durations

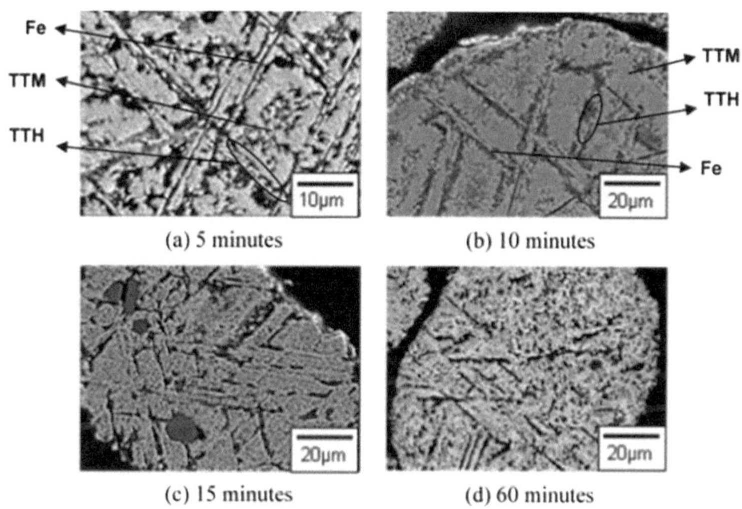

Fig. 2.18 Micromorphology of non-homogeneous ironsands particles reduced in 25 vol% H_2–75 vol% Ar at 900 °C for different durations

2.7 Current Studies on the Deep Reduction and Melt Separation of Ironsands in Electric Furnaces

When considering smelting costs, low-grade, low-cost ores are often prioritized. For instance, during periods of high international iron ore prices, the need to explore alternative, low-cost sources of iron-bearing ores, such as those with a high TiO_2 content, becomes more urgent. However, the high TiO_2 content in these ores can lead to several challenges, including lower ore reducibility, variations in liquid-phase slag temperatures, reduced slag refining capacity, and viscosities of slag that fall outside the optimal range. Despite these challenges, TiO_2-containing slags offer potential benefits. For example, they can form a protective layer of titanium carbide-nitride on the surface of refractory bricks, which helps to mitigate the attack of liquid-phase slag on the refractory material.

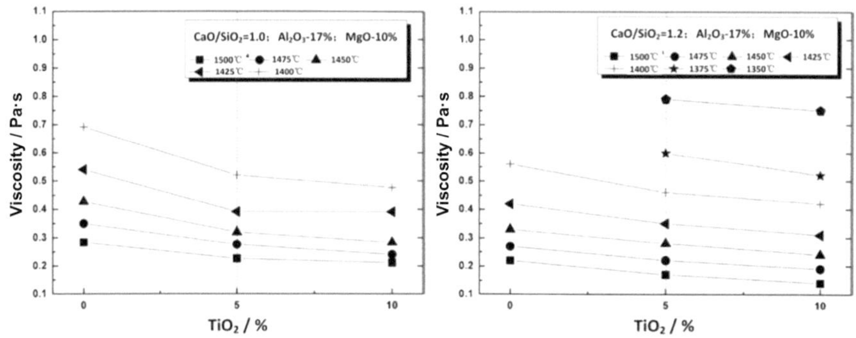

Fig. 2.19 Variation of slag viscosity with TiO_2 content (0–10 wt%)

Regarding the effect of TiO_2 on slag viscosity, current research indicates that a small addition of TiO_2 can reduce slag viscosity, while larger additions have the opposite effect. Under reducing atmosphere conditions, the incorporation of TiO_2 increases the viscosity of the $CaO–SiO_2–Al_2O_3–TiO_2$ slag system. Distinguishing based on TiO_2 content, research by Shankar et al. [94] showed that for low-titanium slag, the viscosity of the $CaO–SiO_2–MgO–Al_2O_3$ tetrameric slag system decreases when 2 wt% TiO_2 is added, within the range of $[(CaO + MgO)/(SiO_2 + Al_2O_3)] = 0.46–0.83$. For medium-titanium slag, Park [95] examined the viscosity of slag systems containing TiO_2 in the range of 0–10 wt% using a viscometer, with results shown in Fig. 2.19. These results indicate that the slag viscosity continues to decrease as the TiO_2 content increases. For medium–high titanium slag, Saito et al. [96] studied the viscosity of $CaO–SiO_2–MgO–Al_2O_3$ slag systems containing 10–20 wt% TiO_2. Their findings revealed that TiO_2 reduces the viscosity of the slag, and the activation energy for viscous flow decreases as the TiO_2 content increases.

Regarding the effect of TiO_2 on the slag structure, the results of Park [95] indicated that the addition of TiO_2 led to the depolymerization of the complex silicate lamellar structure into a simpler silicate configuration. Raman spectroscopy further confirmed that the reduction in viscosity caused by TiO_2 addition primarily affects the dissociation of the silicate and aluminate network structures of the slag. By disrupting the silicate network, TiO_2 addition reduces the slag viscosity.

On the other hand, Nam [97] investigated the properties of low-titanium slag (0–5 wt% TiO_2) from the perspective of cooling crystallization and precipitates, using thermodynamic software (Factsage) for calculations. The experimental results revealed that the slag's liquid phase content increased with the TiO_2 content as the temperature decreased. For four different TiO_2 concentrations, the critical temperature at which viscosity increased sharply occurred when the liquid phase content remained between 40 and 45 wt%. Additionally, the critical temperature was found to be closely related to the magnesium rosmarinite content; when magnesium rosmarinite accounted for 35 wt%, the critical temperature for the viscosity increase was observed. As TiO_2 content increased, both the temperature at which the critical viscosity appeared and the magnesium rosmarinite content decreased.

It can be noted that current research on titanium-containing slags primarily focuses on low and medium titanium slags. For low-titanium slags, due to their minimal impact on viscosity, the focus is more on the phase precipitation behavior during cooling and crystallization. For medium-titanium slags, attention is mostly directed towards changes in viscosity. However, studies on the physical and chemical properties of high-titanium or ultra-high-titanium slags ($TiO_2 > 30$ wt%) remain limited.

Currently, direct reduction processes for sponge iron production, as well as deep reduction and slag-gold separation in electric furnaces, are well-established in industrial production. Tang [98] studied the melt separation of slag and iron phases from pre-reduced high-phosphorus iron ore using a Si–Mo high-temperature resistance furnace. Using iron ore with an 88% pre-reduced degree, Tang varied the time and temperature to determine the conditions for complete separation of slag and iron. The results showed that at 1500 °C, although aggregation of both the metal and slag phases occurred, they could not be fully separated due to factors such as mobility and surface tension. At 1550 °C, the slag and metal phases were clearly separated, although some metal phase remained in the slag. At 1600 °C, the slag phase completely disappeared from the metal phase, and the scanning electron microscope (SEM) microstructure revealed a significant reduction in slag points (Fig. 2.20).

The results shown in Fig. 2.21 demonstrate that the separation of slag and iron melt occurs in four distinct stages. In the first stage (1 min), most of the mineral powder sinters together as the surface begins to melt. After 2 min, the majority of the pores start to disappear, and a volume contraction process occurs, although the initial morphology is still maintained due to the presence of a substantial solid phase. At 3 min, most of the solid phase has transitioned to a liquid phase, forming the initial metal phase. However, numerous pores and slag-phase inclusions remain, with inclusion sizes exceeding 30 μm. By 4 min, the larger pores and inclusions disappear, and the slag-phase inclusions reduce to below 2 μm, resulting in a purer metal phase. Therefore, to achieve complete slag-iron separation, a holding time of 4 min is required after reaching the melt separation temperature.

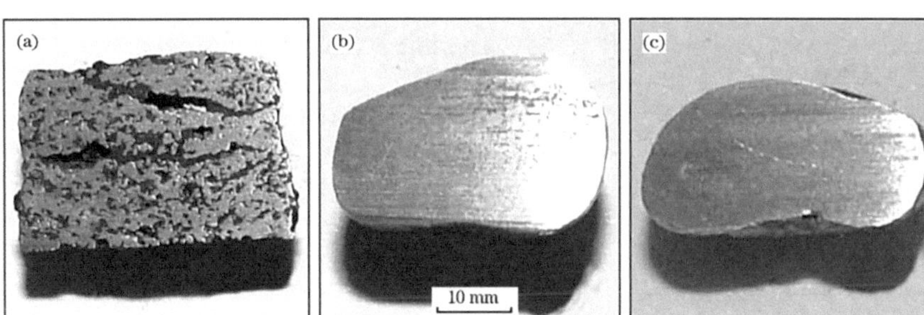

Fig. 2.20 Macroscopic morphology of metal phases after melting and separation for 10 min at different temperatures (**a** 1500 °C; **b** 1550 °C; **c** 1600°C)

Fig. 2.21 Macroscopic morphology of metal phases under different melting time conditions at 1550 °C (**a** 1 min; **b** 2 min; **c** 3 min; **d** 4 min)

In 2014, Park [99] conducted a study on high alumina ores and noted that the slag composition requirements for the direct reduction process are less stringent than those for the blast furnace, as higher Al_2O_3 content adversely affects both sintering and blast furnace smelting. Park focused on the behavior of iron melt separation in pre-reduced high-alumina slag in the presence of graphitic carbon. Figure 2.22 presents in-situ observations from high-temperature confocal electron microscopy, revealing that slag-iron separation during melting can be divided into four stages: (a) the generation of iron and carbon melts at the eutectic temperature, (b) the onset of slag melting, (c) the initiation of slag phase separation from the metal phase, and (d) the completion of slag-iron separation.

Park proposed a different model for the size distribution of iron, charcoal, and slag particles during melt separation compared to that of Ohno [100]. Ohno suggested that the average sizes of iron, charcoal, and slag particles are 150 μm, 25–32 μm, and 25–32 μm, respectively. Under these conditions, the larger iron particles form the backbone of the pellet, with the charcoal and slag particles filling the gaps between them. In contrast, Park found that the average sizes of the particles were 50 μm, 50 μm, and 70 μm for iron, charcoal, and slag, respectively. Based on the densities of these materials—iron (7.6 g/cm^3), charcoal (2.0 g/cm^3), and slag (3.0 g/cm^3)—he concluded that each iron particle would be in contact with six iron particles, two charcoal particles, and three slag particles. In other words, the pellet skeleton is composed of closely packed iron particles in a three-dimensional arrangement, with some iron particles replaced by charcoal and slag particles, as illustrated in Fig. 2.23.

Consequently, Park developed a schematic diagram of the melt separation process, shown in Fig. 2.24. In this diagram: (a) as the slag phase melts, solid iron and charcoal particles move closer together, becoming more compact; (b) the solid iron and charcoal particles are in contact with one another; (c) due to the carburization reaction, liquid iron begins to form; and (d) in the presence of the slag phase, which enhances the wettability of the liquid iron with the charcoal particles, the contact area between liquid iron and charcoal gradually increases.

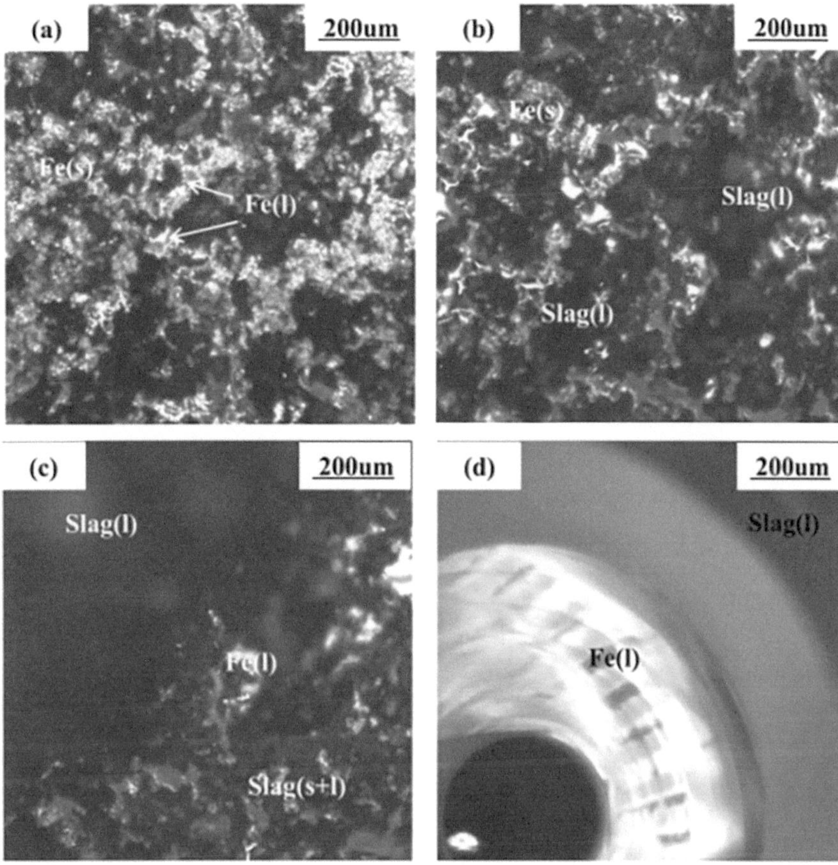

Fig. 2.22 In situ observation of the heating process using high-temperature confocal scanning electron microscopy

2.8 Current Studies on the Utilisation of Titanium-Rich Slag

China is rich in titanium resources, with the world's proven reserves of vanadium-titanium magnetite exceeding 40 billion tonnes. Of this, the reserves in the Panzhihua region alone account for more than 10 billion tonnes, representing approximately 25% of the global total [101]. Currently, during the smelting of vanadium-titanium magnetite in blast furnaces, around 50 wt% of the titanium content is lost to the blast furnace slag. The blast furnace slag stockpiled in the Panzhihua area has surpassed 80 million tonnes and is increasing at a rate of over 3.6 million tonnes per year [102]. Titanium-containing blast furnace slag can be directly utilized as a raw material for products such as concrete, slag cotton, silica-potash fertilizers, and concrete blocks.

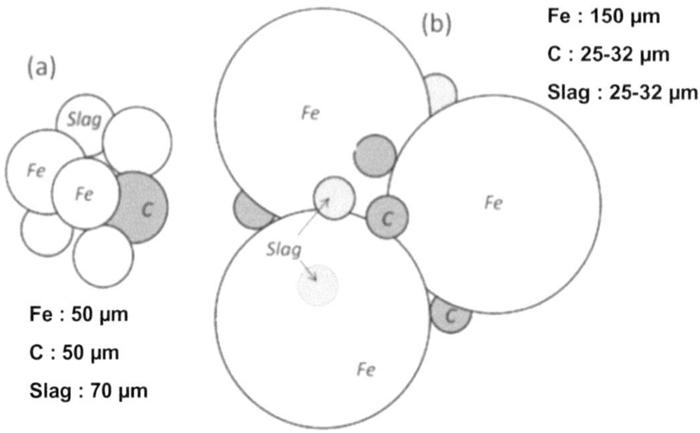

Fig. 2.23 Particle size theory for dual-phase melting and partitioning processes

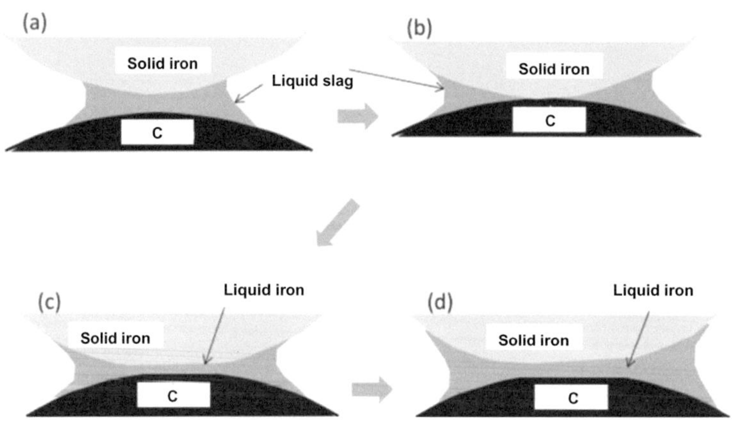

Fig. 2.24 Schematic diagram illustrating the mechanism of the melt separation process

However, despite the simplicity and ease of this method, valuable elements like vanadium and titanium are not effectively utilized, leading to substantial waste.

The oxides in blast furnace slag exist as various silicate minerals, including xanthite, olivine, dicalcium silicate, calcium silicate, wollastonite, and spinel, which are more commonly found in alkaline blast furnace slag. In contrast, acidic blast furnace slag forms different minerals depending on the cooling rate [103]. Some TiO_2 in vanadium-titanium magnetite is closely associated with Fe, which enters the iron ore concentrate after beneficiation. Following blast furnace smelting, the TiO_2 content in the slag exceeds 20 wt%, and the mineral composition of the resulting slag predominantly consists of calcite, titanium-rich crystalline ores, titanium-climbing tremolite, spinel, and titanium carbide [104].

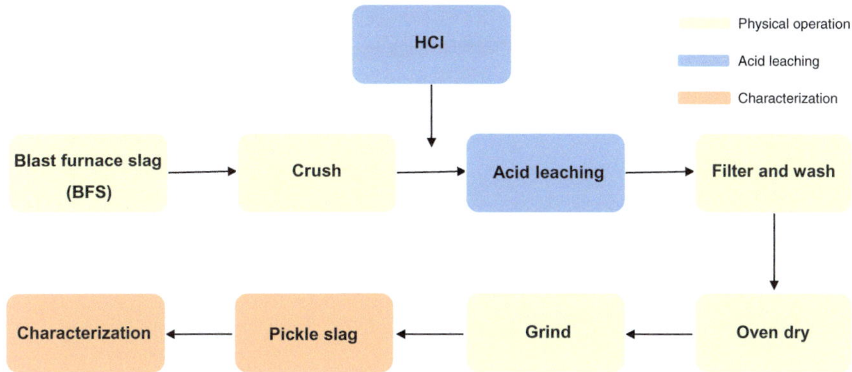

Fig. 2.25 Process flow diagram of hydrochloric acid leaching

For the extraction of titanium from titanium-containing slag, several methods have been developed, including:

Hydrochloric Acid Leaching Method

The hydrochloric acid leaching method involves using hydrochloric acid as a solvent to extract soluble components from solid waste, specifically titanium-containing blast furnace slag, as part of a recycling process. The flowchart of this process is shown in Fig. 2.25.

Research on the physical phase changes during the leaching process has revealed that the phases in the leached material include magnesium–aluminum spinel, titanium-rich minerals, titanium-climbing tremolite, and chalcocite. In the early stage of the leaching process, titanium is primarily extracted from titanium-rich minerals and titanium-climbing tremolite. In the later stages, it is mainly extracted from chalcocite, although some unreacted chalcocite becomes encapsulated by dissolved tremolite and precipitated silicic acid, which slows down the leaching of chalcocite. The microscopic appearance of the leached slag is shown in Fig. 2.26, where the small surface particles in Fig. 2.26c, d are identified as chalcocite [105, 106].

There are several factors influencing the leaching efficiency in the hydrochloric acid leaching process, including raw material particle size, heating temperature, acid-to-slag ratio, hydrochloric acid concentration, and leaching time.

① The leaching behavior of blast furnace slag with particle sizes ranging from 830–2360 μm, 380–830 μm, 180–380 μm, 120–180 μm, and below 120 μm at various temperatures was compared. It was found that when the particle size was in the range of 120–180 μm and the heating temperature was 95 °C, the leaching of TiO_2 and the hydrolysis of $TiOCl_2$ occurred simultaneously. Under these conditions, the leaching slag contained the highest TiO_2 content, with its chemical formula detailed in Eqs. 2.13 to 2.19.

$$CaO + 2HCl = CaCl_2 + 2H_2O \tag{2.13}$$

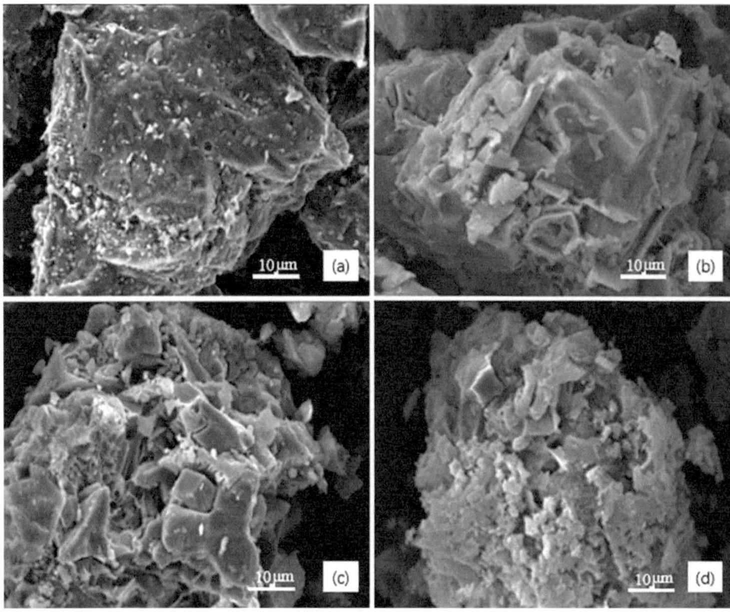

Fig. 2.26 Scanning electron micrographs of blast furnace slag and residue after leaching [105]

$$MgO + 2HCl = MgCl_2 + H_2O \tag{2.14}$$

$$Al_2O_3 + 6HCl = 2AlCl_3 + 3H_2O \tag{2.15}$$

$$Fe_2O_3 + 6HCl = 2FeCl_3 + H_2O \tag{2.16}$$

$$TiO_2 + 2HCl = TiOCl_2 + H_2O \tag{2.17}$$

$$TiOCl_2 + 3H_2O = H_4TiO_4 \downarrow +2HCl \tag{2.18}$$

$$H_4TiO_4 = H_2TiO_3 \downarrow +H_2O \tag{2.19}$$

② Similarly, an acid-to-slag ratio of 1.75:1 resulted in the highest TiO_2 content. However, as the acid amount increased, the acid decomposition rate of TiO_2 also increased, while the presence of excess acid hindered the hydrolysis of $TiOCl_2$. Therefore, to maintain a high TiO_2 content, it is essential to minimize the acid-to-slag ratio. The optimal ratio is found to be 1.5:1.

③ With the increase in hydrochloric acid concentration or leaching time, the TiO_2 content in the leach residue initially increased and then decreased. This

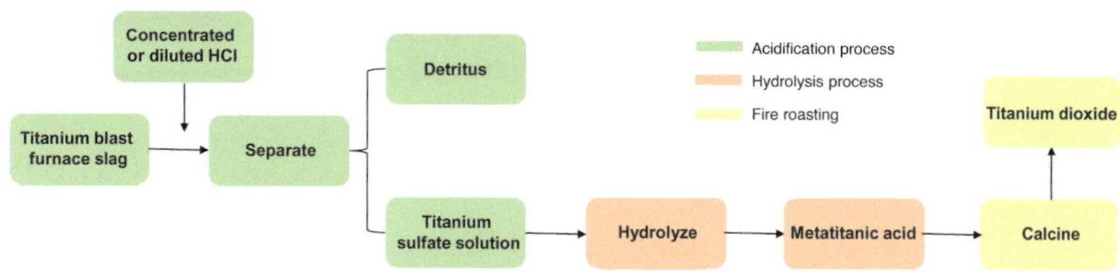

Fig. 2.27 Process flow diagram of sulfuric acid leaching

phenomenon occurs because when the acid concentration exceeds 8 mol/L, the rate of acidolysis of TiO_2 surpasses the rate of hydrolysis of $TiOCl_2$, leading to a decrease in TiO_2 content. Similarly, when the leaching time exceeds 4 h, the Ti in $CaTiO_3$ is also leached, resulting in a reduction of TiO_2 content [106, 107].

Sulphuric acid leaching method

The sulfuric acid leaching method uses sulfuric acid as a solvent to extract soluble components from high-titanium slag. The process flow for this method is shown in Fig. 2.27, and the main reactions involved are detailed in Eqs. 2.20 to 2.23.

Similar to the hydrochloric acid method, the sulfuric acid leaching method also involves factors such as sulfuric acid concentration, liquid-to-solid ratio, and leaching time, which affect leaching efficiency. Yang et al. [108] studied the leaching behavior of blast furnace slag using waste acid and found the following:

The leaching rate increases with higher acid concentration, but the optimal sulfuric acid concentration is found to be between 5 and 10 vol% [109]. Higher acid concentrations can result in excessive leaching of impurities such as aluminum, which can lower the titanium grade.

The titanium leaching rate increases with the raw material's liquid-to-solid ratio. However, when the liquid-to-solid ratio is too high, it can lead to oversaturation of the reaction products in the solution, forming a product film diffusion layer that inhibits the reaction.

The study on the effect of time revealed that the titanium stripping rate decreases with increasing time. This is because reaction products accumulate on the surface of unreacted minerals, hindering contact between reactants and slowing down the reaction.

$$H_2SO_4 + TiO_2 = TiOSO_4 \downarrow + H_2O \tag{2.20}$$

$$H_2SO_4 + CaO = CaSO_4 \downarrow + H_2O \tag{2.21}$$

$$3H_2SO_4 + Al_2O_3 = Al_2(SO_4)_3 + 3H_2O \tag{2.22}$$

$$H_2SO_4 + MgO = MgSO_4 + H_2O \tag{2.23}$$

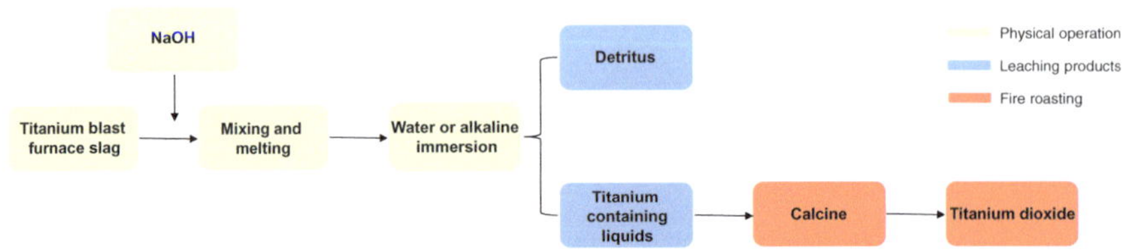

Fig. 2.28 Process flow chart for titanium dioxide preparation via alkali molten salt method

Although the sulfuric acid method for producing titanium dioxide has advantages such as low cost, mature technology, simple equipment, and ease of operation, it comes with drawbacks like high energy consumption, elevated operating costs, and stringent equipment requirements. For instance, the production of 1 ton of titanium dioxide via sulfuric acid leaching generates 6–8 tons of waste acid, which has a negative environmental impact [110]. Nonetheless, this method remains in use, particularly due to its adaptability to raw materials with high calcium and magnesium content, which is typical of titanium-bearing raw materials in China.

Alkali melt salt method

The alkali melt salt method involves the reaction of Ti, Si, and Al oxides in titanium slag with NaOH at high temperatures to generate soluble oxygenated acid salts. This process enables the initial separation of titanium elements. Titanium-containing liquids are obtained through water or alkali leaching, and titanium dioxide is subsequently produced by calcination [111]. The process flow for this method is shown in Fig. 2.28.

Xue [112] and colleagues conducted experiments using the sodium alkali molten salt method to treat high-titanium slag for the preparation of TiO_2. They studied the effects of reaction temperature and the particle size of high-titanium slag on the decomposition rate of the slag, as well as explored the titanium dissolution behavior. The results further confirmed the feasibility of the reaction between high-titanium slag and sodium alkali molten salt, revealing that increasing the temperature or reducing the particle size of the high-titanium slag significantly improved the decomposition rate. However, although this process produces high-grade titanium dioxide, it has several drawbacks: it consumes a large amount of alkali, the cost of sodium salt recovery is high, the process is complex, and improper handling could result in additional environmental pollution.

Titanium tetrachloride as enrichment form

The high-temperature carbonization and low-temperature selective chlorination of $TiCl_4$ technology first involves carbonizing TiO_2 in blast furnace slag to form titanium carbide. This is followed by the selective chlorination of the titanium carbide to produce $TiCl_4$ at low temperatures, utilizing the thermodynamic and kinetic advantages of the titanium carbide chlorination reaction. After chlorination, the slag is

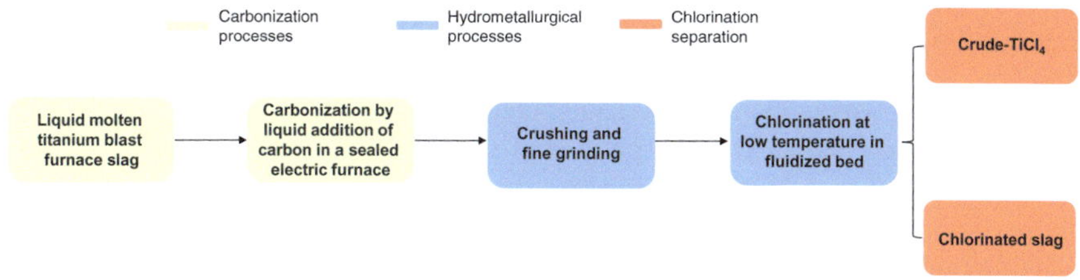

Fig. 2.29 Flow diagram of high-temperature carbonation–low-temperature selective chlorination process

dechlorinated and water-washed, and the resulting material can be used as a raw material for cement production [113]. The process flow is shown in Fig. 2.29.

In this process, carbon and titanium dioxide are initially introduced into an electric furnace for a carbonization reaction to produce titanium carbide and carbon monoxide. The resulting material is then cooled, crushed, and finely ground into granules. These granules are subsequently contacted with chlorine in a fluidized bed for the chlorination reaction, producing crude titanium tetrachloride and chlorinated slag [114]. However, some experimental analyses of TiC chlorination revealed significant chlorination of calcium and magnesium. The primary reason for this is the neglect of the carbon produced in the TiC chlorination reaction. This carbon is highly reactive and readily undergoes carbon-addition chlorination with CaO and MgO. To address this issue, researchers proposed introducing oxygen into the reaction system to facilitate the reaction of the carbon with oxygen to form CO_2. This approach can effectively reduce the chlorination rates of calcium and magnesium, thus increasing the recovery of titanium [115]. While this process is technically feasible, managing the substantial reaction heat generated, addressing local overheating, and controlling the chlorination reaction remain key challenges.

Others

Researchers have discovered that blast furnace slag can be utilized as a photocatalytic degradation agent. Following the pioneering work by Fujishima [116] and colleagues in 1972, which identified the photocatalytic properties of TiO_2, semiconductor materials have gained significant attention in photocatalysis research. Yang [117] observed that titanium-containing blast furnace slag exhibited superior photocatalytic properties in experiments, leading to the development of a new method for treating titanium-containing blast furnace slag and preparing materials with photocatalytic properties. Zhao [118] and others later confirmed that high-titanium slag exhibited a stronger photocatalytic effect on o-nitrophenol, and they demonstrated that titanium-containing blast furnace slag could be used directly as a photocatalyst, eliminating the need for secondary pollution treatment [119].

The carbonitriding-magnetic separation process is a pyrometallurgical method for extracting titanium from blast furnace slag by first carbonitriding the slag and then performing two rounds of magnetic separation. During the first magnetic separation, titanium carbonitride is concentrated around iron beads, which act as carriers, enabling partial separation of the titanium carbonitride. The second round uses strong magnetic separation combined with flotation to separate the remaining titanium carbonitride. However, the process requires fine grinding during the second magnetic separation, which results in high operational costs [120].

Selective enrichment and precipitation methods aim to concentrate the titanium component into a specific enriched phase by altering reaction conditions. The cooling conditions are then controlled to promote the precipitation and growth of the titanium-enriched phase, allowing for selective separation of the titanium-enriched phase from the matrix phase. Common enrichment phases for titanium components include calcite, rutile, and black titanite. While this process offers high capacity for titanium slag, low secondary pollution, and relatively low costs, it suffers from incomplete transformation of titanium-containing minerals, high energy consumption, and additive requirements, leading to lower titanium recovery rates [121, 122].

The carbonation ultra-gravity separation process uses the reducing properties of coke to treat titanium-containing blast furnace slag, with the main products being titanium carbide, yellow feldspar, and spinel. The reduction is followed by ultra-gravity separation technology to isolate the titanium carbide [123].

The electric furnace smelting method is a pyrometallurgical process where a reducing agent, such as anthracite or petroleum coke, is used to reduce the iron oxides in titanium concentrate to metallic iron. This selective iron removal process produces titanium slag with TiO_2 content ranging from 72 to 95 wt%. While this method is effective, it requires a significant initial investment, has a long construction period, and incurs high production costs, especially at low capacity [124, 125].

The reduction rusting method [126, 127] consists of five main processes. Pre-oxidation roasting: The titanium concentrate is subjected to oxidation. Reduction roasting: The titanium concentrate is reduced to metallic iron and TiO_2. Rusting: The reduction products are treated in an aqueous solution containing 0.5 wt% NH_4Cl, with the introduction of oxygen-containing gases to accelerate the rusting process. Separation: The rusted products are rinsed with water. Titanium, the valuable component, remains in the slag, while the iron rusts form a slurry. Calcination: The slag, after the removal of iron rust, is calcined at high temperatures to produce artificial rutile. However, this method is only suitable for processing high-grade ilmenite ore and is not effective for primary ores with high calcium and magnesium content.

The pretreatment-electric furnace combined smelting technology for high-titanium slag is based on electric furnace smelting, with the addition of a rotary kiln pre-reduction process for titanium concentrate. In this setup, high-value iron is reduced to low-value iron in the rotary kiln. After pre-reduction, the high-value iron entering the electric furnace undergoes further reduction, which helps lower the furnace temperature and shortens the smelting time. Additionally, a gas power station is incorporated, allowing the gases produced by both the rotary kiln and the electric furnace to be utilized, thereby improving the overall energy efficiency of the process [128].

2.9 Current Studies on the Titanium–Iron Alloy Smelting Technologies

2.9.1 Properties and Applications of Titanium–Iron Alloy

Ferroalloys are master alloys that contain other elements and are primarily used in the production of cast iron and steel. Ferro-titanium alloy (Fe-Ti) is a type of parent alloy that contains at least 28 wt% titanium. In addition to titanium and iron, it also includes elements such as aluminium, silicon, carbon, phosphorus, sulfur, and manganese. Ferro-titanium, with around 30 wt% titanium, exhibits a silver-grey color, which is prone to oxidation, leading to various colors on its surface due to oxidation.

Ferro-titanium is typically obtained by reducing its corresponding raw material or concentrate. Compared to pure titanium, Fe-Ti offers advantages such as better solubility and a lower cost. The raw materials for Fe-Ti production include ilmenite, leucoxene, chalcocite, and slag concentrates derived from ilmenite [129]. With the increasing availability of titanium scrap, Fe-Ti production is also increasingly reliant on recycled materials.

In iron and steel metallurgy, ferro-titanium alloys are utilized as deoxidizers, degassers, and alloying agents. They help reduce ingot segregation and improve the strength and wear resistance of steel [130, 131]. Moreover, in chemical and energy applications, ferro-titanium serves as an important coating for welding rods and as a hydrogen storage material.

Titanium and iron can form two compounds: $TiFe_2$ and TiFe. TiFe is an unstable compound that only exists in solid alloys. The melting point of TiFe ranges from 1400 to 1500 °C, depending on its chemical composition, and its density is between 5.7 and 6.4 g/cm^3. TiFe is brittle, and TiFe ingots are prone to breaking during cooling. According to the national standard GB3282-87, the common grades and chemical compositions of ferro-titanium are listed in Table 2.5.

Titanium–iron alloy is primarily used as a deoxidizer and alloying agent in the smelting of special steels. It is also employed as an additive in cast iron and welding rods.

Table 2.5 Typical grades and chemical compositions of iron–titanium alloys/wt%

Grades	Ti	Al	Si	P	S	C	Ca	Mg
		≤						
FeTi30-A	25–35	8.0	4.5	0.05	0.03	0.10	0.40	2.5
FeTi30-B	25–35	8.0	5.0	0.06	0.04	0.15	0.40	2.5
FeTi40-A	35–45	9.0	3.0	0.03	0.03	0.10	0.40	2.5
FeTi40-B	35–45	9.0	4.0	0.04	0.04	0.15	0.40	2.5

As a deoxidizer and degassing agent

Titanium exhibits a significantly stronger deoxidizing ability than silicon and manganese. Additionally, its degassing products have a low melting point, making it easier for scattered slag inclusions to polymerize and float to the surface. When titanium is used to deoxidize steel, it reduces segregation in the upper portion of the ingot, thereby improving the quality of the ingot and increasing its yield. Titanium also reacts with nitrogen dissolved in steel to form stable, insoluble iron nitride, enhancing the steel's properties.

As an alloying agent

When used as an alloying agent, the addition of ferro-titanium to steel results in the formation of titanium carbide (TiC) with carbon. This improves the strength of the steel and reduces its susceptibility to intergranular corrosion, particularly in stainless steel, thereby enhancing its corrosion resistance.

Improving the performance of cast iron

High-silicon cast iron, which is predominantly used in the foundry industry, benefits from the addition of titanium. Due to its lower melting point, titanium iron is well-suited for incorporation into cast iron. The addition of titanium aids in the formation of fine-grained graphite during the supercooling process. Furthermore, titanium is sometimes added to high-silicon heat-resistant cast iron to minimize the expansion tendency. The uniformly dispersed titanium carbide and titanium nitride in the softer matrix ensure that the cast iron retains good wear resistance without compromising its machinability.

2.9.2 *Preparation Processes of Titanium–Iron Alloy*

At present, the primary production methods for Titanium–iron alloys include the remelting method, carbon thermal method, electrochemical method, and aluminum thermal method.

Remelting Method

The remelting method primarily utilizes metal iron and metal titanium as raw materials. These materials are melted and recast at high temperatures to obtain Titanium–iron alloys with the required titanium content. Although this method is relatively simple, its application is limited due to the limited availability of scrap titanium, making large-scale production in China unfeasible at present.

Carbon Thermal Method

The carbon thermal reduction method is mainly used to produce high-carbon ferro-titanium. In this process, coke is mixed with titanium-containing raw materials and then subjected to intensive reduction in an electric arc furnace at temperatures exceeding 2000 °C. The chemical reaction involved in the reduction of titanium dioxide with carbon is as follows:

$$TiO_2 + 2C = Ti + 2CO$$
$$\Delta G^{\theta} = 699146 - 342.54T$$

$$TiO_2 + 3C = TiC + 2CO$$
$$\Delta G^{\theta} = 513376 - 328.27T$$

Due to the high carbon content in high-carbon ferro-titanium, it is primarily used as a reducing agent and degassing agent. However, its production and usage remain relatively small at present.

Since the affinity of silicon for oxygen is weaker than that of titanium for oxygen, it is more difficult to reduce TiO_2 with silicon.

The reaction for the reduction of titanium dioxide with silicon is as follows:

$$TiO_2 + Si = Ti + SiO_2$$
$$\Delta G^{\theta} = 8499 + 26.5T$$

If silicon is added during the carbon reduction of TiO_2, titanium and silicon chemically react to form titanium silicides, resulting in silicon-titanium alloys (containing Ti: 20–25 wt%, Si: 20–25 wt%, and C less than 1 wt%). This can reduce the carbon content in the alloy. However, the usage of silicon-titanium alloys is limited, and such alloys are generally not used in smelting processes.

Huang Run, Zhang Jinzhu, and others [132] investigated the physical phase transitions and microstructural changes during the vacuum carbothermal reduction of titanium concentrate at temperatures ranging from 1000 to 1400 °C. Using tools such as Factsage, XRD, and BSE-EDS, they observed that the weight loss of the sample increased with the rise in reduction temperature, indicating that the reduction rate also increased. The main chemical reactions in the carbothermal reduction process are shown in Eqs. 2.24 to 2.30.

$$Fe_3O_4 + C = 3FeO + CO \tag{2.24}$$

$$FeO + C = Fe + CO \tag{2.25}$$

$$FeTiO_3 + FeO = Fe_2TiO_4 \tag{2.26}$$

$$Fe_2TiO_4 + C = Fe + FeTiO_3 + CO \tag{2.27}$$

$$2FeTiO_3 + C = Fe + FeTi_2O_5 + CO \tag{2.28}$$

$$3/5FeTi_2O_5 + C = 3/5Fe + 2/5Ti_3O_5 + CO \tag{2.29}$$

$$2Ti_3O_5 + C = 3Ti_2O_3 + CO \tag{2.30}$$

Liu [133] and colleagues investigated the effects of heating temperature, pressure, and carbon distribution on the reduction process in a thermodynamic simulation of the vacuum carbothermal reduction of Panzhihua titanium concentrate. In their study on the effect of pressure, it was found that as pressure decreased, the total mass of each component volatilized into the gas phase increased, while the content of impurities in the slag phase decreased. This was beneficial for increasing the reduction rate and obtaining a higher purity product. Regarding the effect of heating temperature, it was observed that with constant carbon content, temperatures above 1300 °C led to the volatilization of Mg, SiO, and Mn vapors. When the temperature reached 1700 °C, the content of different components in the gas and slag phases stabilized, with the titanium slag grade reaching up to 88 wt%. Additionally, increasing the carbon distribution resulted in an increase in the reduction rate.

In the study of Fe_2O_3 addition to the carbothermal reduction of titanium-containing blast furnace slag, Kaifei Wang, Guohua Zhang, and others [134] performed thermodynamic calculations using Factsage software, determining the reaction temperature to be 1500 °C. After 4 h at constant temperature, macroscopic analysis of the reduction products, XRD analysis of the slag phase and reduction products, and microstructural analysis of the reduction products were conducted. The results showed that adding Fe_2O_3 powder to the raw material effectively promoted the growth of TiC grains. In a kinetic study of the carbothermal reduction of ilmenite with activated carbon, Gou et al. [135] found that the rate and extent of the reduction reaction increased with temperature, and that, except at 1200 °C, the amount of carbon added had little effect on the reaction rate across different temperatures.

Electrochemical Methods

The electrochemical extraction of titanium metal, known as the FFC-Cambridge process [136, 137], is a versatile technique for the direct conversion of metal oxides to the corresponding metals by cathodic polarization of the oxides in a molten salt electrolyte with calcium chloride as the solute [138]. Titanium–iron alloys can also be extracted from ilmenite using the FFC-Cambridge process [139, 140], and the reduction rate of ilmenite can be enhanced by adding an appropriate amount of calcium oxide to the molten salt electrolyte [141]. Xiong et al. [142] employed the FFC-Cambridge process in the study "The Effect of Calcium Incorporation on the Preparation of Ferrotitanium from Ilmenite Electrochemical Reduction," where ferrotitanium alloys were prepared from ilmenite using a $CaCl_2$-NaCl mixed molten salt as the electrolyte. The study described the duality of the effect of adding CaO on the reduction process: while an appropriate amount of calcium oxide significantly increased the reduction rate of ilmenite, excess CaO hindered the reduction of the intermediate $CaTiO_3$. Subsequently, a new electrochemical method known as the SOM electrochemical method was proposed [143, 144]. This method was later found to reduce preparation costs and energy consumption by shortening electrolysis time and improving electrolysis efficiency [145]. However, prolonged electrolysis time and the production of porous structure products remain significant drawbacks of the electrochemical reduction process, leading the industry to prefer the aluminothermic reduction method for the preparation of Titanium–iron alloys [146].

Aluminothermic Method

Pourabdoli et al. [147] conducted a study on the direct reduction of titanium slag to Titanium–iron alloy. The study involved the aluminothermic reaction of titanium slag and reagents in a high-temperature furnace, investigating the effects of aluminum dosing and the addition of CaO on the composition of the Titanium–iron alloy and the alloy output. The experimental results are shown in Fig. 2.30.

With the increase in Al dosage from 80 to 110 wt%, the yield of Titanium–iron alloy increased from 60 to 66%, while the titanium reduction rate decreased from 57 to 52%. This reduction in titanium recovery is primarily due to the thermal reduction process in the aluminum portion of the reduction, where TiO_2 is converted into TiO and Ti_2O_3. These oxides are more stable and difficult to reduce. TiO, being a strong alkaline compound, readily reacts with Al_2O_3. As the Al dosage increases, the Ti/Al ratio in the generated Titanium–iron alloy decreases. However, with an increase in the amount of CaO, the yield of the Titanium–iron alloy increases significantly. When the CaO/Al_2O_3 ratio reaches 1:1, the yield of the alloy is maximized. Since CaO is more alkaline than TiO, increasing the CaO/Al_2O_3 ratio weakens the reaction between TiO

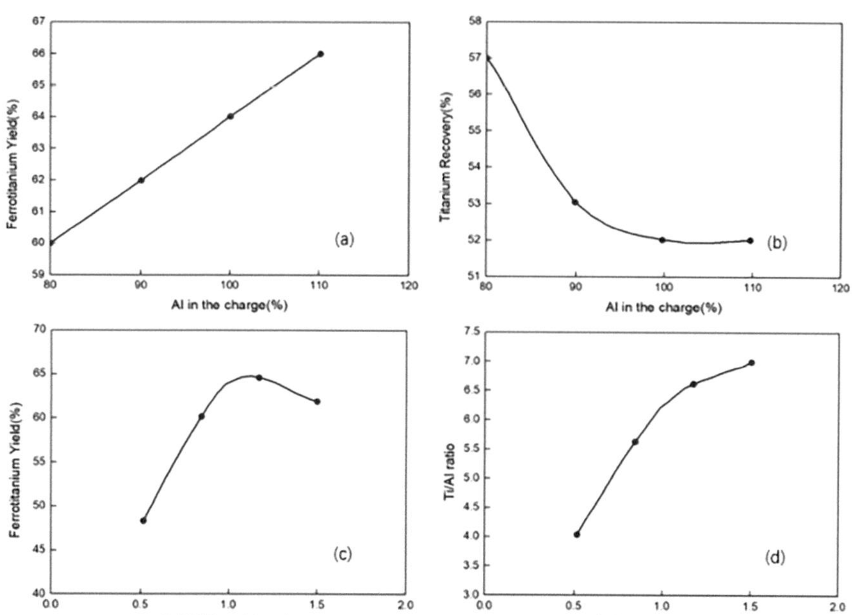

Fig. 2.30 Effect of Al and CaO dosage on the composition of titanium–iron alloy

and Al_2O_3, thereby promoting the reduction reaction. This leads to an increase in the Ti/Al ratio in the alloy. The key chemical equations for the aluminothermic reaction are represented in Eqs. 2.31 to 2.34.

$$3TiO_2 + 4Al = 3Ti + 2Al_2O_3 \quad (2.31)$$

$$Fe_2O_3 + 2Al = 2Fe + Al_2O_3 \quad (2.32)$$

$$3SiO_2 + 4Al = 3Si + 2Al_2O_3 \quad (2.33)$$

$$3FeO + 2Al = 3Fe + Al_2O_3 \quad (2.34)$$

Zhen [148] et al., in their experimental study of the aluminothermal reduction of titanium-containing blast furnace slag, analyzed the influence of reaction temperature, holding time, and CaO addition. They found that the optimal temperature for the aluminothermal reduction reaction was 1600 °C, with the reaction completing after 10 min. The addition of an appropriate amount of CaO reduced the slag viscosity, facilitating the process. Dou et al. [149] employed the aluminothermal self-propagation method to prepare low-oxygen, high-titanium ferroalloys. They studied the thermodynamics of various reaction systems and examined the effects of different aluminum dosages. The results showed that the high-titanium iron alloy primarily consisted of $TiFe_2$, Fe, TiO_2, and Al_2O_3 phases. The microstructure exhibited regions with Al_2O_3 inclusions, Titanium–iron eutectic zones, and titanium matrix zones. The presence of oxide inclusions directly contributed to the high oxygen content and microscopic defects in the alloy. The titanium content in the high-titanium iron alloy ranged from 61.58 to 66.27 wt%, while the iron content varied from 20.53 to 16.15 wt%, and the silicon content ranged from 3.82 to 2.78 wt%. These values conformed to the standards for high-quality high-titanium iron. The oxygen content in the alloy was effectively reduced, with a minimum value of 2.62 wt%, significantly lower than the 12.2 wt% typically achieved by traditional aluminothermal methods.

Zhu [150] and colleagues, in their study on aluminothermal reduction for the preparation of high-titanium iron, performed adiabatic temperature calculations. They concluded that the average value of aluminum reduction was low, making it difficult to maintain energy balance through the heat released by the aluminothermal reaction alone. Thermodynamic calculations indicated that the reduction of TiO_2 and Fe_2O_3 by Al began at approximately 1000 °C, and with the addition of CaO to the TiO_2 and Fe_2O_3 reduction system, the reaction occurred at around 1236 °C.

Moreover, some studies on aluminothermal processes have used potassium chlorate as a heating agent, which increases the consumption of aluminum metal and produces more alumina, resulting in a larger amount of slag. Additionally, the decomposition of potassium chlorate generates oxygen and potassium chloride, which volatilizes with the flue gas at high temperatures, complicating the flue gas treatment process. Although the reaction is a self-oxidation reaction providing heat, the overall cost is high due to the expensive potassium chlorate and metallic aluminum.

The melt preparation of Titanium–iron-silicon alloy from high-titanium blast furnace slag can enhance the transfer processes in the slag, improving the reaction kinetics. This substantially boosts the reduction efficiency, maximizing the recovery of titanium from the blast furnace slag and reducing the TiO_2 content in the residual slag. Efficient separation and extraction of titanium and other valuable metal elements in high-titanium blast furnace slag can be achieved under optimal conditions.

Other methods for preparing Titanium–iron alloys are summarized in Table 2.6.

2.9.3 *Aluminum-Thermal Smelting of Titanium–Iron Alloys*

The raw materials for smelting titanium–iron alloy mainly include: titanium-containing minerals, iron ore, aluminium reducing agent, lime and so on.

Titanium-containing minerals

Common types and properties of titanium-containing minerals are shown in Table 2.7.

China's titanium ores are primarily mined and produced in Guangdong, Guangxi, Hainan, Yunnan, and Panzhihua, Sichuan. The primary product is ilmenite concentrate, with a smaller amount of rutile concentrate. The magnetically separated ilmenite contains the following typical composition: TiO_2 48–51 wt%, TFe 32–38 wt%, FeO < 33 wt%, SiO_2 < 1.5 wt%, P < 0.04 wt%, S < 0.04 wt%, C < 0.04 wt%, and H_2O < 2.0 wt%.

Iron Ore

For the production of ferro-titanium, iron ore is used as an alloy composition adjusting agent. It increases the thermal efficiency of the aluminothermic reaction process and facilitates the adjustment of the iron-to-titanium ratio in the alloy. Typically, iron-rich ores are preferred, with the following composition: TFe < 64 wt%, FeO < 10 wt%, SiO_2 < 7 wt%, P < 0.02 wt%, S < 0.05 wt%, and C < 0.1 wt%.

Aluminium Pellets

Aluminium granules are used in the production of ferro-titanium, and the higher the aluminium content, the better. Typically, A3 grade aluminium is used, which contains more than 98 wt% aluminium, with Fe < 1.1 wt%, Si < 0.02 wt%, and Cu < 0.05 wt%.

Lime

Lime is used as a flux in the furnace charge to improve the fluidity of the slag. Freshly burnt lime, with an effective CaO content > 85 wt%, C > 1.0 wt%, and SiO_2 < 2 wt%, is used in smelting. The processed lime has a particle size of less than 2 mm.

Table 2.6 Summary of iron–titanium preparation results from various studies

Number	Item	Ingredients, methods, content	Literatures
1	Titanium slag	TiO_2:63; FeO:17; Al_2O_3:2.5; MgO:1.2; MnO:2.3; SiO_2:10; CaO:3.5; V_2O_5:0.4; P_2O_5:0.2 (wt%)	[151]
	Reaction method	Ferro-titanium is produced by thermal reduction of titanium slag with aluminum in an AC electroslag furnace	
	Alloying element content	Ti:45–54.5; Fe:22.2–30; Al:6–11.3; Si:9.1–11.5; Mn:1.59–2; C:0.72–0.93; P:0.07–0.08 (wt%)	
2	Rutile and ilmenite	TiO_2:89.2; SiO_2:5.4; Al_2O_3:2.4 TiO_2:50.6; TiFe:32.3; SiO_2:4.1(wt%)	[152]
	Reaction method	The reaction mixture is placed into the closed magnesium oxide lining of a graphite reactor, magnesium powder is placed on the surface of the reactants, and then the magnesium is ignited to induce a self-propagating high-temperature synthesis reaction	
	Alloying element content	Ti:44.87–71.23; Al:1.18–4.43; O0.83–4.37; Fe:18.88–29.01; Si:3.52–24.11(wt%)	
3	High titanium slag	TiO_2:86.03; Fe_2O_3:2.4; Al_2O_3:3.3; SiO_2:5.6; S:0.665(wt%)	[153]
	Reaction method	Reducing agent aluminum powder is mixed with raw materials, the surface is covered with a layer of magnesium powder, and the magnesium powder is ignited to produce a self-propagating reaction	
	Alloying element content	Ti:58.73–60.91; Al:11.29–12.85; Si0.68–0.76; O:3.69–4.39; S:0.49–0.76(wt%)	
4	Blast furnace slag	TiO_2:20.78; Al_2O_3:14.61; MgO:8.57; MnO:0.73; SiO_2:26.48; CaO:26.17; FeO:1.61(wt%)	[154]
	Reaction method	Titanium alloys are prepared from titanium-containing blast furnace slag by the aluminum thermal method in an induction furnace	
	Alloying element content	Ti:38–50; Si:29–34; Al:4–11; Mn:1.6–2.2(wt%)	

Table 2.7 Physical properties of common titanium-bearing minerals

Minerals	Molecular formula	TiO_2 content wt%	Density kg/m^3	Mohs hardness	Color and luster
Chromite	$FeO \cdot TiO_2$	< 53	4560–5240	5–6	Brown or dark brown
Rutile	TiO_2	90–98	4200	6–6.5	Red
Leucite	$TiO_2 \cdot nH_2O$	66–79	3860–4000	5–6	Dark brown to white
Yellow titanium ore	$TiO_2(Al_2O_3) \cdot nH_2O$	80–95	2500–3000	–	Buff
Titanite	$CaO \cdot SiO_2 \cdot TiO_2$	34–42	3400–3600	5–6	Yellow to black
Chalcogenide	$CaO \cdot TiO_2$	58–59	4000	5–6	Unsure

The primary reaction for the preparation of Titanium–iron alloy using the aluminothermic method is as follows:

$$TiO_2 + \frac{4}{3}Al = Ti + \frac{2}{3}Al_2O_3$$
$$\Delta G^\theta = -167472 + 12.1T$$

The reduction of TiO_2 occurs in stages, with some TiO_2 being first reduced to TiO and subsequently to monomeric titanium.

$$2TiO_2 + \frac{4}{3}Al = 2TiO + \frac{2}{3}Al_2O_3$$
$$\Delta G^\theta = -452655 + 14.36T$$

$$2TiO + \frac{4}{3}Al = 2Ti + \frac{2}{3}Al_2O_3$$
$$\Delta G^\theta = -1178588 + 9.92T$$

TiO is strongly basic, and only in a slag containing sufficient CaO does the combination of CaO and Al_2O_3 promote the activity of TiO, facilitating its reduction to Ti, as shown by the following reaction:

$$TiO_2 + \frac{4}{3}Al + \frac{2}{3}CaO = Ti + \frac{2}{3}(CaO \cdot Al_2O_3)$$
$$\Delta G^\theta = -190813 + 12.14T$$

When smelting ferro-titanium using the aluminothermic method, the recovery rate of titanium generally does not exceed 65–75%. Practical experience indicates that the highest titanium recovery rate is achieved when the lime consumption is approximately 20 wt% of the aluminium amount.

The use of aluminium as a reducing agent to produce ferro-titanium in electric furnaces, i.e., the electric-aluminium thermal method, represents the development direction of the aluminothermic method. Compared to the aluminothermic method, refining ferro-titanium in an electric furnace offers the following advantages:

(1) The heat produced by electricity is significantly cheaper than that produced by metal oxidation;
(2) When smelting in an electric furnace, carbon can be used as a reducing agent, and only part of the easily reducible oxides need to be reduced, thus saving a large amount of aluminium;
(3) In an electric furnace, ferro-titanium can be smelted by spreading cheap aluminium chips across the furnace bottom and adding an ignition charge on top. The furnace power ignites the ignition charge, and the resulting slag helps prevent titanium burnout. After the furnace is fully loaded, lime is added, followed by the gradual addition of the basic charge. Once the charge is melted, the refining charge is added. The smelting process, including the furnace sedation (5–10 min), lasts approximately 20 h or more.

Titanium concentrate in the ore portion of the charge can be refined to standard ferro-titanium by replacing 25–100 wt% of calcite concentrate containing 12.5–50 wt% iron ore. The caliche concentrate should be heated to 873–973 °C to prevent the material temperature from falling below 573 °C. The highest economic efficiency is achieved when 25 wt% of ferro-titanium concentrate is replaced, achieving a titanium recovery rate of 77%, with an average titanium content of 27.5–45 wt%. This allows for the production of high-grade ferro-titanium in large quantities. Using calcium-titanium concentrate smelted in an electric furnace, ferro-titanium with a Ti content of 38–45 wt% can be obtained, with a titanium recovery rate of 68–70%.

The Klyuchevsk Ferroalloy Plant of the former Soviet Union employed a two-step melting method to refine ferro-titanium. Calcium-titanium concentrate was added at 50 wt% of the total concentrate before the electric furnace was deactivated. Titanium concentrate was then added and smelted with aluminium. The titanium content of the ferro-titanium refined in this manner ranged from 37.5 to 40.0 wt%, with aluminium consumption of 380 kg/t and a titanium recovery rate of 71%.

In electric furnaces, aluminium powder can also be used to reduce titanium slag to semi-finished products (aluminium-titanium iron) and produce high-quality cement raw materials. Alternatively, ferrosilicon and aluminium powder can be used to reduce semi-finished products to silicon-titanium alloys. After melting titanium iron in the smelting furnace, 70 wt% of the slag is poured into the titanium slag reduction electric furnace. Silica and lime are added, along with aluminium powder for reduction, resulting in silicon-titanium iron and high Al_2O_3 agglomerates.

Furthermore, the former Soviet Union employed induction furnaces for smelting high-titanium iron. This method involves using an induction furnace for remelting metal ferro-titanium scrap and metallic iron. The process is characterized by the use of electric heating slag at the end of the smelting process to maintain a certain temperature, ensuring that metal particles sink completely into the molten metal. This approach increases the recovery of titanium by 8% and reduces aluminium consumption by 70 kg/t.

The composition of industrially produced ferro-titanium is typically: Ti: 25–31 wt%, Si: 3.3–5.0 wt%, P: 0.02–0.03 wt%, with significant amounts of O_2, N_2, and H_2. The average slag content is: TiO_2 11.7–13.3 wt%, SiO_2 < 0.5 wt%, CaO: 10–14 wt%, MgO: 3–4 wt%, FeO: 0.8–2.0 wt%, Al_2O_3: 70–74 wt%, and the slag-iron ratio is 1.3. The production of 1 tonne of ferro-titanium requires 980–1,000 kg of titanium concentrate (containing 42 wt% TiO_2), 420 kg of aluminium, 70 kg of iron ore, 45 kg of titanium metal waste, and 100 kg of lime. The total titanium utilization rate is 66–68%, including burnt losses.

2.10 Distribution of Global Titanium Resources and the Development of Titanium Industry

2.10.1 Global Distribution and Development Trends

Titanium and its alloys are widely used in various industries, including aerospace, petrochemical, shipbuilding, seawater desalination, metallurgy, electric power, biomedicine, sports, and leisure, due to their exceptional properties such as low density, high strength, high specific strength, corrosion resistance, non-magnetic characteristics, and good biocompatibility. The application of titanium and its alloys serve as a key indicator of a country's progress in new materials research and development, as well as its overall national strength [155]. However, the complexity of the titanium smelting process and the high costs involved have limited the widespread adoption of titanium and its alloys. Titanium-containing blast furnace slag, which is rich in titanium resources, presents a promising opportunity for resource extraction. Finding effective methods to extract and utilize the titanium in this slag to develop low-cost, high-performance titanium alloys is a major goal for researchers in this field. The ultimate aim is to enable titanium alloys to penetrate the civilian industrial market, where significant potential exists [156].

Global titanium deposits can be classified into five major categories based on their origin: magmatic deposits, sedimentary deposits, weathering deposits, metamorphic genesis deposits, and metamorphic transformation deposits. These categories can be further divided into 11 subcategories, based on variations in enclosing rocks, protoliths, and metallogenic mechanisms, as shown in Table 2.8 [157]. Currently, the main titanium ores exploited worldwide are ilmenite and rutile, with ilmenite being the predominant type [158, 159]. According to the U.S. Geological Survey, ilmenite accounts for approximately 89% of the global consumption of titanium ore. As of the end of 2017, global titanium ore resources, in terms of TiO_2 equivalent, have surpassed 2 billion tonnes, with reserves amounting to 934 million tonnes. Of these reserves, 873 million tonnes are ilmenite, which represents about 93.47%, as shown in Table 2.9. With the exception of Antarctica, all six other continents are rich in titanium ore resources, spanning 29 countries [160]. Australia holds the largest share of global titanium reserves, followed by China, India, South Africa, Kenya, Brazil, Madagascar, Norway, Canada, Mozambique, Ukraine, the United States, Vietnam, and Sierra Leone [161]. The titanium reserves of these 14 countries account for approximately 97% of the world's total reserves [157].

As shown in Table 2.9, the proven reserves of the top five titanium ore-producing countries account for 78% of the global total. Australia, as the world's largest producer and exporter of titanium, also holds the largest titanium reserves. The country's titanium ores are primarily of the exogenous sedimentary seaside sand type, predominantly found along the east and west coasts of Australia. Rutile deposits on the west coast are located on the Swann Coastal Plain, 1300 km north of Basildon, and on the Scott Coastal Plain, 160 km south of Basildon. These heavy mineral deposits were formed in wave-eroded sills in the underlying Mesozoic bedrock, with the deposits dating back to the Pleistocene epoch due to sea retreat and shoreline elevation. The total amount of heavy mineral deposits in the alluvial sands along the west coast exceeds 59 million tonnes, including significant ilmenite sands.

On the east coast, the heavy mineral deposits extend from the mouth of the Shoalhaven River in New South Wales in the south to Cape Clinton in Queensland in the north, covering a coastline of about 1700 km. These deposits are mainly modern and ancient coastline sands, with some weathered deposits distributed along uplifted seaside sands and coastal sand dunes, as well as sea-intruded sand dune systems just inland from the coast. Most of these deposits were formed in the late Pliocene or early to modern Pleistocene periods. Rutile reserves on the east coast are estimated at about 6.1 million tonnes [157].

Due to the proximity of Australia's ilmenite resources to the coast, approximately 19% of the country's ilmenite and 26% of its rutile resources are unavailable for extraction due to land-use allocations. Mineable deposits are primarily located along the east and west coasts and in the central Murray Basin. In 2017, Australia produced 0.9 million tonnes of ilmenite and 0.45 million tonnes of rutile, with the majority of rutile being exported. Additionally, 51% of the ilmenite was exported, while the remainder was processed to produce synthetic rutile [157].

Japan also boasts a highly developed titanium industry, though it differs from the United States in that it is largely civilian-oriented. Japan's titanium industry includes only two titanium sponge producers: Sumitomo Titanium and Toho Titanium, along with processors such as Kobe Steel, Sumitomo Metals, and Nippon Steel Corporation. Besides the large titanium companies, Japan is home to numerous small, specialized firms focused on developing titanium products for sports, leisure, and medical applications. This vibrant and diverse titanium sector contributes to a wide range of titanium products in Japan [157].

Russia is home to the world's largest titanium sponge plant, Avisma, and the largest titanium fabrication plant, Verkhnyaya Salda. The latter, with its advanced processing capabilities, is a key supplier to Boeing and Airbus. Meanwhile, Kazakhstan and

Table 2.8 Classification scheme of global titanium deposit types [162]

Main category	Subcategory	Typical industrial titanium minerals	Age of mineralization	Typical deposits
Magmatic deposits	Bedrock type	Rutile, ilmenite, vanadium-titanium magnetite	Precambrian, Paleozoic	Tellnes deposit, Norway
	Basic rock type	Calcite, niobium-rutile, niobium-slate titanium ore	From Metazoan to Mesozoic	Powderhorn deposit, Colorado, USA
Hydrodynamic sedimentary deposits	River sand type	Ilmenite, rutile	Cenozoic	Gbangbama deposit, Sierra Leone
	Lakeshore sand type	Ilmenite	Medium—New generation	PortLeyden deposit, New York, USA
	Seaside sand type	Ilmenite, altered ilmenite, rutile	Medium—New generation	RichardsBay deposit, South Africa
Protolithic weathered deposits	Alkaline protolithic	Titanium ore	Cenozoic	Tapira deposit, Brazil
	Basaltic protolithic	Ilmenite	Cenozoic	Roseland deposit, Virginia, USA
	Sandy protolith type	Altered ilmenite	Cenozoic	TrailRidge deposit, Florida, USA
Metamorphic deposits	Titanium-bearing magmatic protoliths	Rutile	Precambrian	Piampaludo deposit, Italy
	Titanium-bearing sedimentary protoliths	Rutile	Precambrian	Evergreen deposit, Colorado, USA
	Titanium ore primary rock type	Rutile	Precambrian, Paleozoic	Dinning deposit, Maryland, USA

Ukraine have minimal domestic demand for titanium and export nearly all of their titanium sponge to the U.S. and Europe. Europe, on the other hand, lacks titanium sponge producers and has limited titanium melting and casting capacity. However, European countries excel in advanced technologies, including superplastic forming, forging, welding, and precision casting for titanium and its alloys [156].

Table 2.9 Global titanium ore reserves by major producing countries (in TiO_2 equivalent) [157]

Country	2017 Production/10^4t			Reserves/10^4t			Global share/%	Mineable life/a	
	Chromite	Rutile	Total	Chromite	Rutile	Alloy		Chromite	Rutile
Australia	90	45	135	25,000	2900	27,900	29.86	277	64
China	80		80	22,000		22,000	23.54	275	
India	20	2	22	8500	740	9240	9.89	425	370
South Africa	130	6.5	136.5	6300	830	7130	7.63	48	127
Kenya	37.5	8	45.5	5400	1300	6700	7.17	144	162
Brazil	5		5	4300		4300	4.60	860	
Madagascar	14		14	4000		4000	4.28	285	
Norway	26		26	3700		3700	3.96	142	
Canada	47.5		47.5	3100		3100	3.32	65	
Mozambique	55	0.7	55.7	1400	88	1488	1.59	25	125
Belarus	35	9	44	590	250	840	0.90	16	27
USA	10		10	200		200	0.21	20	
Vietnam	30		30	160		160	0.17	5	
Sierra Leone		16	16		49	49	0.05		3
Senegal	30	1	31			NA	NA		
Other	9	1.5	105	2600	400	2640	2.83	288	
Total	619	89.7	708.7	87250	6197	93,447	100		

In 2017, global production of titanium ore raw materials, measured in TiO_2 equivalent, reached 7.08 million tonnes. Of this, the top ten suppliers accounted for approximately 79% of the total global supply, with the top six suppliers dominating more than 60% of the market share, as shown in Table 2.10. Among these, Rio Tinto, Iluka, and Tronox control the majority of titanium ore resources and raw material supply in Canada, Australia, and South Africa, respectively. Currently, although there are numerous small- and medium-sized titanium ore production enterprises in China, no Chinese company ranks among the top ten globally. Furthermore, the output and quality of domestic production do not meet the needs of downstream industries, leading to the continued import of large quantities of high-quality titanium raw materials from countries such as Vietnam and Australia.

2.10.2 China's Titanium Resources and Industrial Development

China's titanium resources currently rank first in the world, with a total of 142 titanium deposits distributed across 20 provinces and regions. The main origins are

Table 2.10 Major global suppliers of titanium ore raw materials

Provider	Company headquarters	Mineral products	Production/10^4t	Market share/%
Rio Tinto	UK	Rutile	170	25.0
Iluka	Australia	Titanium concentrate, rutile	110	16.2
Tronox	USA	Titanium concentrate, rutile, titanium slag	75	11.0
Kronos	Norway	Titanium concentrate	40	5.9
Kenmare	Irish	Titanium concentrate, rutile	33	4.9
Cristal mining	Australia	Titanium concentrate, rutile	27	4.0
IREL	India	Titanium concentrate, rutile	25.5	3.8
V. V. Mineral	India	Titanium concentrate, rutile	22	3.2
Irshansky GOK	Belarus	Titanium concentrate	19	2.8
CRL	Australia	Titanium concentrate, rutile	17	2.5

Sichuan, Hebei, Hainan, Hubei, Guangdong, Guangxi, Shanxi, Shandong, Shaanxi, Henan, and other provinces, with reserves and distribution shown in Table 2.11. Ilmenite accounts for 98% of the total titanium resource reserves in China, while rutile makes up only 2%. The country's titanium deposits are relatively diverse, comprising both primary and secondary ores, with primary vanadium-titanium magnetite being the dominant industrial type. In the case of ilmenite-type titanium resources, primary ores account for 97%, while alluvial ores account for 3%. For rutile-type titanium resources, the vast majority consists of low-grade primary ores, which represent 86% of the total rutile reserves in the country, with alluvial ores making up the remaining 14%. While China possesses abundant titanium resources and is the world's largest consumer of titanium, its deposits are primarily composed of vanadium-titanium magnetite, which generally has lower titanium purity compared to high-quality titanium sands found abroad. As a result, China relies heavily on imports of titanium concentrates. Vanadium-titanium magnetite is typically used first for smelting iron and steel, with the tailings subsequently processed into titanium concentrates. Therefore, the output of titanium concentrates is significantly influenced by the iron and steel market [158].

Titanium deposits primarily consist of ilmenite and rutile deposits. Based on their formation, they can be categorized into primary and secondary deposits. Furthermore, according to their genesis, titanium deposits are divided into five main types:

Table 2.11 Titanium resource distribution in major production regions of China [163]

Region	Titanium ore type	Reserves*10^4t	Ratio/%
Sichuan	Vanadium and titanium magnetite	87,349	86.36
He'nan	Rutile mine	5000	4.94
Hainan	Ilmenite sand ore	2556	2.53
Hebei	Vanadium and titanium magnetite	2031	2.0
Yunan	Ilmenite sand ore	1146	1.13
Guangxi	Ilmenite sand ore	708	0.7
	Rutile mine	0.3	
Guangdong	Vanadium and titanium magnetite	1062	1.77
	Ilmenite sand ore	629	
	Rutile mine	91.1	
	High-titanium rutile sand ore	11	
Hubei	Rutile mine	565	0.56

magmatic deposits, volcanic sedimentary deposits, metamorphic deposits, residual accumulation (weathered crust) deposits, and sand deposits. In China, the most significant type of titanium deposit is the late-basaltic and ultra-basal magma crystallization of differentiated and penetrative vanadium-titanium magnetite rock deposits. The second category consists of seashore sedimentary ilmenite and rutile (symbiotic or associated) sand deposits. Thirdly, weathering of titanium-rich geological bodies gives rise to residual ilmenite and rutile sand deposits. The fourth category includes regional metamorphism of titanium-rich basalts or ancient metamorphic rocks, leading to the formation of titanium minerals. Lastly, river alluvial or lakeshore sedimentary ilmenite and rutile sand deposits form the fifth category. Currently, the main industrial types of titanium deposits are classified as magmatic, alluvial, and metamorphic deposits [159].

Although China has relatively abundant titanium resources, most of these resources are primary ores with low titanium content, which primarily exist in the form of rock-mined titanium magnetite. Rutile ores, which are easier to mine and utilize, account for only 2%. The country's titanium ore purification technology is still underdeveloped, with limited equipment capacity. As a result, the titanium concentrates produced typically have low grades, and the demand for high-grade and high-quality titanium ores is largely met through imports. In 2017, China's net imports of titanium concentrates reached 3,065,100 tonnes, marking a 20.91% increase from the previous year. The main import sources were Mozambique, Kenya, Australia, Senegal, and Vietnam. Currently, the domestic titanium industry faces challenges related to the homogenization and overcapacity of low-end and medium-end titanium products. However, the demand for high-end titanium materials has been increasing rapidly, with the chemical industry accounting for the largest share of this demand (43.40%) [158]. This indicates that China's need for titanium resources remains robust, and it is crucial to ensure the security and stability of its resource supply [156].

References

1. Mitchel G. An account of a new semi-metallic substance, called menacane, and its ores. Trans R Irish Acad. 1806:11–30.
2. Thomson TXVI. Chemical analysis of a black sand, from the river dee in Aberdeenshire. Philos Mag. 1810;35(142):98–105.
3. Thomson TVII. Chemical analysis of a black sand, from the River Dee in Aberdeenshire; and of a copper ore, from Arthrey in Stirlingshire. Earth Environ Sci Trans R Soc Edinb. 1812;6:253–68.
4. Weaver TI. On the gold-workings formerly conducted in the county of Wicklow, Ireland. Lond Edinburgh, Dublin Philos Mag J Sci. 1835;7(37):1–6.
5. Weaver T. IV. On Irish tin ore. Lond Edinburgh, Dublin Philos Mag J Sci. 1841;19(121):27–31.
6. Smith A. XXVIII. On Irish tin ore. Lond Edinburgh, Dublin Philos Mag J Sci. 1841;18(115):134–6.
7. Forbes DI. Researches on the mineralogy of South America. Lond Edinburgh Dublin Philos Mag J Sci. 1865;29(193):1–10.
8. Davidson J. 1. Laboratory note—analysis of titaniferous iron sand from North Berwick. Proc R Soc Edinburgh. 1875;8:523–5.
9. Wright C A, Luff A. LX.—Second report to the chemical society on "researches on some points in chemical dynamics". J Chem Soc Trans. 1878;33:504–45.
10. Monckton HW, Herries R. On some bagshot pebble-beds and pebble gravel. Proc Geol Assoc. 1889;11(1):13–23.
11. Jukes-Browne A. The geology of Devizes: with remarks on the grouping of cretaceous deposits. Proc Geol Assoc. 1892;12(6):254–6.
12. Cope T. Note on the titaniferous-iron sand of Porth Dinlleyn. Proc Liverpool Geol Sot. 1902:208–19.
13. Mackenzie GC. The magnetic iron sands of Natashkwan, County of Saguenay, Province of Quebec. Ottawa: Government Printing Bureau; 1912. p. 145.
14. Heskett J. The utilisation of titaniferous iron ore in New Zealand. J Iron Steel Inst. 1920;101:201.
15. Scott A. Hafnium and New Zealand sand. Nature. 1923;111(2792):598–598.
16. Scott A. CV.—Isolation of the oxide of a new element. A correction. J Chem Soc Trans. 1923;123:881–4.
17. Thorpe T. Hafnium and titanium. Nature. 1923;111(2782):252–3.
18. Bishop L. A note on the composition of a Maori compost from Taranaki. J Polynesian Soc. 1924;33(4):317–20.
19. Inouye K. Iron resources of Japan. J Geogr (Chigaku Zasshi). 1924;36(5):7–10.
20. Wylie A. The ironsands of New Zealand. New Zealand: Government Printer; 1937.
21. Gibbs HS. The vanadium content of Taranaki ironsand. Wellington: Victoria University of Wellington; 1937.
22. Kirkaldy J, Wooldridge S. Notes of the geology of the country around Haslemere and Midhurst. Proc Geol Assoc. 1938;49(2):135–47.
23. Hull AJ. An investigation on the extraction of titanium and vanadium present in taranaki ironsand. Melbourne: Victoria University College; 1939.
24. Hutton CO. The titaniferous ironsands of Patea, with an account of the heavy residues in the underlying sedimentary series. Department of Scientific and Industrial Research; 1940.
25. Hutton CO. The ironsands of Fitzroy, New Plymouth. Government Printer; 1945.
26. Hutton CO. Vanadium in the Taranaki titaniferous iron-ores. New Zealand J Sci Tech. 1945;27:15–6.
27. Sudo T. Geology and mineralogy of the tertiary iron-sand deposits in Japan. J Geol Soc Japan. 1942;49(558):335–55.
28. Modriniak N. A preliminary geophysical survey of the Whakatane District. N Z J Sci Technol. 1945;26B:327–31.

29. Mason B. The utilization of New Zealand ironsands as a source of iron, titanium and Vanadium. J Sci Technol. 26:227–38.
30. Ongley P. Possible New Zealand base metal industries. NZ Eng. 1947;2(12):1236.
31. Fleming C. Magnetic ironsand-ores west of Wanganui. NZ J Sci Technol. 1946;27:347–65.
32. Finch J. The Wanganui Wangaehu ironsands. J Sci Technol. 1947.
33. Beck A. Ironsand deposits at Waitara. New Plymouth; 1947.
34. Okada J. Beneficiation test of titaniferous pig iron ore. Bull Res Inst Min Dress Metall Tohoku Univ. 1950;5(2):1–6.
35. Martin WRB. The iron and titanium ores of New Zealand: Occurrence, composition, attempts to use and status today with special reference to ironsands. NZ Eng. 1955;10(10):317.
36. Nicholson D, Cornes J, Martin W. Ilmenite deposits in New Zealand. NZ J Geol Geophys. 1958;1(4):611–6.
37. Nicholson D, Fyfe H. Borehole survey of North Island ironsands from New Plymouth to Kaipara Harbour. NZ J Geol Geophys. 1958;1(4):617–34.
38. McDougall J. Ironsand deposits offshore from the west coast, North Island, New Zealand. NZ J Geol Geophys. 1961;4(3):283–300.
39. Hirst J, Kennedy E. Singatoka iron sands. Government Press; 1962.
40. Maung MT. Beneficiation of ironsands from Raglan, New Zealand. Dunedin: University of Otago; 1962.
41. Ross D. Surface magnetic permeability measurements on some Taranaki ironsand deposits. NZ J Geol Geophys. 1963;6(2):197–208.
42. Wright J. Iron-titanium oxides in some New Zealand ironsands. NZ J Geol Geophys. 1964;7(3):424–44.
43. Wright J, Lovering J. Electron-probe micro-analysis of the iron–titanium oxides in some New Zealand ironsands. Mineral Mag J Mineral Soc. 1965;35(272):604–21.
44. Wright J. Heating experiments on New Zealand ironsands and the presence of pseudobrookite. NZ J Geol Geophys. 1967;10(3):659–65.
45. Wright J. Pseudobrookite in New Zealand ironsands and related notes. NZ J Geol Geophys. 1968;11(2):546.
46. Gow AJ. Petrographic studies of ironsands and associated sediments near Hawera, South Taranaki. NZ J Geol Geophys. 1967;10(3):675–96.
47. Jessop A. Steelmaking from New Zealand ironsands (retroactive coverage). In: Proceedings. 1971. pp. 439–50.
48. Pajalich W. Sand and gravel. Commodity Data Summaries. 1976:146–7.
49. McAdam G, ÓBrien D, Marshall T. Rapid reduction of New Zealand ironsands. Ironmak Steelmaking. 1977;4(1):1–9.
50. Mcdma G, GD M. Instability of titanium-rich ironsands in reducing gases. Ironmakg Steelmakg. 1974;1(3):138–50.
51. McConnell A, Ade G, Cooper R, et al. Symposium on Taharoa ironsands project. NZ Eng. 1974;29(5):129.
52. Christie AB. Sedimentology of pleistocene and recent ironsands, West Coast of the North Island, New Zealand: a study of grain size distributions of some recent ironsands and the sedimentology. Victoria University of Wellington, 1975.
53. Foley Fisher N. The Waipipi ironsands project—aspects of mineralogy and beneficiation. Unpublished MSc thesis, lodged in the Library, University of Otago, New Zealand, 1975.
54. Ingram J, Cooper R. Steelmaking from Ironsands in New Zealand. Metals Australia. 1975;7(1):12–3.
55. Watson JL. Magnetic assessment of Waipipi ironsands. NZ J Sci. 1979;22:87–93.
56. Watson J, Falconer A. The application of magnetic assessment to ironsands production. In: New Zealand conference. 1980. pp. 205–13.
57. Graham IJ, Watson JL. Compositional, structural, and magnetic variations in Waipipi titanomagnetites. NZ J Geol Geophys. 1980;23(4):447–54.
58. Low HF. Mineralogical-beneficiation relationships for Waipipi ironsands. Dunedin: University of Otago; 1981.

59. Carter L. Ironsand in continental shelf sediments off western New Zealand—a synopsis. NZ J Geol Geophys. 1980;23(4):455–68.
60. Falconer AJ. The application of magnetic assessment to ironsands production. Dunedin: University of Otago; 1979.
61. Watson J, Low H. The role of titanomagnetite composites in gravity separation of New Zealand ironsands. In: Process mineralogy II: applications in metallurgy, ceramics and geology. 1982. pp. 135–49.
62. Watson JL. Conceptual model of a drum magnetic separator to beneficiate beach sands. J Metals. 1983:20–40.
63. Smith PN. Metallurgical-operations evaluation for Waipipi ironsands. Dunedin: University of Otago; 1982.
64. Hukkanen E, Walden H. The production of vanadium and steel from titanomagnetites. Int J Miner Process. 1985;15(1–2):89–102.
65. Stokes S, Nelson CS, Healy TR, et al. The Taharoa ironsand deposit. Mineral deposits of New Zealand. Australas Inst Min Metall Monograph. 1989;13:105–9.
66. Lawton DC, Hochstein MP. Geophysical study of the Taharoa ironsand deposit, west coast, North Island, New Zealand. NZ J Geol Geophys. 1993;36(2):141–60.
67. Bates C, Lawrence B. An integrated flat products steel works based on the use of New Zealand ironsands. In: Proceedings of international symposium on exploitation and utilization of vanadium-bearing titanomagnetite. 1989. pp. 73–86.
68. MacKenzie KJ, Berezowski R, Cardile C. The formation of ferrites from spinel ironsands. Mater Sci Forum. 1988:893–7.
69. Mangkusubroto K. Indonesia's ocean resource developments. In: First ISOPE ocean mining symposium. 1995.
70. Pearce CI, Henderson CMB, Telling ND, et al. Fe site occupancy in magnetite-ulvospinel solid solutions: A new approach using X-ray magnetic circular dichroism. Am Mineral. 2010;95(4):425–39.
71. Akimoto S-i. Thermo-magnetic study of ferromagnetic minerals contained in igneous rocks. J Geomagn Geoelectr. 1954;6(1):1–14.
72. Néel L. Some theoretical aspects of rock-magnetism. Adv Phys. 1955;4(14):191–243.
73. Akimoto S-i, Katsura T, Yoshida M. Magnetic properties of $TiFe_2O_4$-Fe_3O_4 system and their change with oxidation. J Geomagn Geoelect. 1957;9(4):165–78.
74. O'reilly W, Banerjee S. Cation distribution in titanomagnetites (1–x) Fe_3O_4-xFe_2TiO_4. Phys Lett. 1965;17(3):237–8.
75. Wechsler BA, Lindsley DH, Prewitt CT. Crystal structure and cation distribution in titanomagnetites ($Fe_{3x}Ti_xO_4$). Am Mineral. 1984;69(7–8):754–70.
76. Kakol Z, Sabol J, Honig J. Cation distribution and magnetic properties of titanomagnetites $Fe_{3x}Ti_xO_4(0\leq x<1)$. Phys Rev B. 1991;43(1):649.
77. Hamdeh H, Barghout K, Ho J, et al. A Mössbauer evaluation of cation distribution in titanomagnetites. J Magn Magn Mater. 1999;191(1–2):72–8.
78. Bosi F, Halenius U, Skogby H. Crystal chemistry of the magnetite-ulvospinel series. Am Mineral. 2009;94(1):181–9.
79. Kasama T, McEnroe SA, Ozaki N, et al. Effects of nanoscale exsolution in hematite–ilmenite on the acquisition of stable natural remanent magnetization. Earth Planet Sci Lett. 2004;224(3–4):461–75.
80. Gehring A, Mastrogiacomo G, Fischer H, et al. Magnetic metastability in natural hemo-ilmenite solid solution ($y\approx0.83$). J Magn Magn Mater. 2008;320(23):3307–12.
81. Wei X, Lu XG, Zou XL, et al. Phase transitions, micro-morphology and its oxidation mechanism in oxidation of ilmenite ($FeTiO_3$) powder. Trans Nonferr Metals Soc China. 2013;23(8):2439–45.
82. Zhang L, Zhang L, Wang M, et al. Dynamic oxidation of the Ti-bearing blast furnace slag. ISIJ Int. 2006;46(3):458–65.
83. McAdam G, Dall R, Marshall T. Direct reduction of New Zealand ironsand concentrates. 1. Gaseous reduction. New Zealand J Sci. 1969;12(4):649.

84. McAdam G, Dall R, Marshall T. Direct reduction of New Zealand ironsand concentrates. 2. Solid reductants. New Zealand J Sci. 1969;12(4):669.
85. Park E, Ostrovski O. Effects of preoxidation of titania-ferrous ore on the ore structure and reduction behavior. ISIJ Int. 2004;44(1):74–81.
86. Park E, Ostrovski O. Reduction of titania-ferrous ore by carbon monoxide. ISIJ Int. 2003;43(9):1316–25.
87. Wang YM, Yuan ZF. Reductive kinetics of the reaction between a natural ilmenite and carbon. Int J Miner Process. 2006;81(3):133–40.
88. Kucukkaragoz CS, Eric RH. Solid state reduction of a natural ilmenite. Miner Eng. 2006;19(3):334–7.
89. Wang YM, Yuan ZF, Guo ZC, et al. Reduction mechanism of natural ilmenite with graphite. Trans Nonferr Metals Soc China. 2008;18(4):962–8.
90. Paunova R. Thermodynamic study of the reduction of titanium magnetite concentrate with solid carbon. Metall Mater Trans B-Proc Metall Mater Proc Sci. 2002;33(4):633–8.
91. Schwebel GL, Filippou D, Hudon G, et al. Experimental comparison of two different ilmenites in fluidized bed and fixed bed chemical-looping combustion. Appl Energy. 2014;113:1902–8.
92. Longbottom RJ, Ostrovski O, Zhang JQ, et al. Stability of cementite formed from hematite and titanomagnetite ore. Metall Mater Trans B-Proc Metall Mater Proc Sci. 2007;38(2):175–84.
93. Park E, Ostrovski O. Reduction of titania-ferrous ore by hydrogen. ISIJ Int. 2004;44(6):999–1005.
94. Shankar A, Gornerup M, Lahiri AK, et al. Experimental investigation of the viscosities in CaO-SiO_2-MgO-Al_2O_3 and CaO-SiO_2-MgO-Al_2O_3-TiO_2 slags. Metall Mater Trans B-Proc Metall Mater Proc Sci. 2007;38(6):911–5.
95. Park H, Park JY, Kim GH, et al. Effect of TiO_2 on the viscosity and slag structure in blast furnace type slags. Steel Res Int. 2012;83(2):150–6.
96. Saito N, Hori N, Nakashima K, et al. Viscosity of blast furnace type slags. Metall Mater Trans B-Proc Metall Mater Proc Sci. 2003;34(5):509–16.
97. Nam SJ, Kang YB, Jung SM, et al. Feasibility of BF hearth protection using spinel formation by slag composition control. ISIJ Int. 2013;53(10):1779–85.
98. Tang HQ, Ma L, Wang JW, et al. Slag/metal separation process of gas-reduced oolitic high-phosphorus iron ore fines. J Iron Steel Res Int. 2014;21(11):1009–15.
99. Park JO, Jeong IH, Jung SM, et al. Metal-slag separation behaviors of pellets consisted of iron, graphite and CaO-Al_2O_3 based slag powders. ISIJ Int. 2014;54(7):1530–8.
100. Ohno T, Sarukawa K, Tokieda K, et al. Morphology of a TiO_2 photocatalyst consisting of anatase and rutile crystalline phases. J Catal. 2001;203(1):82–6.
101. Lu P. Research progress and prospect analysis of comprehensive utilization of Panzhihua blast furnace slag industrialization. Iron Steel Vanadium Titanium. 2013;34(03):33–8.
102. Peng Y. Progress in titanium extraction technology from Panzhihua blast furnace slag. Titan Ind Prog. 2005;03:44–8.
103. Wang XL. Iron and steel metallurgy (ironmaking part). Beijing: Metallurgical Industry Press; 1991.
104. Gao Y, Gui YL, Song CY, et al. Current status and prospects of comprehensive utilization of high-titanium blast furnace slag. Multipurpose Util Min Resources. 2019;01:6–10.
105. Xiong Y, Li C, Liang B, et al. Hydrochloric acid leaching of naturally cooled titanium-bearing blast furnace slag. Chinese J Nonferr Metals. 2008;03:557–63.
106. Jia F, Liu R, Yang M. Direct hydrochloric acid leaching of high-titanium blast furnace slag. J Nanjing Normal Univ (Eng Technol Ed). 2012;12(02):43–9.
107. Wang ZJ, Zhang LH, Wang HB, et al. Preparation of synthetic rutile by direct hydrochloric acid leaching of Panxi ilmenite under atmospheric pressure. Nonferrous Metals. 2007;04:108–11.
108. Yang DJ, Liu DJ, Xu CH. Study on leaching Panzhihua blast furnace slag with titanium white waste acid. Sichuan Chem Industry. 2007;02:44–6.
109. Zhong BN, Xue TY, Hu GP, et al. Enrichment of low-grade reduced titanium slag by phosphoric acid activation roasting-acid leaching. Chin J Process Eng. 2013;13(03):378–84.

110. Liu J. Research progress on preparation of titanium-rich raw materials for fluidized chlorination from Panzhihua titanium resources. China Nonferr Metall. 2018;47(06):49–53.
111. Wang Q, Zhang QW. Preparation of rutile TiO_2 by molten salt method from titanium-rich slag. Appl Chem Industry. 2008;(02):140–141+145.
112. Xue TY, Qi T, Wang LN, et al. Basic research on preparation of TiO_2 from high-titanium slag by sodium alkali molten salt method. In: Proceedings of the 2008 national conference on metallurgical physical chemistry, Guiyang, Guizhou, China; 2008. p. 5.
113. Li XH, Pu JT. Latest progress in comprehensive utilization of high-titanium blast furnace slag in Panzhihua. Iron Steel Vanadium Titanium. 2011;32(02):10–4.
114. Zhang RL. Method for producing titanium tetrachloride from titanium-bearing blast furnace slag: China, CN87107488.5. 1987–10–27.
115. Peng Y. Thermodynamic analysis of low-temperature selective chlorination of carbonized Panzhihua blast furnace slag. Titan Ind Prog. 2005;06:45–9.
116. Fujishima A, Honda K. Electrochemical photolysis of water at a semiconductor electrode. Nature. 1972;238(5358):37–8.
117. Yang H, Xue XX, Zuo L, et al. Photocatalytic degradation of methylene blue by titanium-bearing blast furnace slag catalyst. Chin J Process Eng. 2004;03:265–8.
118. Zhao N, Yang H, Xue XX, et al. Experimental study on degradation of o-nitrophenol by high-titanium slag as photocatalytic material. J Chinese Ceramic Soc. 2005;02:202–5.
119. Qiu ST, Zhang MB, Li JX, et al. Current status and prospects of comprehensive utilization of titanium-bearing blast furnace slag resources. Iron & Steel. 2016;51(07):1–8.
120. Yi XX, Li YW, Yang DB. Study on titanium extraction from Panzhihua titanium-bearing blast furnace slag after carbonization and nitridation by magnetic separation. Min Metall. 2008;03:46–9.
121. Sui ZT, Guo ZZ, Zhang L, et al. Green separation technology of titanium components from titanium-bearing blast furnace slag. J Mater Metall. 2006;02:93–7.
122. Xu RZ, Zhang JL, Chang ZY, et al. Research progress on selective enrichment and precipitation of titanium components in high-titanium blast furnace slag. Iron Steel Vanadium Titanium. 2017;38(06):6–12.
123. Gao QR, Song B, Yang ZB, et al. Study on carbonization of titanium-bearing blast furnace slag and supergravity separation of titanium carbide. Nonferr Metals Sci Eng. 2017;8(02):1–7.
124. Yu DQ. Research and practice on electric furnace smelting recovery of titanium-containing waste slag in titanium production. Light Metals. 2001;08:45–7.
125. Yang RM. Discussion on design optimization of titanium slag electric furnace smelting. Nonferr Metals Des. 2009;36(04):58–61.
126. Li TL, Li WC, Yang YX, et al. Ten years of practice in producing synthetic rutile by reduction-rusting method. Rare Metals and Cemented Carbides. 1992;04:46–51.
127. Hu KJ, Xi G, Yao J, et al. Current status of synthetic rutile production by reduction-rusting method and feasibility analysis of adopting this process in Panzhihua. Titan Ind Prog. 2006;04:17–22.
128. Wang XD, Zheng RN, Wang Q. New process of pretreatment-electric furnace combined smelting of high-titanium slag. Nonferr Metals Des. 2018;45(02):82–5.
129. Panigrahi M, Shibata E, Iizuka A, et al. Production of Fe–Ti alloy from mixed ilmenite and titanium dioxide by direct electrochemical reduction in molten calcium chloride. Electrochimica Acta. 93:143–51.
130. Konovalov KN, Krasnoryadtsev NN, Pashchenko VE, et al. Melting stainless steel with different brands of ferrotitanium. Metallurgist. 1978;21(8).
131. Shutkin NI, Goncharenko MS. The smelting of titanium-containing stainless steel without using ferrotitanium. 1959;3(10):424–5.
132. Huang R, Lyu XD, Wu QH, et al. Vacuum carbothermal solid-state reduction of Panzhihua titanium concentrate. J Chongqing Univ. 2019;42(04):56–62.
133. Liu C, Huang R, Yuan Y, et al. Thermodynamic simulation study on vacuum carbothermal reduction of Panzhihua titanium concentrate. Iron Steel Vanadium Titanium. 2015;36(04):13–8.

134. Wang KF, Zhang GH, Wang L, et al. Effect of Fe_2O_3 addition on carbothermal reduction of titanium-bearing blast furnace slag. Chin J Process Eng. 2018;18(06):1276–82.
135. Gou HP, Zhang GH, Hu XJ, et al. Kinetic study on carbothermic reduction of ilmenite with activated carbon. Trans Nonferr Metals Soc China. 2017;27(08):1856–61.
136. Schwandt C, Doughty GR, Fray DJ. The FFC-Cambridge process for titanium metal winning. Key Eng Mater. 436:13–25.
137. Hussain P, Bhagat R, Dashwood R. Precursor preparation for Ti-Al-V-Y alloy via FFC cambridge process. Appl Mech Mater. 330:62–7.
138. Mohanty J. Electrolytic reduction of titania slag in molten calcium chloride bath. JOM. 64(5):582–4.
139. Qi C, et al. Preparation of ferrotitanium alloy from ilmenite by electrochemical reduction in chloride molten salts. JOM. 2016;68(2):668–74.
140. Ma M, Wang DH, Hu XH, et al. A direct electrochemical route from ilmenite to hydrogen-storage ferrotitanium alloys. Chem-A Eur J. 2010;12(19):5075–81.
141. Ersin K. Hysteretic features of Ising-type segmented nanostructure with alternating magnetic wires. J Alloy Compd. 2016;676:337–46.
142. Li X, Hua Y, Xu C, et al. Effect of CaO addition on preparation of ferrotitanium from ilmenite by electrochemical reduction in $CaCl_2$-NaCl molten salt. Nonferr Metals. 2016;676:383–9.
143. Suput M, Delucas R, Pati S, et al. Solid oxide membrane technology for environmentally sound production of titanium. Min Process Extract Metall Trans. 2008;117(2):118–22.
144. Ye XS, Lu XG, Li CH, et al. Preparation of Ti–Fe based hydrogen storage alloy by SOM method. Int J Hydrogen Energy. 2010;36(7):4573–9.
145. Zou XL, Lu XG, Zhou ZF, et al. Environmentally friendly electrochemical method for the direct extraction of titanium from high-titanium slag. Adv Mater Res. 361:655–9.
146. Wang Z, Zhang J, Liu Z, et al. Production of ferrotitanium alloy from titania slag based on aluminothermic reduction. J Alloy Compd. 2019;810: 151969.
147. Gao ZX, Cheng GJ, Yang H, et al. Preparation of ferrotitanium using ilmenite with different reduction degrees. Metals 2019;9(9).
148. Zhen YL, Zhang GH, Zhou GZ. Experimental study on aluminothermic reduction of titanium-bearing blast furnace slag. Iron Steel Vanadium Titanium. 2014;35(05):40–4.
149. Dou ZH, Zhang TA, Zhang HB, et al. Preparation of low-oxygen ferrotitanium alloy by aluminothermic self-propagating method. J Central South Univ (Sci Technol). 2012;43(06):2108–13.
150. Niu LP, Zhang TA, Zhang HB, et al. Thermodynamics and kinetics of aluminothermic reduction for preparing ferrotitanium. Chinese J Nonferr Metals. 2010;20(S1):425–8.
151. Pourabdoli M, Raygan S, Abdizadeh H, et al. A new process for the production of ferrotitanium from titania slag. Can Metall Q. 2013;46(1):17–23.
152. Dou ZH, Wang C, Fan SG, et al. Al control in high titanium ferro with low oxygen prepared by thermite reaction. In: 6th International symposium on high-temperature metallurgical processing, 2016. pp 11–17.
153. Cheng C, Dou ZH, Zhang TA, et al. Sulfur distribution in preparation of high titanium ferroalloy by thermite method with different CaO additions. Rare Met. 2019;38(08):793–9.
154. Huang R, Bai C, Lv X, et al. Preparation of titanium alloy from titania-bearing blast furnace slag. In: 3rd International symposium on high-temperature metallurgical processing. Hoboken, NJ, USA: John Wiley & Sons, Inc.; 2012. pp. 75–83.
155. Zhou L, Zhao YQ, Wang XD. Strategic research on the development of titanium alloy materials and applications in China. Beijing: Chemical Industry Press; 2012.
156. Sun LF. Fundamental research on the preparation of Ti-Si-(Al) alloy by aluminothermic reduction of titanium-bearing blast furnace slag. Kunming: Kunming University of Science and Technology; 2017.
157. Sun RB, Wang QS, Yuan CH, et al. Analysis of the global titanium resource situation. China Min Mag. 2019;28(06):1–6+12
158. Wu X, Zhang J. Distribution and characteristics of titanium resources in China. Titan Ind Prog. 2006;06:8–12.

159. Wang LP, Wang H, Gao Q, et al. Distribution and production status of titanium resources in China. Rare Met. 2004;01:265–7.
160. Zhang FL, Liu SW, Hu YD, et al. Current status and future development suggestions for the titanium industry in China. Modern Min. 2015;31(04):1–4.
161. Liu YQ, Ding H, Xie D. Titanium resources and environmental effects of sponge titanium processing in China. Earth Sci Front. 2014;21(05):281–93.
162. Zhao YM. Types, distribution, and main geological characteristics of rutile deposits. Min Deposits. 2008;04:520–30.
163. Sun YZ, Zhang Q. Progress and future work on chloride-process titanium dioxide production technology in China. Titan Ind Prog. 2001;03:6–10.

Chapter 3
Mineralogical Properties of Ironsands

Ironsands, also known as titanomagnetite based on its primary mineral phase, is a titanium-rich iron ore typically formed through rapid cooling of volcanic lava. These ores are predominantly found in coastal regions such as New Zealand, Indonesia, and South Africa, where they have undergone prolonged weathering, marine abrasion, and atmospheric oxidation. This process results in the formation of dense, rounded, and nearly spherical black sand particles [1]. Compared with conventional iron ores, ironsands possess distinct advantages such as abundant reserves and ease of exploitation. However, their low grade and poor reducibility necessitate comprehensive mineralogical characterization to support efficient recovery of iron and titanium values [2–7].

3.1 Room-Temperature Characterization of Ironsands

3.1.1 Chemical Composition

The ironsands studied in this book were sourced from a coastal mine on East Java Island, Indonesia. The chemical composition of the ironsands was determined through chemical synthesis titration, and the results are presented in Table 3.1.

From Table 3.1, it can be seen that the total iron grade of the ironsands exceeds 55 wt%, meeting the standard for second-grade iron ore for ironmaking (first-grade > 58 wt%; second-grade > 55 wt%). The mass fractions of Fe^{3+} and Fe^{2+} ions were calculated to be 32.61 wt% and 23.02 wt%, respectively, indicating that the iron oxides primarily exist in the form of magnetite. The TiO_2 content is 11.41 wt%, which is slightly lower than the TiO_2 content of vanadium and titanium concentrates from the Panxi region of China (12.88 wt%). The combined mass fraction of calcium, silicon, magnesium, and aluminium oxides is 11.85 wt%, and the ratio of (CaO + MgO) to (SiO_2 + Al_2O_3) is 0.58, classifying the ores as semi-self-melting acidic

Z. Wang et al., *Smelting and Utilization Technology of Ironsands*,
https://doi.org/10.1007/978-981-95-2868-4_3

Table 3.1 Chemical composition of ironsands/wt%

TFe	FeO	TiO_2	SiO_2	MgO	Al_2O_3	CaO	MnO	V_2O_5
55.63	29.60	11.41	4.13	3.74	3.38	0.60	0.50	0.48
NaO	K_2O	PbO	ZnO	C	P	S	Cl	Cauterize
0.020	0.017	0.001	0.005	0.10	0.031	0.013	0.009	3.14

ores. The vanadium oxide content is 0.48 wt%, indicating a relatively low effective use value.

The harmful elements potassium, sodium, and zinc are present in low concentrations. When projected based on an ore ratio of 1:1.4, the alkali load per tonne of iron in the ironsands is 518 g, and the zinc load is 70 g, both of which are below the industry standards for furnaces ($Na_2O + K_2O < 3$ kg; ZnO < 150 g). The sulphur and phosphorus contents are 0.013 wt% and 0.031 wt%, respectively, both of which meet the standard for first-grade ore (sulphur < 0.1 wt%; phosphorus < 0.1 wt%).

In conclusion, it can be observed that the ironsands contain low levels of harmful elements, and its iron and titanium components are valuable for industrial applications.

3.1.2 *Physical Properties*

In order to further understand the physical composition of ironsands, the raw ironsands was analysed by X-ray diffraction, and the results are shown in Fig. 3.1.

As shown in Fig. 3.1:

(1) The primary physical phase in the ironsands used in this study is titanomagnetite, abbreviated as TTM, which is a solid solution system of magnetite and titanomagnetite spinel. It can be represented as $(1 - x)Fe_3O_4 - xFe_2TiO_4$ or

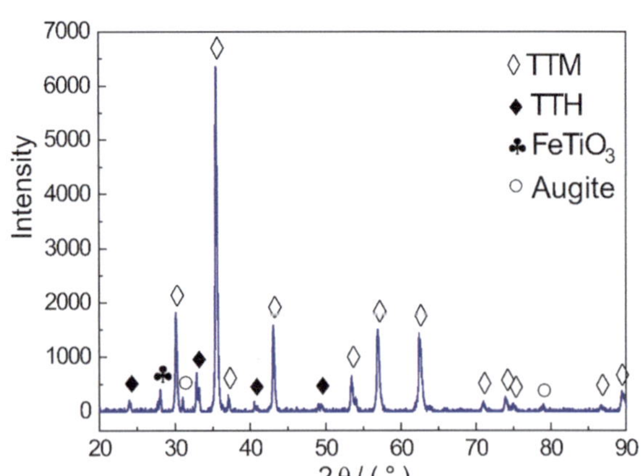

Fig. 3.1 X-ray diffraction pattern of ironsands

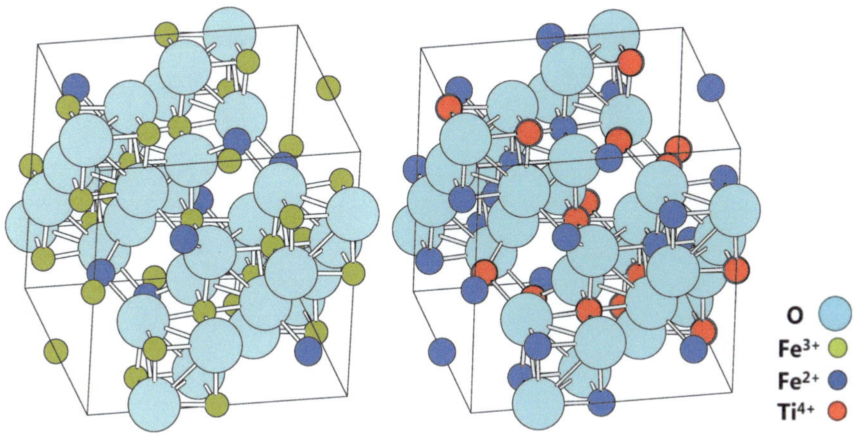

Fig. 3.2 Schematic crystal structure of the two solid solution components in TTM (magnetite and titanomagnetite spinel)

$Fe_{3-x}Ti_xO_4$, where x indicates the solid solubility of titanomagnetite spinel. The crystal structure of titanomagnetite exhibits a cubic crystal system containing 32 oxygen atoms in one unit cell, with the general formula $A_8B_{16}O_{32}$, where A and B represent tetrahedral and octahedral vacancies, respectively [7–16]. The formation of TTM is believed to result from the replacement of Fe^{3+} by Ti^{4+} in the B vacancies of magnetite (Fe_3O_4). Simultaneously, to maintain valence equilibrium, Fe^{3+} in the A vacancies is reduced to Fe^{2+}. The substitution process of these ions is complex, and using X-ray magnetic circular dichroism, Carolyn et al. [3] proposed a new model for the formation of TTM, which will not be repeated here. The crystal structures of its two solid solution components are depicted in Fig. 3.2.

(2) The secondary phase in the ironsands is titanomagnetite, abbreviated as TTH, which is a solid solution system of hematite and ilmenite. It can be represented as $(1-y)Fe_2O_3 - yFeTiO_3$ or $Fe_{2-y}Ti_yO_3$, where y indicates the solid solubility of ilmenite. The crystal structure of TTH is a triclinic crystal system, with $[FeO_6]$ octahedra and $[TiO_6]$ octahedra connected in a co-prismatic form based on a specific solid-solution ratio. Both titanium and iron are six-coordinate, while oxygen is coordinated in four-fold geometry. TTH initially precipitated due to the oversaturation of elemental titanium during the rapid cooling of volcanic lava. The subsequent development of TTM occurred through oxidation by dissolved oxygen in the atmosphere and water over an extended period, resulting in the formation of the stable TTH phase at room temperature and pressure. The crystal structure of TTH is shown in Fig. 3.3.

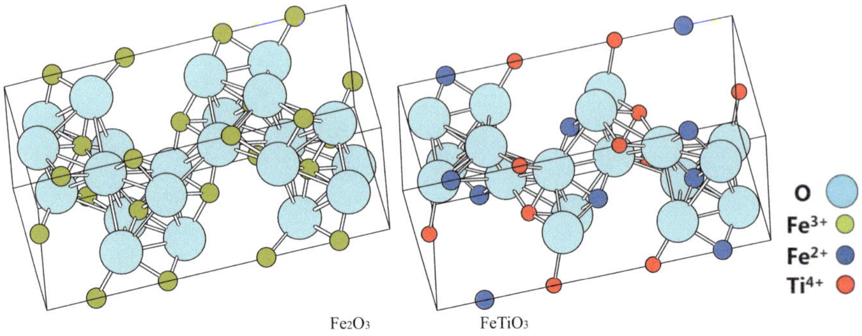

Fig. 3.3 Schematic crystal structure of the two solid solution components in TTH (hematite and ilmenite)

(3) Characteristic peaks of ilmenite appear in the range of 25–30 degrees, indicating that the raw ironsands also contain a small amount of pure ilmenite. Additionally, oxides of silicon, magnesium, aluminium, and calcium are predominantly present in the ironsands in the form of chain pyroxene.

3.1.3 Particle Size Distribution

The laser particle size analyzer (LPSA) was employed to measure the particle size distribution of the ironsands, and the results are presented in Fig. 3.4. From Fig. 3.4, it is evident that the particles of the ironsands are distributed within the range of 90–420 μm, with a more concentrated distribution between 130 and 250 μm. This range constitutes 69.12 wt% of the total ore particles, and it is neither within the ideal size range for nuclear particle formation nor suitable for the sintering process. In this particle size range, for sintering purposes, the ore particles are not within the optimal nuclear particle size nor suitable for adhesion, which means that if the ironsands are not pre-treated and are directly used as an ingredient in the sintering process, they will significantly affect the ore granularity mixing and sintering air permeability. For the pelletizing process, it is essential that the mineral particles below 76 μm comprise over 70–80 wt%, indicating that the raw ironsands must be milled before they can be used as an ingredient in the pelletizing process.

In conclusion, ironsands can be directly added to the sintering process without particle size grinding, but the raw ore size does not fall within the optimal range. Therefore, it can only be added in small quantities. However, when used as a pelletizing ingredient, the ironsands must undergo grinding and refining to reach the desired pelletizing size range.

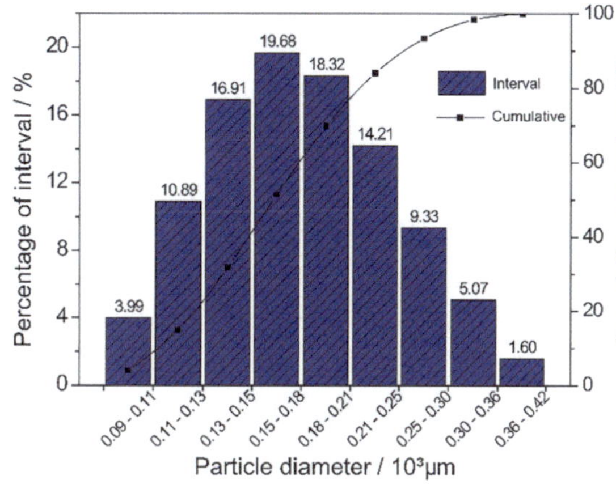

Fig. 3.4 Particle size distribution of raw ironsands particles

3.1.4 Microscopic Morphology and Structural Analysis

Through scanning electron microscopy (SEM) and energy dispersive spectroscopy (EDS), the micro-morphology of the raw ironsands can be obtained, and the results are shown in Fig. 3.5.

Through SEM and EDS analyses of the raw ironsands, two main forms of microscopic morphology were observed. The first type consists of smooth, homogeneous particles with a relatively even distribution of elements. The primary phase of these particles is titanomagnetite (TTM), as shown in particle A in Fig. 3.5. The second type consists of non-homogeneous particles, with laminar phases precipitated at specific angles on a substrate, where the substrate remains TTM, and the laminar phases correspond to titanium hematite (TTH), as shown in particle B in Fig. 3.5 and the locally enlarged C-plot. To reduce interfacial energy during crystal growth, crystals preferentially grow in specific directions, a phenomenon known as the anisotropy of

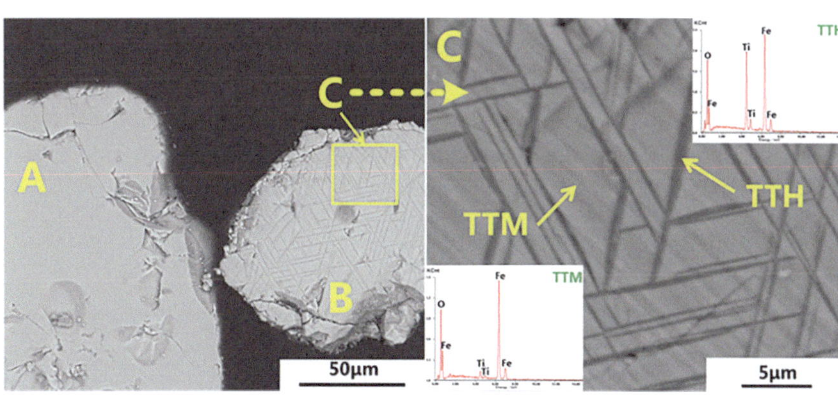

Fig. 3.5 Micro-morphology and energy dispersive spectroscopy of raw ironsands

crystal growth. The lamellar TTH shows a regular growth direction, with the angles between the intersecting lamellae close to 60° and 120°, indicating that the gradual oxidation of the ironsands leads to the melting and growth of TTH along specific crystal faces on the TTM substrate.

From the EDS results, the iron and titanium elements are evenly distributed in the A particles, while in the B particles, the lamellar TTH phases that have melted from the substrate exhibit higher titanium content and lower iron content compared to the basal TTM phases. Quantitative EDS analyses of these three different phases (homogeneous particles, non-homogeneous particles substrate, and non-homogeneous particles with lamellar structure) yielded the atomic percentages of iron, titanium, and oxygen, from which the solid solubility x and y were calculated. The results are shown in Table 3.2.

As shown in Table 3.2, the atomic percentages of titanium and oxygen in the lamellar structure of non-homogeneous particles are significantly higher than in other parts, which is primarily due to the prolonged oxidation caused by dissolved oxygen in the atmosphere and water after the formation of the ironsands. This leads to the melting and further growth of the lamellar TTH, resulting in the enrichment of titanium and oxygen elements from the basal part of the non-homogeneous particles. As a result, the lamellar content of non-homogeneous particles is notably higher than in other areas. Moreover, comparing the basal portion of the homogeneous ironsands particles with that of the non-homogeneous particles, although both are composed of TTM, they show different solid solubilities of magnetite and ilmenite spinel, as well as varying elemental contents of iron, titanium, and oxygen. This difference is also attributed to the phase transformation and elemental enrichment during the oxidation process, leading to lower titanium content in the basal part of the non-homogeneous particles, which corresponds to a lower ilmenite solid solubility and x value.

In summary, due to factors such as cooling rate, crystal solubility, and the atmospheric oxidation environment, the ironsands minerals primarily consist of the following solid solution phase systems:

(1) Magnetite-titanium-iron spinel solid solution system (Fe_3O_4–Fe_2TiO_4), known as titanomagnetite (TTM);
(2) Hematite-ilmenite solid solution system (Fe_2O_3–$FeTiO_3$), referred to as titanium hematite (TTH).

Additionally, a pseudobrookite-subpseudobrookite solid solution system (Fe_2TiO_5–$FeTi_2O_5$), known as pseudobrookite (PSB), exists, which represents a higher oxidation state compared to TTM and TTH. This solid solution system was

Table 3.2 Atomic percentages of the three distinct phases in raw ironsands particles

Physical state	Fe/at.%	Ti/at.%	O/at.%	x or y
Homogenized particles	43.02	4.07	44.10	$x \approx 0.26$
Non-homogeneous granular substrate	49.94	2.03	45.10	$x \approx 0.12$
Layered structure of inhomogeneous particles	31.92	14.36	50.73	$y \approx 0.62$

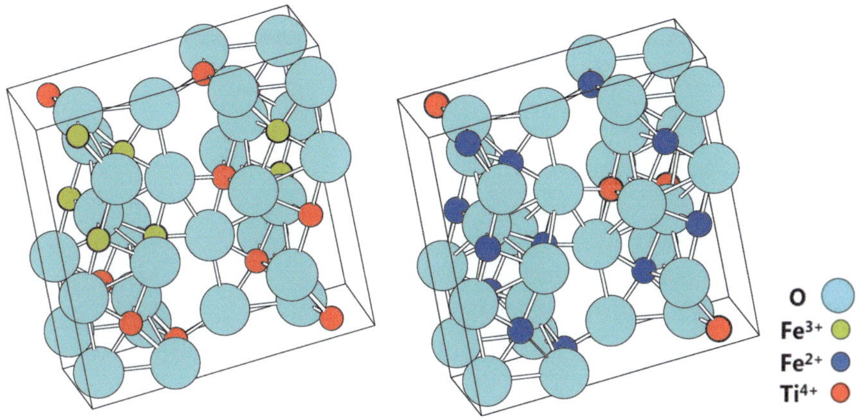

Fig. 3.6 Schematic crystal structure of the two solid solution components in PSB (pseudobrookite and subpseudobrookite)

not present in the raw ironsands but was detected during the pyrolysis and pre-oxidation experiments conducted in this study. The crystal structure of this solid solution system is depicted in Fig. 3.6.

3.2 High-Temperature Characteristics of Ironsands

The Gibbs free energy change ($\triangle G_r$) of a chemical reaction under varying temperature and pressure conditions is influenced by the chemical potentials of the reactants and products involved. The chemical potential in the standard state can be derived from the Gibbs free energy of formation, as shown in Eq. 3.1. On the other hand, the chemical potential of a group element can be obtained from the chemical potential in the standard state, the temperature, and the activity of the group element, as expressed in Eq. 3.2. Based on the chemical potential of each component, the Gibbs free energy change for the chemical reaction can be calculated, as shown in Eqs. 3.3 and 3.4 [17, 18].

$$\mu_i^\theta = DeltaG_f^\theta \tag{3.1}$$

$$\mu_i = \mu_i^\theta + RT \ln a_i \tag{3.2}$$

$$Delta\ G_r(n, T, p) = \left(\sum_1^M n_i \mu_i\right)_{prod} - \left(\sum_1^N n_i \mu_i\right)_{react} \tag{3.3}$$

$$Delta\ G_r(n,T,p)=\sum_{1}^{M+N} n_i\mu_i^\theta+\sum_{1}^{M+N} n_iRT\ln a_i$$
$$=Delta\ G_r^\theta(n,T,P)+RT\sum_{1}^{M+N}\ln a_i^{n_i} \quad (3.4)$$

where

μ_i	Chemical potential of the group element *i*;
a_i	The activity of group element *i*;
T	Kelvin temperature, K;
R	Ideal gas constant, $J{\cdot}mol^{-1}{\cdot}K^{-1}$;
p	Pressure, Pa;
ΔG_r	Gibbs free energy change for the reaction, $J{\cdot}mol^{-1}$;
n	Coefficient of stoichiometry;
M	The number of product species after the reaction;
N	The number of reactant species before the reaction;
θ	Standard state.

When the reaction reaches a steady state, the Gibbs free energy of the chemical reaction changes to zero, at which point Eq. 3.4 can be expressed as Eq. 3.5 [19].

$$Delta\ G_r^\theta(n,T,P)=-RT\sum_{1}^{M+N}\ln a_i^{n_i} \quad (3.5)$$

Using the FactSage 6.4 thermodynamic computation model (CRCT and Ftoxid) and the associated oxide database (FACT oxide database), it is possible to calculate the standard state Gibbs free energy for chemical reactions under different pressure and temperature conditions [20]. In this section, the conditions of 1 bar atmospheric pressure, a temperature range of 100–1400 °C, and pure solid matter activity of 1 are used as benchmarks to calculate the temperature dependence of the minimum oxygen partial pressure required for the reactions that occur during the generation process of the three substances (hematite, ilmenite, and pseudobrookite) in the pyrolysis process. The results are shown in Fig. 3.7, with the three reactions labeled as a, b, and c, respectively.

From Fig. 3.7, it is apparent that the minimum oxygen partial pressure required for reactions a, b, and c increases gradually as the temperature rises from 100 to 1400 °C. Furthermore, below 1400 °C, all three reactions can occur at partial oxygen pressures lower than the atmospheric level. The minimum oxygen partial pressure curves for the oxidative formation of ilmenite (reaction c) intersect the curves for the oxidative formation of hematite (reaction a) and ilmenite (reaction b) at 546.2 °C and 1157.6 °C, respectively.

In the early stages of non-isothermal pyrolysis, while the partial pressure of oxygen required for the oxidative formation of ilmenite ($FeTiO_3$) is the lowest, suggesting that, thermodynamically, the reaction should occur preferentially, the

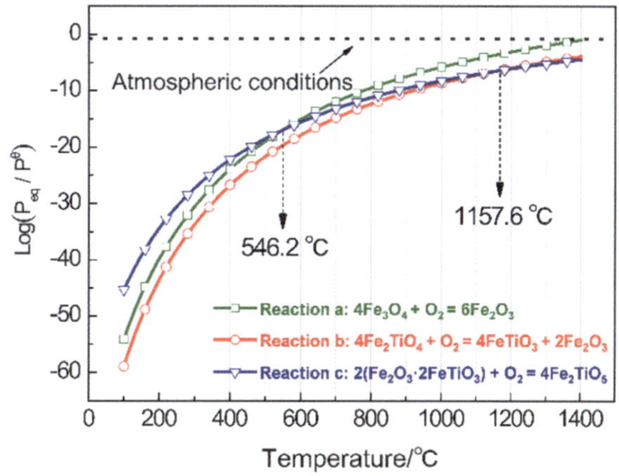

Fig. 3.7 Minimum oxygen partial pressure required for reaction as a function of temperature

oxidative formation of ilmenite is limited by the low content of its precursor, ferrotitanium spinel (Fe_2TiO_4). Therefore, during this stage, the oxidation of magnetite (Fe_3O_4) to hematite (Fe_2O_3) predominates, and the oxidation of ilmenite ($FeTiO_3$) from ferrotitanium spinel (Fe_2TiO_4) occurs secondarily. As the temperature increases and the reaction time extends, with further oxidation and reduction of magnetite (Fe_3O_4) in the titanomagnetite (TTM) and more ilmenite spinel (Fe_2TiO_4) formation, the oxidation of ilmenite spinel (Fe_2TiO_4) to ilmenite ($FeTiO_3$) (reaction b) becomes more significant. When the temperature reaches 1157.6 °C, reaction (c), which leads to the oxidative formation of pseudobrookite (Fe_2TiO_5), requires the lowest partial pressure of oxygen among the three reactions. Additionally, hematite (Fe_2O_3) and ilmenite ($FeTiO_3$) have been produced in large quantities through reactions (a) and (b), providing sufficient reactants for the formation of pseudobrookite (Fe_2TiO_5), allowing for the smooth production of pseudobrookite (Fe_2TiO_5) in the later stages of pyrolysis.

Thermogravimetric analysis (TGA) is a technique used to study the mass change of a substance during a reaction at a constant temperature and atmosphere. This allows for the characterization of the rate of mass change, as well as the start and end temperatures of the reaction, through the first- and second-order derivatives of the thermogravimetric curve.

In complex pyrolytic oxidative weight gain reactions, it can be challenging to determine the onset temperatures of sub-reactions or distinguish different reaction phases from thermogravimetric curves (TG) alone, as several sub-reactions may occur simultaneously. Thus, in this section, the first-order derivative of the thermogravimetric curve (DTG) and the second-order derivative (DDTG) are introduced to characterize the mass gain/loss rate of ironsands during the pyrolysis process and to differentiate the overlapping sub-reactions, as illustrated in Eqs. 3.6–3.8 [21, 22].

$$\alpha_{i,T} = \frac{m_{i,T} - m_{i,0}}{m_0} = \frac{\Delta m_{i,T}}{m_0} \tag{3.6}$$

$$DTG = \frac{d(m_T/m_0)}{dT} = \sum_{i=1}^{N} \frac{d\left(m_{i,T}/m_0\right)}{dT} = \sum_{i=1}^{N} \frac{d\left(\left(m_{i,0} + \Delta m_{i,T}\right)/m_0\right)}{dT} = \sum_{i=1}^{N} f_i \frac{d\alpha_{i,T}}{dT} \tag{3.7}$$

$$DDTG = \frac{d^2(m_T/m_0)}{dT^2} = \sum_{i=1}^{N} f_i \frac{d^2\alpha_{i,T}}{dT^2} \tag{3.8}$$

where m_0 and m_T represent the mass of the system at the initial moment and at temperature T, respectively; $m_{i,0}$ and $m_{i,T}$ represent the initial mass of histone i and the mass of histone i at temperature T, respectively; and $\alpha_{i,T}$ and $f_i(m_{i,0}/m_0)$ denote the rate of mass change of histone i and the initial mass fraction, respectively.

When a subreaction begins, i.e., when a new histone i starts to form, the rate of change of $\sum fi(d\alpha_{i,T}/dT)$ in Eq. 3.7 increases rapidly, implying that $\sum fi(d^2\alpha_{i,T}/dT^2)$ in Eq. 3.8 is near its extremum. Therefore, the onset of a new subreaction results in a point of maximum slope on the DTG curve and a point of maximum value on the DDTG curve. Mathematically, the location of an independent change point is more easily identified by the maximum of the DDTG curve rather than the point of maximum slope on the DTG curve. As such, the DDTG curve is selected to determine the number of subreactions and their corresponding temperatures during the pyrolysis process of the ironsands.

The equipment used for the non-isothermal pyrolysis experiments was a thermogravimetric analyzer (model: BJHJ HCT-3). Initially, 10 mg of ironsands was weighed and placed in a pre-weighed high-alumina crucible (4 mm in diameter × 4 mm in height). The crucible containing the ironsands was then placed on the analytical balance of the thermogravimetric analyzer. The ironsands were heated from room temperature to 1200 °C at a rate of 5 °C/min, with air continuously flowing into the reactor at a rate of 60 mL/min to maintain constant atmospheric conditions. The thermogravimetric analyzer was connected to a computer control system that recorded the absolute mass of the ironsands sample every 2 s. Upon reaching a predetermined temperature, the thermogravimetric analyzer automatically cooled to room temperature. The TG baseline for the pre-experiment was established by conducting a similar experiment without ironsands, under the same conditions. The experimental TG curve was then corrected by subtracting the pre-experiment TG baseline to yield a more accurate TG curve. The DTG and DDTG curves were computed mathematically from the TG curve [23–25], as shown in Fig. 3.8 (The blue solid line indicates the change in the weight gain fraction of the ironsands with temperature, the red dashed line represents the first-order derivative of the TG curve, and the green dotted line represents the second-order derivative of the TG curve. The results exhibit good reproducibility across three replicate experiments, with uncertainty fluctuations within 5%.)

As shown in Fig. 3.8, multiple maxima appear on the DDTG curves, indicating the presence of five distinct subreaction steps. New subreactions are initiated at

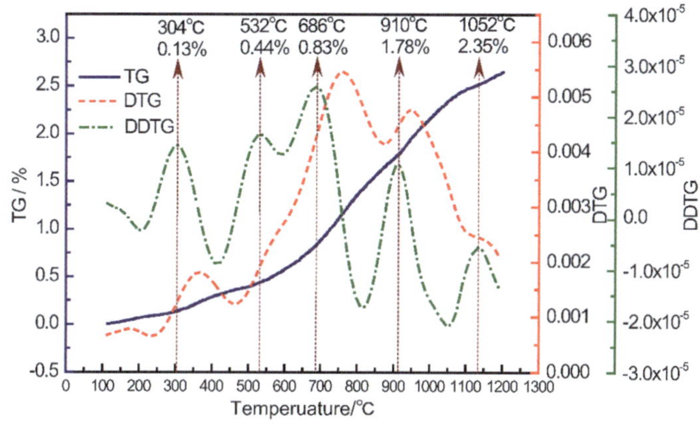

Fig. 3.8 Thermogravimetric analysis of non-isothermal pyrolysis of ironsands

temperatures of 304 °C, 532 °C, 686 °C, 910 °C, and 1052 °C during the pyrolysis of ironsands, with corresponding rates of weight gain of 0.13 wt%, 0.44 wt%, 0.83 wt%, 1.78 wt%, and 2.35 wt%. The two subreactions indicated by the initial temperature peaks at 304 °C and 532 °C correspond to the conversion of magnetite (Fe_3O_4) to hematite (Fe_2O_3) in the TTM phase, which is also confirmed by the thermodynamic analysis presented in Fig. 3.7. Additionally, in the XRD patterns shown in Fig. 7.2, peak intensities corresponding to TTH along the (116) and (104) crystallographic directions are observed. A significant increase in peak intensity between 300–400 °C and 500–600 °C further substantiates the occurrence of the aforementioned subreactions.

In the subsequent stage, the maximum peak of the five extreme points occurs at 686 °C, indicating that the subreaction occurring at this temperature exhibits the fastest rate of weight gain among the five reactions. This is attributed to the fact that, between 600 and 700 °C, TTH develops and grows along the (113), (024), and (012) crystal directions, accompanied by the initial appearance of PSB along the (101) crystallographic direction, as shown in Fig. 7.2. Consequently, this results in a peak on the DDTG curve at 686 °C, where the weight gain rate of ironsands is highest.

The extreme point on the DDTG curve at 910 °C marks the initiation of another subreaction. Due to the oxidation of TTM in the early stages of the reaction, a significant amount of titanite-iron spinel material phase is generated and accumulated. This accumulation provides a sufficient supply of reactants for the formation of ilmenite, as depicted in Fig. 3.7. As a result, the growth of ilmenite along the (111) crystallographic direction leads to a further increase in the mass of ironsands at 910 °C.

The oxidation of PSB at 1052 °C represents the final subreaction. Ultimately, at 1200 °C, the mass increase of ironsands is approximately 2.65 wt%, compared to the theoretical weight gain of 3.28 wt%. This indicates that only about 80.8 wt% of

Fe^{2+} was oxidized to Fe^{3+}, suggesting that some unoxidized TTM remains in the raw ironsands.

3.3 Summary

The ironsands used in this study contain low levels of harmful elements, and the iron and titanium components hold significant industrial value. The primary phase of the material is titanomagnetite (TTM), which has a cubic crystal structure. TTM is a solid solution system of magnetite and ilmenite-spinel. The secondary phase is titanohematite (TTH), which exhibits a trigonal crystal structure and is a solid solution system of hematite and ilmenite. TTH initially precipitates during the rapid cooling of volcanic lava due to the high solubility of titanium. Subsequently, TTH undergoes further development and growth under long-term oxidation by atmospheric and dissolved oxygen in water. Ultimately, the stable TTH phase is formed under standard temperature and pressure conditions.

Ironsands can be directly incorporated into the sintering process of ironmaking without the need for particle size reduction; however, the raw ore size is not within the optimal range. If used as a pellet feed, it must be milled and refined to achieve a suitable pelletable size. Scanning electron microscopy (SEM) and energy dispersive X-ray spectroscopy (EDS) analyses of the raw ironsands reveal two primary micromorphological forms. One form consists of homogeneous particles with a smooth surface and a relatively uniform element distribution, with TTM as the dominant phase. The other form consists of non-homogeneous particles that exhibit a laminar phase precipitated at a specific angle, with the substrate remaining TTM, while the laminar phase is identified as TTH.

In the initial stages of non-isothermal pyrolysis, the oxidation of magnetite (Fe_3O_4) to form hematite (Fe_2O_3) predominates, while the oxidation of ilmenite spinel (Fe_2TiO_4) to form ilmenite ($FeTiO_3$) plays a secondary role. As temperature increases and the reaction time extends, magnetite (Fe_3O_4) is further oxidized and reduced within the TTM phase, leading to the formation of additional ilmenite spinel (Fe_2TiO_4). At this stage, the oxidation of ilmenite spinel (Fe_2TiO_4) to ilmenite ($FeTiO_3$) gradually becomes the dominant reaction. When the temperature reaches 1157.6 °C, the oxygen partial pressure required for the oxidation of Fe_2TiO_5 is the lowest among the three reactions, allowing for the smooth generation of Fe_2TiO_5 in the late stage of the pyrolysis process.

References

1. Bowles JA, Jackson MJ, Berquo TS, et al. Inferred time- and temperature-dependent cation ordering in natural titanomagnetites. Nat Commun. 2013;4.

2. Jung SM. Effects of $CaO/CaCO_3$ on the carbothermic reduction of titanomagnetite ores. Metall Mater Trans B-Proc Metall Mater Proc Sci. 2015;46(3):1162–74.
3. Pearce CI, Henderson CMB, Telling ND, et al. Fe site occupancy in magnetite-ulvospinel solid solutions: a new approach using X-ray magnetic circular dichroism. Am Mineral. 2010;95(4):425–39.
4. Bosi F, Halenius U, Skogby H. Crystal chemistry of the magnetite-ulvospinel series. Am Mineral. 2009;94(1):181–9.
5. Longbottom RJ, Ostrovski O, Zhang JQ, et al. Stability of cementite formed from hematite and titanomagnetite ore. Metall Mater Trans B-Proc Metall Mater Proc Sci. 2007;38(2):175–84.
6. Hu T, Lv XW, Bai CG, et al. Reduction behavior of Panzhihua titanomagnetite concentrates with coal. Metall Mater Trans B-Proc Metall Mater Proc Sci. 2013;44(2):252–60.
7. Zboril R, Mashlan M, Petridis D. Iron(III) oxides from thermal processes-synthesis, structural and magnetic properties, Mossbauer spectroscopy characterization, and applications. Chem Mater. 2002;14(3):969–82.
8. Wright J. Iron-titanium oxides in some New Zealand ironsands. NZ J Geol Geophys. 1964;7(3):424–44.
9. Ayyub P, Multani M, Barma M, et al. Size-induced structural phase transitions and hyperfine properties of microcrystalline Fe_2O_3. J Phys C: Solid State Phys. 1988;21(11):2229.
10. Katsura T. Generalized titanomagnetite in Hawaiian volcanic rocks. Pacific Sci. 1964;XVIII:223–8.
11. Pascal C, Pascal J, Favier F, et al. Electrochemical synthesis for the control of γ-Fe_2O_3 nanoparticle size. Morphology, microstructure, and magnetic behavior. Chem Mater. 1999;11(1):141–7.
12. Cannas C, Gatteschi D, Musinu A, et al. Structural and magnetic properties of Fe_2O_3 nanoparticles dispersed over a silica matrix. J Phys Chem B. 1998;102(40):7721–6.
13. Dormann J, Viart N, Rehspringer J, et al. Magnetic properties of Fe_2O_3 particles prepared by sol-gel method. Hyperfine Interact. 1998;112(1):89–92.
14. O'Reilly W. Applications of rock and mineral magnetism. In: Rock and mineral magnetism. 1984. pp. 194–212.
15. Gillot B. Infrared spectrometric investigation of submicron metastable cation-deficient spinels in relation to order-disorder phenomena and phase transition. Vib Spectrosc. 1994;6(2):127–48.
16. Goodenough J, Loeb A. Theory of ionic ordering, crystal distortion, and magnetic exchange due to covalent forces in spinels. Phys Rev. 1955;98(2):391.
17. Li N, Hu P, Zhang XH, et al. Effects of oxygen partial pressure and atomic oxygen on the microstructure of oxide scale of ZrB_2-SiC composites at 1500 degrees. Corros Sci. 2013;73:44–53.
18. Kazanskii V. Chemical potential of mass transfer in colloidal bodies. J Eng Phys. 1966;10(3):233–4.
19. Kim MS, Lee HG, Kang YB. Determination of Gibbs free energy of formation of MnV_2O_4 solid solution at 1823 K (1550 °C). Metall Mater Trans B. 2014;45(1):131–41.
20. Hekmat-Ardakan A, Ajersch F. Thermodynamic evaluation of hypereutectic Al-Si (A390) alloy with addition of Mg. Acta Mater. 2010;58(9):3422–8.
21. Wang YM, Yuan ZF, Guo ZC, et al. Reduction mechanism of natural ilmenite with graphite. Trans Nonferr Metals Soc China. 2008;18(4):962–8.
22. Gillot B, Jemmali F, Rousset A. Kinetics and mechanism of ferrous spinel oxidation studied by electrical conductivity and thermogravimetry. J Mater Sci. 1986;21(12):4436–42.
23. Sanders JP, Gallagher PK. Thermomagnetometric evidence of γ-Fe_2O_3 as an intermediate in the oxidation of magnetite. Thermochim Acta. 2003;406(1–2):241–3.
24. Sanders J, Gallagher P. Kinetics of the oxidation of magnetite using simultaneous TG/DSC. J Therm Anal Calorim. 2003;72(3):777–89.
25. Li KY, Huang XY, Fleischmann C, et al. Pyrolysis of medium-density fiberboard: optimized search for kinetics scheme and parameters via a genetic algorithm driven by Kissinger's method. Energ Fuel. 2014;28(9):6130–9.

Chapter 4
Magnetic Separation of Ironsands

Relative to typical hematite and magnetite ores, ironsands are disadvantaged by both lower grades of valuable elements and inferior reduction kinetics. To address these issues, this book proposes a two-fold pretreatment strategy: (1) physical enrichment via grinding and magnetic separation to enhance iron and titanium oxide content; and (2) chemical modification via pre-oxidation to improve reduction behavior. This chapter focuses on the grinding dissociation behavior, optimization of magnetic separation parameters, and the associated force mechanisms. Chapter 7 further details the effects of pre-oxidation treatment on improving reducibility and overall metallurgical performance [1].

4.1 Fine Grinding Magnetic Separation Methodology of Ironsands

In this section, grinding and magnetic separation experiments were conducted on ironsands, utilizing X-ray diffraction (XRD), scanning electron microscopy (SEM), and laser particle size analysis (LPSA) to examine the microscopic dissociation characteristics at different grinding stages. The effects of particle size and magnetic field intensity on the grade and recovery of the concentrate were studied, alongside the force mechanisms involved in magnetic separation under fine-grained conditions. These results provide valuable insights for further enhancing the grade of ironsands suitable for both iron and titanium extraction processes.

The dry ball milling method is particularly effective for processing large quantities of hard-to-grind ores. In this study, a dry ball mill was employed for grinding experiments on raw ironsands. A 500 g sample of raw ironsands was taken from each of four copies, with grinding durations of 10, 20, 30, and 40 min. After grinding, subsamples were collected, and particle size distributions were analyzed using LPSA,

Z. Wang et al., *Smelting and Utilization Technology of Ironsands*,
https://doi.org/10.1007/978-981-95-2868-4_4

while SEM was used to observe the microscopic dissociation and morphology of the ironsands at different milling times.

Preliminary experiments showed that when the magnetic field strength exceeded 200 mT, the grade and recovery of the concentrate remained virtually unchanged. Therefore, a magnetic field strength range of 40–200 mT was selected for further investigation.

For each powder sample obtained at different ball milling times, 10 g of ore was mixed with a consistent amount of water to maintain a constant feeding concentration. The magnetic separator was adjusted to the predetermined magnetic field strength, with the rotational speed of the magnetic drum and water flow velocity kept constant. Once the magnetic separator reached a stable operating condition, slurry was added to the sluice every 30 s using a rubber-tipped burette. After the experiment, the concentrate collected under various magnetic separation conditions was dried, weighed, and sampled. The total iron (TFe) and titanium dioxide (TiO_2) content were then determined through chemical analysis.

4.2 Microstructural Analysis of Particle Dissociation During Ball Milling

The particle size and micro-dissociation morphology of ironsands with different ball milling times are shown in Fig. 4.1.

As shown in Fig. 4.1, as the grinding time increases, the proportion of ore particles smaller than 0.076 mm gradually increases, although this increase slows down, indicating that grinding efficiency decreases over time. SEM analysis of the raw ironsands reveals three main forms of mineral morphology: greyish-white titanium-iron solid solution (useful minerals); greyish-black SiO_2-dominated vein monoliths; and concretions composed of both veins and useful minerals. After 10 min of grinding, a small portion of the mineral grains dissociate, but most of the particles remain intact, and the dissociation is insufficient to lower the concentrate grade after magnetic separation. After 20 min of grinding, most of the mineral grains dissociate, allowing for a concentrate grade above 59 wt% under weak magnetic separation conditions. However, for the 30-min grind, the grains are primarily present in the form of inclusions, making dissociation more difficult. As the particle size reduces, milling efficiency significantly decreases. Although further size reduction occurs after 40 min of grinding, the improvement in concentrate grade after magnetic separation is minimal.

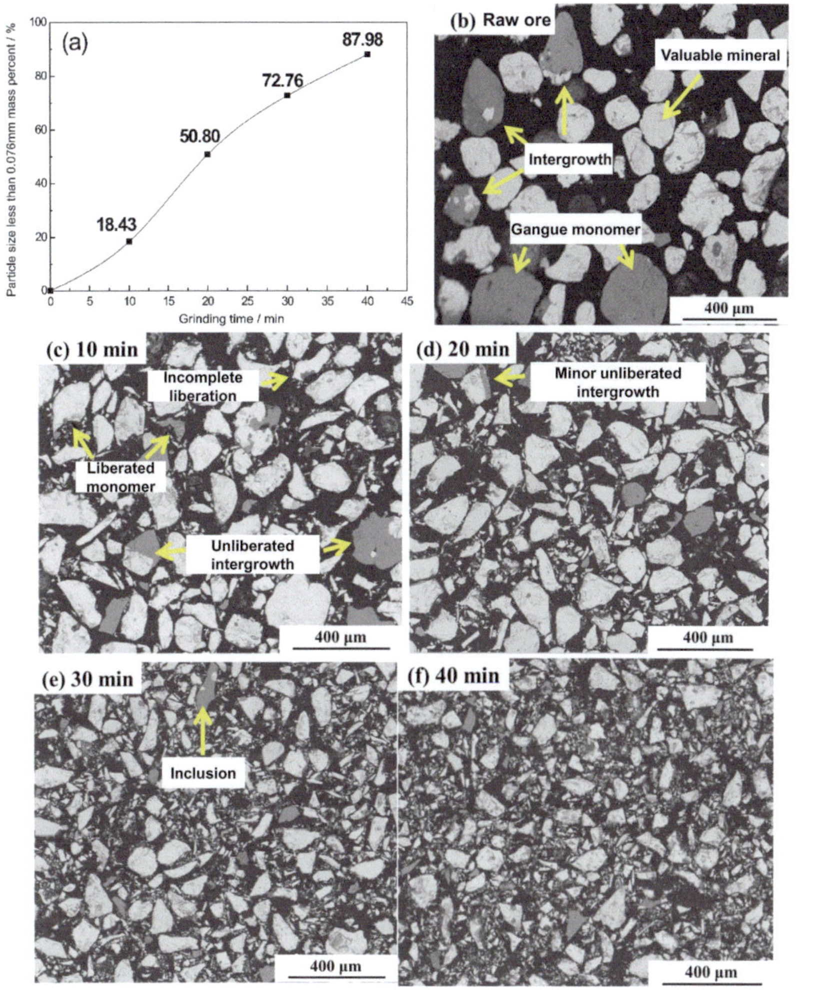

Fig. 4.1 Particle size distribution and microscopic morphology of ironsands under different grinding durations. **a** Particle size of ironsands at varying grinding times; **b** SEM image of raw ironsands; **c** SEM image after 10 min of grinding; **d** SEM image after 20 min of grinding; **e** SEM image after 30 min of grinding; **f** SEM image after 40 min of grinding

4.3 Process Parameter Optimization for Magnetic Separation

The recovery of TFe and TiO_2 from the concentrate of ironsands after fine grinding and magnetic separation was calculated following drying, weighing, and chemical analysis, using Eq. 4.1:

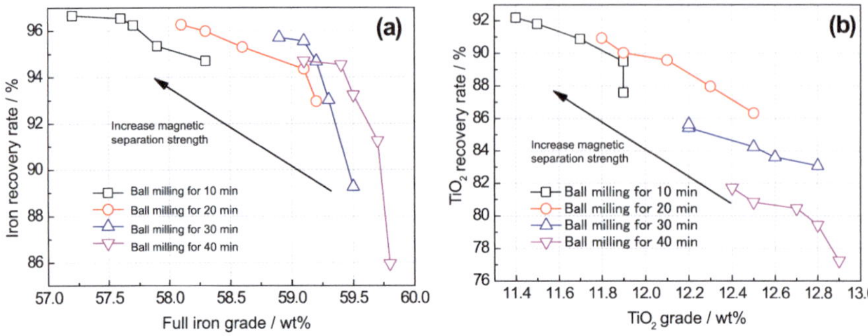

Fig. 4.2 Total Fe (TFe) and TiO_2 grade and recovery of ironsands concentrate under different magnetic separation conditions: **a** TFe grade and recovery; **b** TiO_2 grade and recovery

$$R_i = \frac{m_{\text{concentrate}} \times i\%_{\text{concentrate}}}{m_{\text{raw}} \times i\%_{\text{raw}}} \times 100\% \tag{4.1}$$

where R_i is the magnetic separation recovery of component i, %; $m_{\text{concentrate}}$ and m_{raw} are the mass of raw ore and concentrate, g, respectively; $i\%_{\text{raw}}$ and $i\%_{\text{concentrate}}$ are the mass fraction of component i in raw ore and concentrate, respectively.

The grade and recovery of TFe and TiO_2 in the concentrate, based on the chemical analysis results, are shown in Fig. 4.2.

From Fig. 4.2a, as the magnetic field strength increases, the TFe% in the concentrate decreases gradually, while the R_{TFe} increases. This is due to incomplete dissociation of certain ore particles, which are carried into the concentrate with the tailings under the influence of the progressively stronger magnetic field. As a result, the TFe% of the concentrate decreases while the RTFe increases. Additionally, under the same magnetic field conditions, as the particle size decreases, the TFe% of the concentrate gradually increases, while the R_{TFe} decreases. This is because fine-grained ironsands exhibit a higher degree of dissociation, facilitating the separation of vein minerals from useful minerals, thereby increasing the TFe%. The finer the particle size, the higher the water resistance, making it more difficult for the particles to reach the surface of the drum, thus decreasing R_{TFe}.

When the magnetic separation intensity increased from 40 to 200 mT, the TFe wt% of the concentrate from the 10-min ball-milled sample decreased from 58.3 to 57.2 wt%, a drop of 1.1%. In contrast, the TFe wt% of the concentrate from the 40-min ball-milled sample only decreased by 0.7 wt%. This suggests that coarser particles are more strongly affected by the magnetic field force (the force mechanism in the magnetic field will be discussed in Sect. 4.4). Therefore, weak magnetic separation should be applied to coarser grains to maintain a higher TFe wt% in the concentrate, while stronger magnetic separation should be employed for finer grains to increase R_{TFe} while maintaining an acceptable TFe wt%.

In Fig. 4.2b, it can be observed that after ball milling and magnetic separation, the TiO_2 mass fraction in the concentrate increased compared to the TiO_2 mass fraction of the raw ore (11.41 wt%). The TiO_2 mass fraction is higher under weak magnetic separation conditions and fine-grain sizes. As mentioned earlier, Ti and Fe coexist in ironsands in the form of solid solutions, such as titanomagnetite and titanohematite. Consequently, only a small number of fine-grained titano-iron solid solution particles are removed into the tailings due to the lower magnetic field strength, while most of the particles, due to the presence of both iron and titanium, increase the TiO_2 mass fraction and recovery, while simultaneously enhancing the grade and recovery of iron in the concentrate and reducing the chalcopyrite mass fraction and recovery.

As the grinding time increases and the magnetic field strength decreases, the TFe wt% and TiO_2 wt% of the concentrate after magnetic separation gradually increase, while the recovery decreases. After considering all factors, the optimal conditions for obtaining a higher grade concentrate while ensuring better recovery are grinding for 20 min at 80 mT, 30 min at 120 mT, and 40 min at 160 mT.

4.4 Mechanical Analysis of Magnetic Separation Forces

During the process of magnetic separation, the ore particles are subjected to a combination of forces, including gravity, buoyancy, water resistance, and magnetic field force. These forces can be expressed by Eqs. 4.2–4.5 as follows:

$$G_{\text{total}} = \frac{\pi d^3}{6} \cdot (\rho_{\text{ore}} - \rho_{\text{water}}) \cdot g \tag{4.2}$$

$$F_{\text{resistance}} = \psi \left(1 + \frac{2k}{\sqrt{\text{Re}}}\right)^2 \cdot \rho_{\text{water}} \cdot d^2 \cdot \nu^2 \tag{4.3}$$

$$\text{Re} = \frac{\rho_{\text{water}} d \nu}{\mu_{\text{water}}} \tag{4.4}$$

$$f_{\text{magnetic}} = \mu \cdot \chi \cdot V \cdot H \cdot gradH \tag{4.5}$$

where

d	maximum diameter of a single ore grain, m;
ρ_{ore}, ρ_{water}	density of ore grains and magnetic separation medium (water), kg/m^3;
ψ, k	resistance coefficient constant;
Re	Reynolds number;
μ_{water}	viscosity coefficient of water, Pa·s;
ν	velocity of the movement of mineral particles, m/s;

μ magnetic permeability of the water medium, N/A^2;
χ magnetization rate of the mineral grains;
H magnetic field strength, A/m;
$gradH$ magnetic field gradient at the location of the ore grain, A/m^2.

Thus, under the same magnetic separation conditions, the force acting on an ore particle in the magnetic field is primarily dependent on its maximum diameter. The magnetic separation tube in the separator is specified as φ60 mm × 1000 mm. The time required for an ore particle to travel from the center to the edge of the magnetic separation tube is t = 2.4 s. Assuming uniform particle velocity, the speed is $v = 0.025$ m·s^{-1}. By using the parameters outlined in Table 4.1, we can derive the relationship between the force on the ironsands particle and the maximum diameter, as shown in Fig. 4.3.

From Fig. 4.3, it is observed that as the diameter of the ironsands particles increases, the combined forces of gravity and buoyancy (henceforth referred to as gravity), water resistance, and magnetic field force all exhibit an increasing trend. For particles with diameters ranging from 0 to 30 μm, water resistance dominates, with relatively smaller contributions from gravity and magnetic field force. In this size range, water resistance becomes the primary force acting on the mineral particles. As

Table 4.1 Force parameters of ironsands mineral grains

Grain density/ (kg·m^{-3})	Specific magnetization of ore particles/ (m^3·kg^{-1})	Magnetic permeability of aqueous media μ/ (N·A^{-2})	Magnetic field strength H/ (A·m^{-1})	Drag coefficient ψ;k	Viscosity of water μ/(Pa·s)
4070	3740×10^{-7}	$4\pi \times 10^{-7}$	6.4×10^4	0.11;4.53	10^{-3}

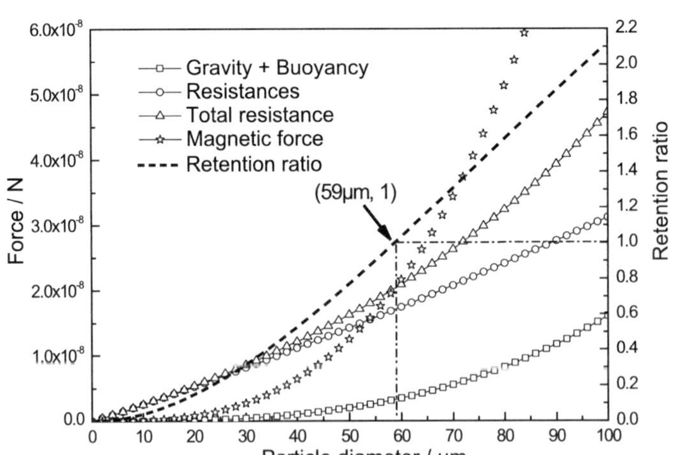

Fig. 4.3 Relationship between the force exerted on ironsands mineral grains and particle diameter

the particle diameter reaches 30 μm, the magnetic field force increases rapidly, while the rate of increase for gravity and water resistance slows down. When the particle diameter reaches 59 μm, the magnetic field force and the total resistance become equal, resulting in a retention ratio of 1. At this point, the ore particles are just able to reach the surface of the drum. For particles larger than 59 μm, the increase in gravity accelerates, but the magnetic field force surpasses the total resistance, allowing the ore particles to reach the surface of the drum. Therefore, the magnetic field force becomes the dominant force for larger particles.

This analysis indicates that excessive grinding can negatively impact the beneficiation efficiency. From a mechanical perspective, it also explains why the recovery rate for fine-grained ironsands (as shown in Fig. 4.2) is lower compared to coarse-grained ironsands, with the recovery of coarse particles being more significantly influenced by the magnetic field force.

4.5 Summary

With an increase in grinding time and a decrease in magnetic field strength, the grades of total iron (TFe) and titanium dioxide (TiO_2) in the concentrate of ironsands after magnetic separation gradually increased. However, the recovery rate showed a decreasing trend. Considering both grade and recovery, optimal conditions were found to be grinding for 20 min at 80 mT, 30 min at 120 mT, and 40 min at 160 mT, as these parameters ensured a higher grade of concentrate while maintaining a better recovery rate.

As the grinding time increased, the degree of dissociation of the ore particles also improved. However, after 30 min of grinding, the particles predominantly existed as inclusions, making them more difficult to dissociate. Consequently, the increase in grade after magnetic separation became less significant.

During the magnetic separation process, ore particles are primarily subjected to gravity, buoyancy, water resistance, and magnetic field force. In the case of ironsand particles under the experimental magnetic separation conditions, when the diameter of the particles was between 0 and 30 μm, water resistance was the dominant force. As the diameter increased to 59 μm and above, the magnetic field force exceeded the total resistance, with the interception ratio rising above 1. This allowed the particles to reach the surface of the drum. This phenomenon helps explain the lower recovery rate for fine-grained ironsands, whereas coarse-grained ironsands are more significantly influenced by magnetic field forces.

Reference

1. Wang ZY, Zhang JL, Liu ZJ, et al. Microscopic liberation and force analysis of fine-grained magnetic separation of ironsand during ball milling. J Northeastern Univ (Nat Sci). 2016;37(05):711–5.

Chapter 5
Sintering Processes for Ironsands

In China, sintered ore is the primary iron-bearing material used in blast furnaces. To assess the suitability of ironsands in sintering, this chapter investigates the influence of ironsands addition on sintering performance. The study employs single-factor experiments examining the effects of ironsands content and particle size on sintering quality and process parameters. The goal is to optimize the incorporation of ironsands into the sintering process, ensuring both productivity and sinter quality.

5.1 Basic Properties of Ironsands for Sintering

Chemical composition

The chemical composition of the ironsands was analyzed through chemical analysis, and the results are presented in Table 5.1.

As shown in Table 5.1, the total iron (Fe) content of the ironsands is 55.63 wt%, which is relatively high compared to typical titanium-bearing rock ores. The FeO content is also elevated, indicating that the iron is primarily present in the form of magnetite (Fe_3O_4). The SiO_2 content is 4.13 wt%, placing it within the medium range for silica. Additionally, the TiO_2 content is relatively high at 11.41 wt%.

Physical phase composition

The physical phase composition of the ironsands was determined using X-ray diffraction (XRD) analysis, and the results are shown in Fig. 5.1.

From the XRD diffraction pattern in Fig. 5.1, it is evident that the dominant mineral phase in the ironsands is titanomagnetite. Titanomagnetite is a solid solution with the chemical formula $(1-x)Fe_3O_4 - xFe_2TiO_4$, a product formed by the solid solution of magnetite and titanomagnetite spinel in specific proportions. Due to the small difference in the ionic radius and atomic mass between Fe and Ti atoms, Ti atoms can substitute for Fe atoms in the crystal lattice, leading to homogeneity. This

Z. Wang et al., *Smelting and Utilization Technology of Ironsands*,
https://doi.org/10.1007/978-981-95-2868-4_5

Table 5.1 Chemical composition of ironsands/wt%

TFe	FeO	SiO_2	CaO	MgO	Al_2O_3	TiO_2	K_2O	Na_2O	ZnO	Cl	S
55.63	29.6	4.13	0.6	3.74	3.38	11.41	0.017	0.020	0.005	0.009	0.013

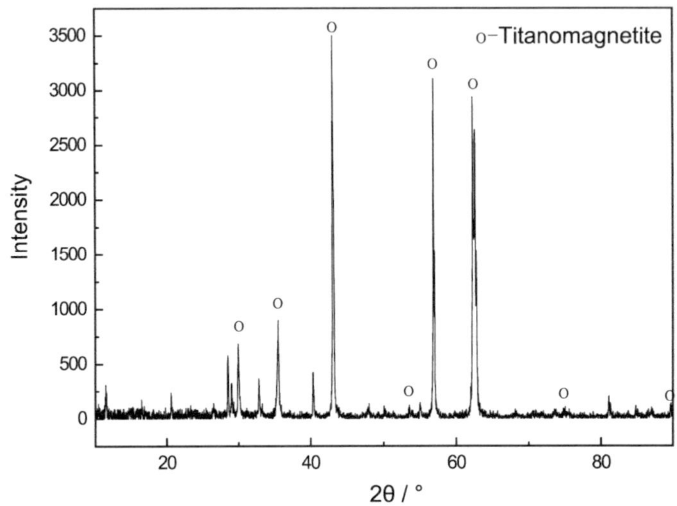

Fig. 5.1 XRD diffraction analysis of ironsands

phenomenon is the primary reason why Fe and Ti are not easily separated during the beneficiation of titanomagnetite [1–3].

Particle size distribution

The particle size, specific surface area, and other factors influence the sintering process, including the second mixing stage of granulation and the adhesion of mineral powders. Therefore, it is essential to investigate the particle size distribution and morphology of the ironsands.

Scanning Electron Microscopy (SEM) was used to analyze the morphology of the ironsands, and the results are presented in Fig. 5.2. A Laser Particle Size Analyzer (LPSA) was employed to characterize the particle size distribution, and the results are shown in Fig. 5.3.

From Fig. 5.2, it is observed that, compared with the vanadium-titanium magnetite from the Chengde rock ore type, the original grain-grade ironsands are mostly spherical or ellipsoidal in shape, with more rounded corners and a smaller specific surface area. This is attributed to the long-term scouring of the ores by seawater, which has smoothed the edges of the ore particles. Consequently, the ball formation and adhesion performance of the original grain-grade ironsands are not as effective as those of the titanium minerals found in rock ores.

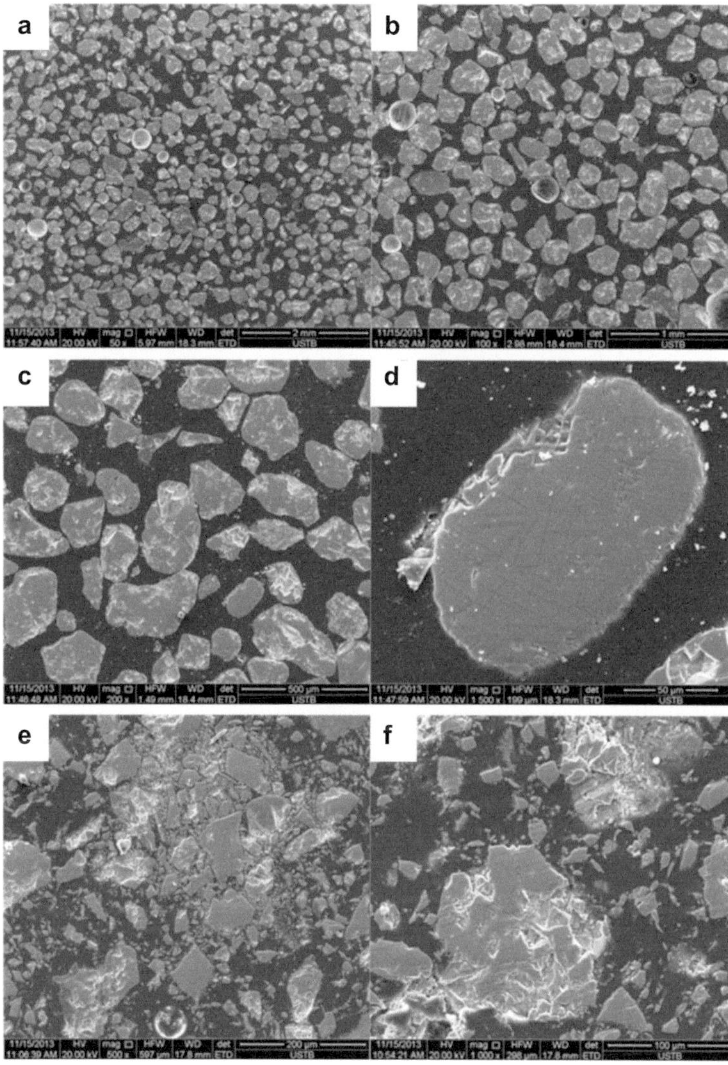

Fig. 5.2 Comparison of micro-morphology between raw-grained ironsands and Chengde vanadium-titanium magnetite. **a**–**d** Ironsands at different multiples of the original grain size; **e**, **f** Chengde vanadium-titanium magnetite at different multiples

From Fig. 5.3, the particle size distribution of the original granular ironsands is concentrated in the 70–115 mesh range, with the largest proportion (19.68 wt%) falling within the 80–100 mesh range. The proportion of particles in the 70–80 mesh and 100–115 mesh ranges are 18.32 wt% and 16.91 wt%, respectively. Furthermore, 68.21 wt% of the particles are larger than 100 mesh, while 31.79 wt% fall within the 100–200 mesh range. Notably, no particles smaller than 200 mesh were detected.

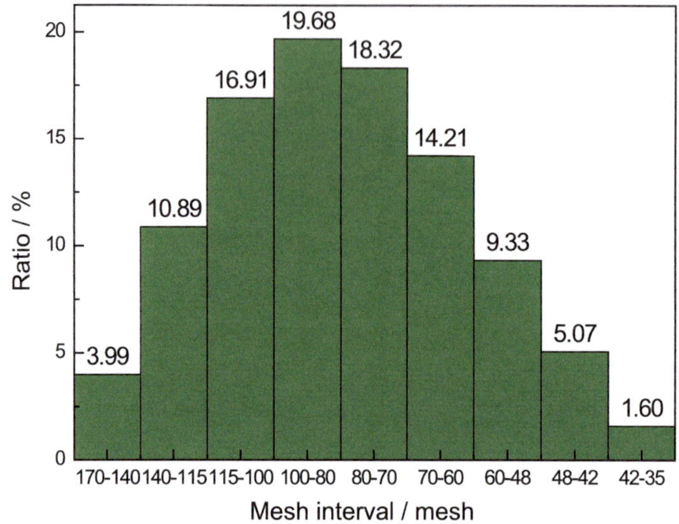

Fig. 5.3 Particle size distribution range of raw-grained ironsands

For mineral powders with a particle size smaller than 200 mesh comprising 60–65 wt%, granulation can be enhanced to improve the permeability of the mixture, thereby boosting both the sintered mineral yield and quality. The sinter granulation particles consist of core particles and an adhesion layer. In this context, returned ore particles of around 3–5 mm diameter can serve as effective core particles. The adhesion layer powder corresponds to particles smaller than 100 mesh, with intermediate grain sizes between 100 and 18 mesh being the primary contributor to reduced permeability in the mixture. Given that 68.21 wt% of the ironsands fall into the intermediate grain size range, this factor plays a crucial role in determining the permeability and efficiency of the sintering process.

Therefore, in addition to exploring the application of raw ironsands at the primary grain size level in the sintering process, the study also investigates the milling of ironsands to refine the particle size. This refinement is expected to enhance the role of ironsands as an effective adhesive powder in sintering. To this end, the grinding properties of ironsands were examined to assess the feasibility and impact of utilizing refined ironsands in the sintering process.

Grinding characteristics

Ball milling experiments were conducted on ironsands using 2 kg of ironsands per group. The milling times were set to 5, 10, 15, 20, 25, 30, and 35 min. After each milling period, 100 g of the milled sample was taken and sieved for 1 h using a vibrating sieve machine to determine the proportion of particles smaller than 200 mesh. The results are shown in Fig. 5.4.

From Fig. 5.4, it can be observed that as the ball milling time increases, the proportion of particles smaller than 200 mesh also increases. During the early stages

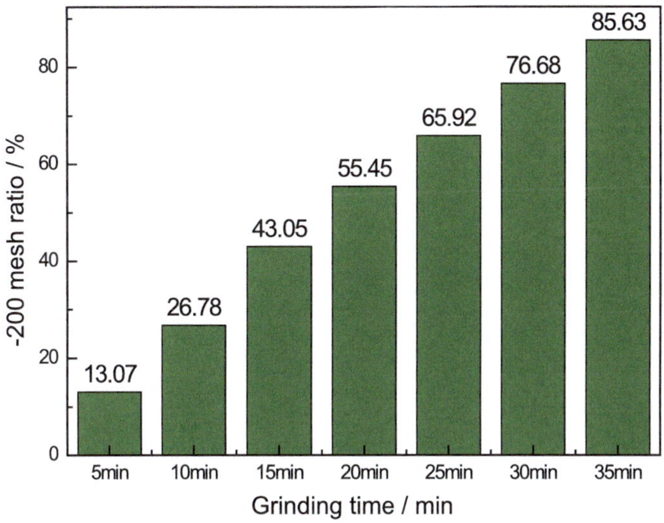

Fig. 5.4 Relationship between ball milling time and proportion of ironsands passing through 200 mesh

of milling, the proportion of particles smaller than 200 mesh increases significantly with a slight increase in milling time, from 13.07 to 43.05 wt% after just 10 min of milling. In the middle and later stages of milling, the increase in the proportion of particles smaller than 200 mesh becomes more gradual, with approximately a 10 wt% increase for every 5 min of milling. When the milling time reaches 25 min, the proportion of particles smaller than 200 mesh reaches 65 wt%. At this point, the material meets the standard for enhanced granulation and permeability of the material layer, as indicated in previous studies [4].

Based on these findings, the sintering cup experiments utilized the original granular ironsands and the ironsands milled for 25 min (fine-grained ironsands) as experimental materials. Additionally, considering a balance between grinding costs and sintered ore quality, ironsands milled for 10 min (intermediate-grained ironsands) were selected as the third experimental ingredient.

5.2 Experimental Method for Ironsands Sintering

Sintering cup experimental equipment

The sintering experiments using ironsands were conducted in a sintering cup. The sintering laboratory is equipped with several key systems, including the main control system, the sintering system, material boxes, a fan, a sintering mixing granulator, a sintering drop strength tester, a sintered ore horizontal reciprocating mechanical sieve, a sintering drum experimental machine, and a sieve machine for post-sintering.

Figure 5.5 shows some of the equipment and testing devices used in the sintering cup experiment. The main parameters of the equipment are provided in Table 5.2.

The sintering cup experimental process consists of five main stages: ingredient preparation, primary mixing, secondary mixing, sintering, and finished product inspection. The experimental procedure is illustrated in Fig. 5.6.

Charge mixture

Before conducting the sintering cup experiments, the preparation of the mixture is essential. This includes the incorporation of mineral powders, binders, fuels, and re-mineralization agents. The amount of ironsands added to the mixture varies, and the dosing bin used for this process is shown in Fig. 5.7.

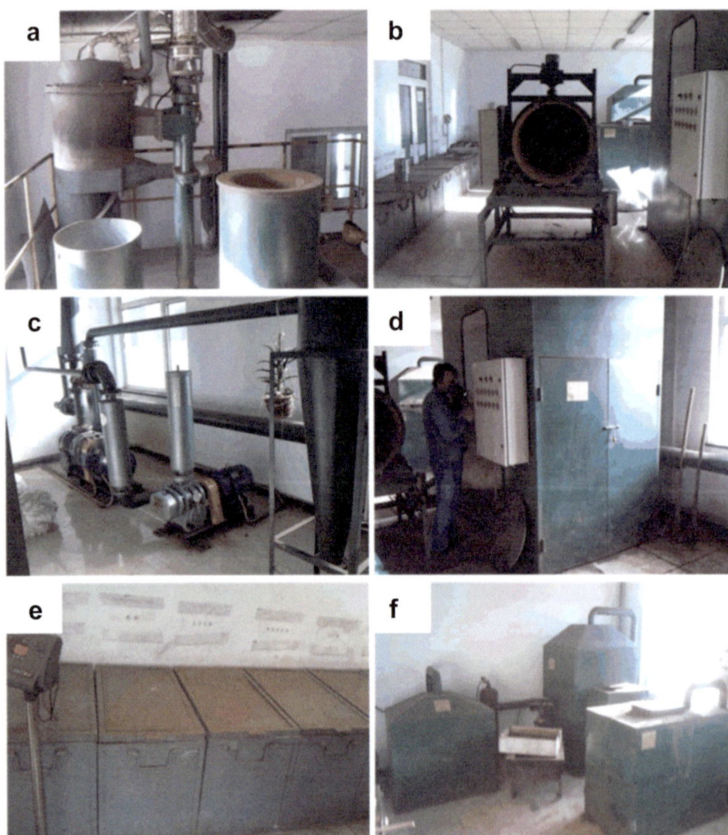

Fig. 5.5 Major equipment and instruments in the sintered cup laboratory. **a** Sintering main system; **b** sintering mixing and granulating machine; **c** blower; **d** sintering drop strength test machine; **e** sintering bins; **f** mechanical sieve, rotary drum test machine, post-drum sieving machine

Table 5.2 Sintered cup main equipment parameters

Installation	Parameters
Sintered post-drum sifter	Sieve hole size 6.3 × 6.3 mm; sieve surface inclination angle − 45° to + 45°; reciprocating speed 20 times/min; motor power 1.1 kW
Sintering drum test machine	Specification φ1000 × 750 mm; loading capacity 7.5 kg; motor power 1.5 kW
Sinter ore horizontal reciprocating mechanical screen	Specification 800 × 500 mm; Reciprocation speed 60 times/min; Motor power 1.5 kW
Sintered mixing and granulating machine	Specification φ600 × 1,400 mm; Rotating speed 7–30 r/min; Swinging inclination − 18° to + 25°; Motor power 4 kW
Sintered drop strength tester	Experimental drop height 2000 mm; motor power: lifting 1.1 kW, pouring 0.75 kW, panning 0.75 kW
Sintering cup system	Sintering cup specification φ300 × 850 mm; motor power: crusher 4 kW, sintering cup 1.5 kW

Primary mixing

The purpose of the primary mixing stage is to ensure the even distribution of all components, promoting uniformity and stability in the sintered ore composition. During mixing, water is added to wet the materials and aid in granulation, which improves the permeability of the sintered mixture and supports the smooth progression of sintering. In the primary mixing process, the weighed raw materials are placed on the mixing plate, where the appropriate amount of water is added to meet the requirements for sintered mixture granulation.

In accordance with the experimental procedure, 127 kg of sintered raw materials were weighed for primary mixing. The mixture was stirred alternately in the left and right pouring directions three times. After the preliminary mixing, water was gradually added to the sintered mixture in increments. After each water addition, the larger particles generated during the mixing process were crushed and mixed again. This process was repeated until a total of 8 wt% water was incorporated into the mixture. The primary mixing stage is considered complete when the mixture begins to form small agglomerates.

Secondary mixing

The objective of secondary mixing is to form granules from the primary mixed sintered material. After the primary mixing is completed, the material is transferred into the secondary mixing machine for pelletizing. The secondary mixing is carried out in a cylindrical blender, as shown in Fig. 5.8. The cylindrical mixer used in this experiment has a diameter of 0.6 m, with a rotational speed of 18.6 r/min and a tilt angle of − 5°. The mixing time for this stage is 5 min.

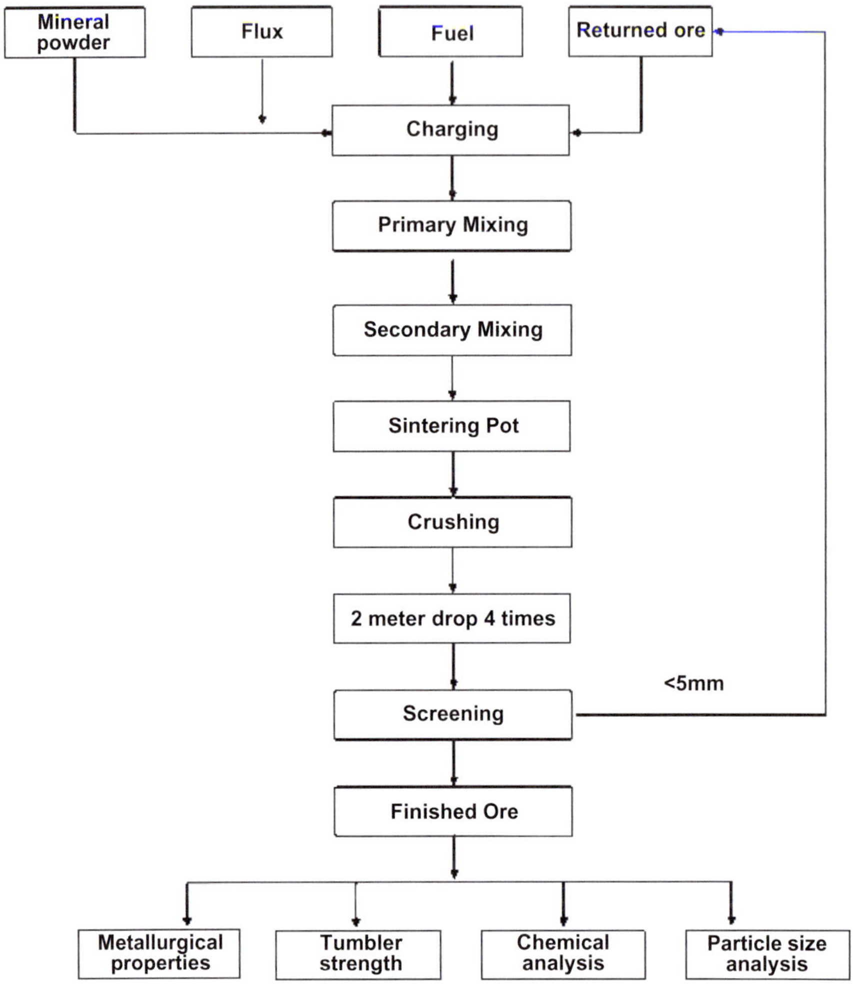

Fig. 5.6 Flow chart of the sintered cup experiment

Fig. 5.7 Charge container

Fig. 5.8 Cylinder blender

Sintering

In the first step, 3 kg of sintered ore with a particle size range of 10–16 mm is used as the base material. This is placed at the bottom of the sintering cup, followed by the addition of the sintered mixture after the secondary mixing. Once the material is loaded, the igniter is activated and automatically moved to the top of the sintering cup when the ignition temperature reaches 1050 °C. The ignition time is controlled at 188 s, with the ignition negative pressure set to 8000 Pa and the sintering negative pressure set to 16,000 Pa. The sintering process concludes when the temperature reaches its peak and begins to decrease.

The sintered ore is then cooled under bottom pumping conditions, with the cooling negative pressure maintained at 12,000 Pa. The cooling process continues until the exhaust gas temperature drops to 250 °C, at which point the pumping is stopped. The sintering cup will automatically discharge the sintered cake, which is then transferred to a single-roll crusher for crushing, producing the final sintered ore product.

Certain parameters in this sintering cup experiment are designed to simulate the operating conditions of two 550 m^2 sintering machines at Shougang Jingtang. The experimental parameters are listed in Table 5.3.

Table 5.3 Sintered cup experimental parameters

Parameter	Set value	Parameter	Set value
Negative ignition pressure/Pa	8000	Ignition time/s	188
Negative sintering pressure/Pa	16,000	Crusher running time/s	70
Negative cooling pressure/Pa	12,000	Pouring waiting time/s	70
Ignition temperature/°C	1050	Sintering thickness/mm	850
Cooling temperature/°C	250	First mixing water volume/L	8

Finished product inspection

① At the end of the sintering process, the material is automatically turned over, and the time taken for sintering is recorded.

② The weight of the sintered ore is measured to determine the quality of the sintered product.

③ The sintered material is placed in a falling test machine, with a fall height of 2 m. The material is subjected to four repeated drops.

④ After the falling test, the sintered ore is placed into a sieve machine for screening. The sieving is done for six particle size categories: > 40 mm, 40–25 mm, 25–16 mm, 16–10 mm, 10–5 mm, and < 5 mm. The ore finer than 5 mm is classified as cold return ore, while the ore larger than 5 mm is categorized as finished ore.

⑤ For the finished ore, a sample weighing 7.5 kg is taken from the 40–25 mm, 25–16 mm, and 16–10 mm size categories, and placed into a rotary drum for further testing. The drum rotation time is 8 min, with a rotational speed of 25 r/min, completing a total of 200 revolutions. After the test, the sample is removed and sieved through a 6.3 mm post-drum sieve machine. The percentage of the material that passes through the sieve is used to determine the drum index.

The sinter cup system used in this experiment is shown in Fig. 5.9, while the equipment used for the finished sintered ore inspection is displayed in Fig. 5.10.

⑥ The yield rate of sintered ore, particle size composition, vertical sintering speed, sintering loss rate, utilization coefficient, drum index, and other relevant parameters can be calculated using the formulas presented in Eqs. 5.1 to 5.6. Based on the results obtained from the experiment, each of these indices can be derived through the following calculation methods.

$$\text{Yield} = \frac{W_{\text{finished ore}} - W_{\text{bedding}}}{W_{\text{sintered ore}} - W_{\text{bedding}}} \times 100\% \tag{5.1}$$

$$\text{Vertical sintering speed} = \frac{H_{\text{layer}}}{T_{\text{Sintering}}} \tag{5.2}$$

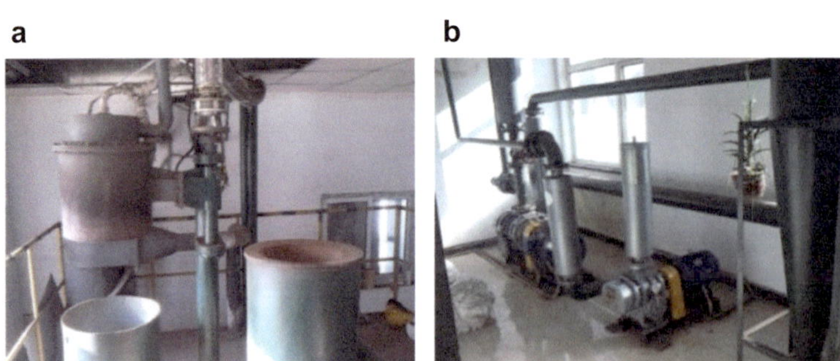

Fig. 5.9 Sintered cup system. **a** Sintering cup master system; **b** sintering extractor fan

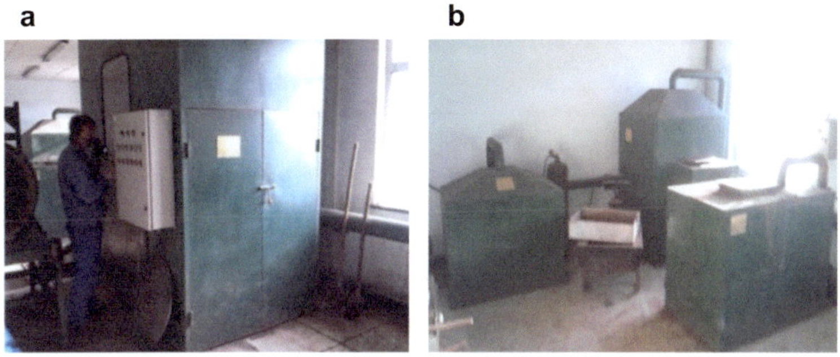

Fig. 5.10 Finished ore inspection equipment. **a** Sintered drop strength tester; **b** mechanical sieves, rotary drum testers, post-drum sifters

$$\text{Burn rate} = \frac{W_{\text{charge}} - \left(W_{\text{sintered ore}} - W_{\text{bedding}}\right)}{W_{\text{charge}}} \times 100\% \quad (5.3)$$

$$\text{Utilization factor} = \frac{W_{\text{finished sinter}} - W_{\text{bedding}}}{T_{\text{sintering}} \times S_{\text{sintering}}} \left(\text{t/m}^2 \times \text{h}\right) \quad (5.4)$$

$$\text{Drum index} = \frac{W_{>6.3\,\text{mm after drum}}}{W_{\text{sinter in the drum}}} \times 100\% \quad (5.5)$$

$$\text{Fuel consumption} = \frac{W_{\text{fuel additions}}}{W_{\text{finished sinter}}} (\text{kg/t}) \quad (5.6)$$

5.3 Experimental Design for Ironsands Sintering

In the sintering cup experiments, three types of ironsands—original, intermediate, and fine granularity—were tested at dosages of 2 wt%, 5 wt%, and 8 wt%, respectively. To ensure the accuracy of the experiments and minimize the impact of occasional errors, parallel experiments were conducted. In total, 18 experimental groups, with varying additions of ironsands, and one reference sample group were included, as shown in Table 5.4.

The experimental design consisted of a baseline sample and three treatment programs. The baseline sample served as the control experiment, which did not include any ironsands. Programs I, II, and III included 2 wt%, 5 wt%, and 8 wt% ironsands, respectively. In Program I and II, ironsands replaced Massaje ore, while in Program III, ironsands replaced Massaje ore, Peruvian raw ore, and steel slag. The chemical compositions of the various sintering ingredients used in the sintering cup experiments are presented in Table 5.5.

Due to varying moisture content in the ores, sintering batching calculations were based on the dry material mass. The actual batching quantities for each scheme are listed in Table 5.6.

Table 5.4 Overall experimental program of sintering cups for ironsands

Number of experiments	Reference frame	Dosed with 2 wt%	Dosed with 5 wt%	Dosed with 8 wt%
Primary grade ironsands	–	Twice	Twice	Twice
Intermediate grade ironsands	–	Twice	Twice	Twice
Fine grade ironsands	–	Twice	Twice	Twice

Table 5.5 Sintered cup experimental ingredient chemical composition/wt%

Raw material	TFe	SiO_2	CaO	MgO	Al_2O_3	TiO_2
Ironsands	55.63	4.13	0.6	3.74	3.38	11.41
Baking powder	63.31	6.3	0.24	0.22	1.14	–
Baka powder	64.76	2.88	0.19	0.23	1.4	–
Australian powder	61.68	3.64	0.22	0.21	2.1	–
Yandi fines	57.23	5.59	0.22	0.21	1.36	–
Massajeti	56.97	4.72	0.41	0.29	2.68	–
Peruvian raw ore	55.9	10.2	2.2	2.79	1.51	–
Afrikaans powder	64.28	4.48	0.2	0.19	1.78	–
Peruvian pelletizing powder	69.43	1.43	0.53	0.74	0.28	–
Steel slag	20.5	12.98	35.65	11.41	1.61	–
Galvanized iron sheet	72	0.75	0.25	0.08	0.27	–
Limestone	–	0.61	53.15	1.36	–	–
Dolomite	–	1.62	30.16	21.5	–	–
Blast furnace return ore	55.8	5.5	10.3	1.6	1.59	–
Dusting ash	47	4.2	3.75	1.58	1.63	–

The dry and wet ingredient batching percentage schemes used in the sintering cup experiments are shown in Tables 5.7 and 5.8, respectively.

To determine the moisture content of each batch, 1 kg of the mixed wet material was weighed, dried, and re-weighed. This provided the moisture content for each experimental group. The mass of the wet material was recorded before and after loading into the sintering cup to calculate the amount of wet material in the sintering cup. After sintering, the mass of the sintered ore was weighed following single-roll crushing, allowing for the determination of the sintered ore mass post-crushing, as shown in Table 5.9.

Table 5.6 Actual weight of dry ingredients for sintered cup experiment/kg

Name	Reference frame	Option 1	Option 2	Option 3
Brazilian sintering powder	7.22	7.23	7.24	7.25
Brazil	13.66	13.67	13.69	13.72
Australian powder	16.91	16.92	16.94	16.98
Yandi fines	9.78	9.79	9.80	9.82
Massajeti	3.50	2.10	0	0
Peruvian raw ore	1.60	1.60	1.60	0
Indonesian ironsands	0	1.32	3.31	5.31
Afrikaans powder	2.94	2.94	2.95	2.95
Peruvian pelletizing powder	4.15	4.15	4.16	4.16
Steel slag	2.28	2.28	2.28	1.88
Iron oxide scale	0.99	0.99	0.99	1.00
Desulfurization slag iron	0.73	0.73	0.73	0.73
Sintering return ore	33.04	33.07	33.12	33.18
Blast furnace return ore	11.33	11.34	11.35	11.37
Limestone	5.47	5.44	5.39	5.20
Lime	5.66	5.67	5.68	5.69
Dust removal ash	2.68	2.68	2.68	2.68
Coke powder	3.95	3.95	3.95	3.95
Blast furnace gas ash	1.13	1.13	1.14	1.14
Total	127	127	127	127

5.4 Room-Temperature Characteristics of Sinter Ore Containing Ironsands

5.4.1 Chemical Composition

The chemical composition of the finished sintered ores obtained from the sinter cup experiments, under different ironsands particle sizes and dosing conditions, was analyzed using titrimetric chemical analysis. The results are presented in Table 5.10.

Total iron content

The total iron (TFe) content in sintered ore is a crucial index for assessing its quality. In the blast furnace ironmaking process, every 1 wt% increase in the grade of iron-containing charge can reduce the coke ratio by 2 wt% and increase the furnace output by 3 wt%. As shown in Fig. 5.11, the total iron (TFe) content of the sintered ore remains stable at approximately 55.56 wt%, with only minimal fluctuation of less than 0.3 wt%. According to the metallurgical industry standards for high-quality iron ore, a fluctuation range of $\pm$ 0.4 wt% for TFe content is considered acceptable.

Table 5.7 Sintered cup experimental dry ingredient batching program/wt%

Variety	Reference frame	Option 1	Option 2	Option 3
Brazilian sintering powder	10.6	10.6	10.6	10.6
Baka powder	19.6	19.6	19.6	19.6
Australian powder	24.1	24.1	24.1	24.1
Yandi fines	13.5	13.5	13.5	13.5
Massajeti	5.0	3.0	0	0
Peruvian raw ore	2.4	2.4	2.4	0
Indonesian ironsands	0	2.0	5.0	8.0
Afrikaans powder	4.4	4.4	4.4	4.4
Peruvian pelletizing powder	5.8	5.8	5.8	5.8
Steel slag	3.4	3.4	3.4	2.8
Iron oxide scale	1.5	1.5	1.5	1.5
Limestone	6.8	6.8	6.8	6.8
Desulfurization slag iron	1.0	1.0	1.0	1.0
Dusting ash	1.9	1.9	1.9	1.9

Note Molten white ash, fuel coke powder and blast furnace gas ash are not included in the calculation of dry material ratio

Therefore, it is reasonable to incorporate a certain proportion of ironsands in the sintering process while maintaining the TFe content within this acceptable range.

FeO content

From Fig. 5.12, it is observed that when the proportion of ironsands in the sintering process is 0 wt% (baseline sample), the FeO content in the sintered ore is 4.56 wt%. When the proportion of ironsands ranges from 2 to 8 wt%, the FeO content fluctuates between 4.45 wt% and 8.48 wt%, with an average of 6.23 wt%. This increase in FeO content is due to the addition of ironsands, which is a form of vanadium-titanium magnetite. The introduction of ironsands into the sintering mixture increases the magnetite phase content, leading to an upward trend in FeO content in the sintered ore.

The FeO content in sintered ore is closely related to several factors, including the mineral type of the iron ore powder, the solid fuel ratio, the sintering atmosphere, and the permeability of the sintered layer. For sintering processes using magnetite powder as the main iron source, the FeO content tends to be higher because magnetite (Fe_3O_4) contains a higher proportion of Fe^{2+} ions. In contrast, sintering with enriched ore, which is primarily composed of hematite (Fe_2O_3), typically results in lower FeO content. Additionally, when the amount of solid fuel used in the sintering process is larger, a more pronounced reducing atmosphere is generated in the sintered material layer. This local reduction leads to the reduction of higher-valent iron oxides, thereby increasing the FeO content of the sintered ore [5].

Table 5.8 Wet material dosing scheme for sintered cup experiments/wt%

Name	Reference frame	Option 1	Option 2	Option 3
Brazilian sintering powder	5.68	5.69	5.70	5.71
Brazil	10.75	10.76	10.78	10.80
Australian powder	13.31	13.32	13.34	13.37
Yandi fines	7.70	7.71	7.72	7.73
Massajeti	2.76	1.65	0.00	0.00
Peruvian raw ore	1.26	1.26	1.26	0.00
Indonesian ironsands	0.00	1.04	2.61	4.18
Afrikaans powder	2.31	2.31	2.32	2.32
Peruvian pelletizing powder	3.27	3.27	3.28	3.28
Steel slag	1.79	1.80	1.80	1.48
Iron oxide scale	0.78	0.78	0.78	0.79
Desulfurization slag iron	0.57	0.57	0.57	0.57
Sintering return ore	26.01	26.04	26.08	26.12
Blast furnace return ore	8.92	8.93	8.94	8.95
Limestone	4.31	4.28	4.24	4.09
Lime	4.46	4.46	4.47	4.48
Dusting ash	2.11	2.11	2.11	2.11
Coke powder	3.11	3.11	3.11	3.11
Blast furnace gas ash	0.89	0.89	0.90	0.90
Total	100	100	100	100

To ensure good reducibility, the FeO content in the sintered ore should not exceed 11 wt%. For high-quality sintered ore, the FeO content should be kept below 9 wt%. In the sintering cup experiment, the FeO content of the sintered ore obtained was consistently below 9 wt%, indicating that the addition of a certain proportion of ironsands does not cause the FeO content to exceed the relevant standards.

Analysis of binary basicity

From Fig. 5.13, the basicity of the sintered ore in the baseline sample is $R = 2.16$. For the sintered ores obtained with the addition of ironsands, the average basicity is $R = 2.13$. This indicates that the basicity of the sintered ore remains relatively stable, with only minimal fluctuation following the addition of ironsands [6].

Sintered ore basicity ($R = CaO/SiO_2$) is a key quantitative index influencing the phase composition of sintered ores. In China, the majority of sintered ores are produced based on the theory of composite calcium ferrate formation and high basicity sintering. The composite calcium ferrate phase has high strength, second only to hematite, and also exhibits good reducibility. To ensure that the main binding phase in sintered ore is calcium ferrate, it is essential to increase the basicity of the sintered mixture. This allows for sufficient calcium oxide in the mixture to combine

Table 5.9 Moisture content and material weight per group for sintering cup experiments/kg

Number	Moisture content	Total before adding cup	Remainder after filling the cup	Mass in cup	Weight of sinter ore after thermal crushing
Reference frame	6.6	138.04	19.72	118.32	103.27
Primary particle size -2%-1	7.1	141.17	30.44	110.73	96.88
Primary particle size -2%-2	6.8	141.56	30.31	111.25	96.09
Primary particle size -5%-1	6.9	141.84	26.88	114.96	101.52
Primary particle size -5%-2	7.1	140.69	27.61	113.08	99.26
Primary particle size -8%-1	6.9	140.51	27	113.51	101.43
Primary particle size -8%-2	6.9	140.53	27.52	113.01	99.06
Intermediate particle size -2%-1	7	140.77	27.48	113.29	100.25
Intermediate particle size -2%-2	7	141.43	28.67	112.76	97.84
Intermediate particle size -5%-1	6.9	138.84	22.78	116.06	101.69
Intermediate particle size -5%-2	7	141.89	26.96	114.93	100.30
Intermediate particle size -8%-1	6.8	140.69	24.54	116.15	102.51
Intermediate particle size -8%-2	6.6	140.21	24.21	116	102.38
Fine particle size -2%-1	6.9	141.38	27.95	113.43	100.72
Fine particle size -2%-2	6.9	139.98	27.41	112.57	97.42

(continued)

Table 5.9 (continued)

Number	Moisture content	Total before adding cup	Remainder after filling the cup	Mass in cup	Weight of sinter ore after thermal crushing
Fine particle size -5%-1	7.1	141.26	28.28	112.98	99.90
Fine particle size -5%-2	7	141.05	28.16	112.89	100
Fine particle size -8%-1	6.8	140.79	26.32	114.47	100.40
Fine particle size -8%-2	6.9	139.92	24.82	115.1	101.58

Note The table labels are organized as follows: "grain size—dosage of ironsands—parallel sample number." For example, "intermediate grain size – 2% – 1" refers to "10-min milled ironsands—2 wt% dosage—parallel sample 1#". "Original grain size" denotes unprocessed ironsands, while other labels follow the same pattern for clarification

with iron ore powder, forming calcium ferrate complexes that provide both strength and reducibility [7, 8]. High basicity sintered ores typically require a basicity of $1.8 < R < 2.2$, and the sintered ores containing ironsands fall within this range.

MgO content

As illustrated in Fig. 5.14, the average MgO content in sintered ore is 1.89 wt% under various ironsands dosing conditions. In comparison, the MgO content in the reference sample is 1.79 wt%. The addition of ironsands increases the MgO content in the sintered ore by 5.6%, primarily due to the higher MgO content present in the ironsands.

MgO can have differing effects on the sintered ore depending on its concentration. Initially, divalent Mg^{2+} ions can replace divalent Fe^{2+} ions in magnetite, reducing the lattice defects of magnetite, stabilizing it, and decreasing the likelihood of secondary hematite formation. This, in turn, reduces the low-temperature reduction pulverization of the sintered ore. However, excessively high MgO content can lead to the formation of other complex compounds within the sintered ore. These compounds can generate internal stresses due to differences in crystallization ability, which could result in the formation of microcracks. This phenomenon could exacerbate pulverization of the sintered ore.

Al_2O_3 content

As shown in Fig. 5.15, the addition of ironsands to the sintering process results in a decrease in Al_2O_3 content in the sintered ore. The Al_2O_3 content in the reference sample is 1.95 wt%, whereas the average Al_2O_3 content in the sintered ore with added ironsands is 1.86 wt%. This represents a reduction of 4.6 wt%.

A high Al_2O_3 content in sintered ore significantly affects its quality, especially contributing to increased chalking during low-temperature reduction. This

Table 5.10 Chemical composition of sintered ore containing ironsands under varying particle sizes and dosing levels/wt%

	TFe	FeO	SiO_2	CaO	MgO	Al_2O_3	TiO_2	C
Reference frame	55.68	4.56	5.36	11.48	1.79	1.95	0.178	0.06
Primary particle size -2%	55.41	5.21	5.39	11.56	1.94	1.87	0.303	0.05
Primary particle size -2%	55.44	5.03	5.30	11.76	1.84	1.86	0.361	0.04
Primary particle size -5%	55.44	5.96	5.35	11.58	1.92	1.90	0.361	0.07
Primary particle size -5%	55.32	6.03	5.26	11.57	1.84	1.86	0.584	0.04
Primary particle size -8%	55.18	6.68	5.34	11.48	1.89	1.88	0.858	0.08
Primary particle size -8%	55.42	6.03	5.25	11.41	1.93	1.90	0.894	0.02
Intermediate particle size -2%	55.73	6.90	5.44	11.14	1.96	1.86	0.500	0.06
Intermediate particle size -2%	55.97	7.33	5.61	11.27	1.93	1.89	0.468	0.03
Intermediate particle size -5%	55.42	6.68	5.40	10.94	2.10	1.92	0.739	0.11
Intermediate particle size -5%	55.40	6.97	5.39	11.16	2.07	1.86	0.767	0.03
Intermediate particle size -8%	55.79	4.96	5.27	11.12	1.87	1.88	0.935	0.05
Intermediate particle size -8%	55.81	4.45	5.31	10.84	2.26	1.86	0.956	0.03
Fine particle size -2%	55.30	4.88	5.23	11.46	1.80	1.80	0.389	0.02
Fine particle size -2%	55.67	4.89	5.19	11.52	1.68	1.78	0.462	0.03

(continued)

Table 5.10 (continued)

	TFe	FeO	SiO_2	CaO	MgO	Al_2O_3	TiO_2	C
Fine particle size -5%	55.83	8.48	5.33	11.14	1.81	1.85	0.895	0.05
Fine particle size -5%	55.63	6.68	5.25	11.71	1.69	1.80	0.648	0.03
Fine particle size -8%	55.52	7.47	5.30	11.36	1.83	1.86	0.889	0.11
Fine particle size -8%	55.69	7.54	5.38	11.18	1.71	1.83	0.878	0.13

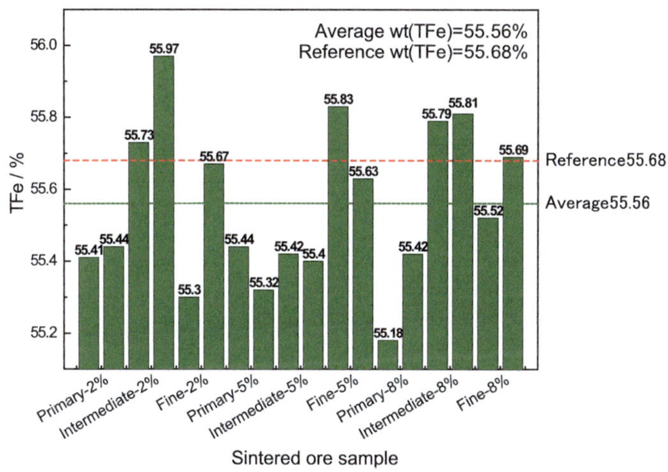

Fig. 5.11 Total iron (TFe) content of sintered ores containing ironsands under varying ironsands particle sizes and dosing levels

is primarily due to the fact that alumina contributes to stress concentration during the reduction of hematite, leading to the expansion and crack propagation in the sintered ore. Therefore, the incorporation of ironsands helps mitigate the negative effects of Al_2O_3 on sinter quality.

TiO_2 content

From Fig. 5.16, it can be observed that the TiO_2 content in the sintered ore increases linearly as the ironsands ratio rises from 0 to 8 wt%. Specifically, the TiO_2 content in the sintered ore is 0.18 wt% when the proportion of ironsands is 0 wt% (base sample), and it approaches 1 wt% when the ironsands proportion reaches 8 wt%.

The TiO_2 content in the sintered ore significantly influences its low-temperature reduction pulverization performance. Therefore, it is crucial to control the TiO_2 content, as excessive amounts can negatively impact the performance. However, some blast furnaces in China, due to the need for furnace protection during specific

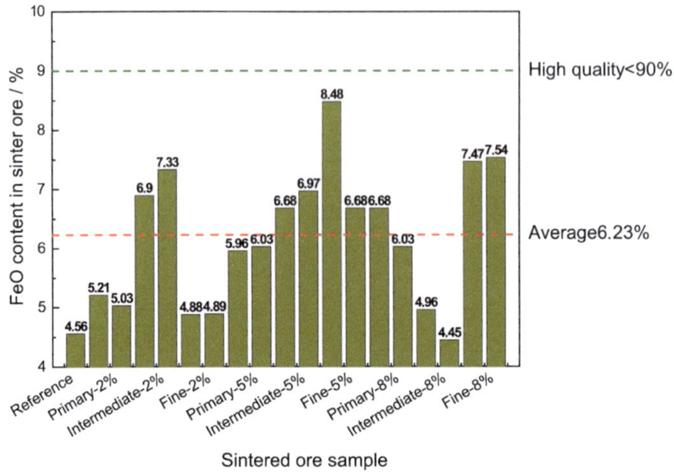

Fig. 5.12 FeO content of sintered ores containing ironsands under varying ironsands particle sizes and dosing levels

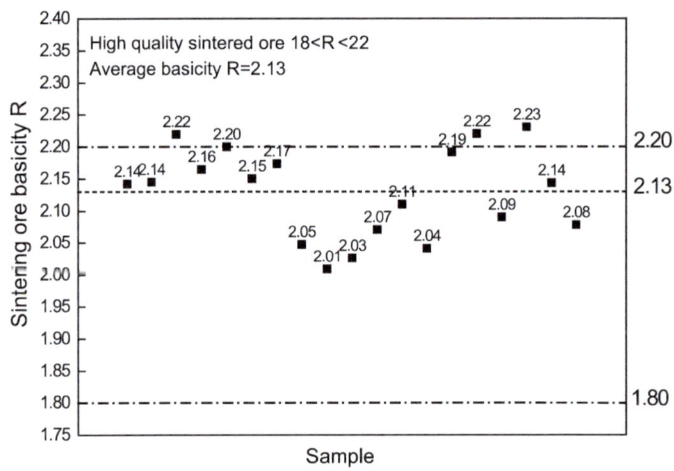

Fig. 5.13 Basicity of sintered ores containing ironsands under varying ironsands particle sizes and dosing levels

operational phases, intentionally incorporate a certain amount of titanium dioxide in sintered or pelletized ores to enhance the furnace's longevity.

Moreover, the TiO_2 content in the returned ore is notably high. Specifically, the TiO_2 content in the sintered returned ore with a particle size smaller than 1 mm is the highest. This is attributed to the top-down pumping mineralization process in the sintering operation, where the temperature within the sintered material layer varies significantly. The mixture at the top of the sintered material layer remains

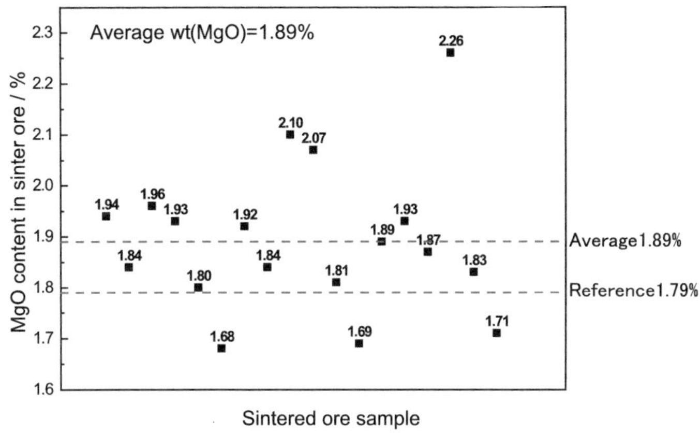

Fig. 5.14 MgO content of sintered ores containing ironsands under varying ironsands particle sizes and dosing levels

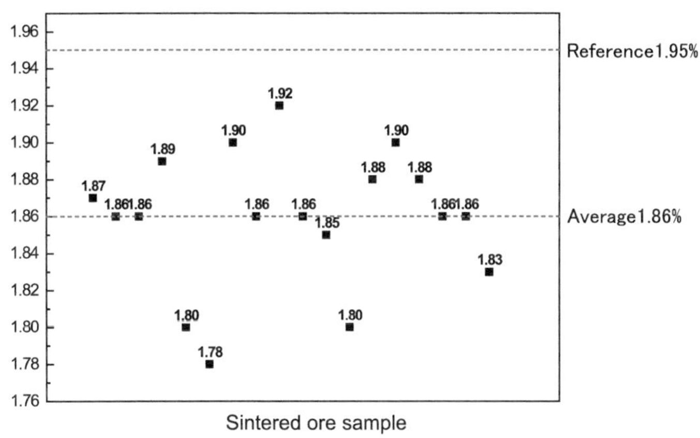

Fig. 5.15 Al_2O_3 content of sintered ores containing ironsands under varying ironsands particle sizes and dosing levels

incompletely mineralized due to insufficient heat supply, resulting in the sintered ore at the top remaining in bulk form. Consequently, this portion of the sintered ore is primarily returned to the ore, along with fine-grained seabird ores. Additionally, as the particle size of ironsands is predominantly < 1 mm, the sintered ore with particle sizes smaller than 1 mm exhibits the highest content of ironsands, which leads to the increased TiO_2 content in the returned ore.

Harmful element content

An analysis of the harmful elements, phosphorus (P) and sulfur (S), in the sintered ore revealed that the addition of ironsands to the sintering process could reduce the levels of these harmful elements to some extent. This reduction has a positive impact

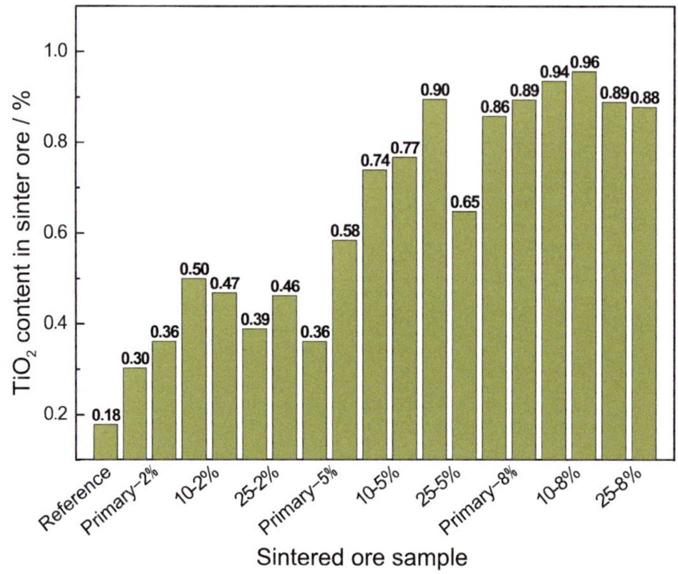

Fig. 5.16 TiO_2 content of sintered ores containing ironsands under varying ironsands particle sizes and dosing levels

on both blast furnace smelting and converter steelmaking, as the P and S contents in ironsands are lower compared to other types of added iron ores.

Overall, the inclusion of 2 to 8 wt% of ironsands in the sintering process does not have a significant detrimental effect on the main components of the sintered ore. In fact, the alumina content and harmful element levels are improved under these conditions.

5.4.2 *Microstructure*

The microstructure of sintered ores plays a vital role in determining their metallurgical properties, which vary based on the raw materials and sintering process parameters. The primary iron-containing minerals include magnetite (Fe_3O_4), hematite (α-Fe_2O_3 and γ-Fe_2O_3), and wüstite (Fe_xO), while the bonding phase minerals comprise SFCA, wollastonite, dicalcium silicate, tricalcium silicate, ferro-peridotite, and calcium-ferric peridotite.

Under different sintering conditions, these mineral types exhibit significant differences in the microstructure of the sintered ore, which in turn affects the ore's metallurgical properties, such as reducibility and mechanical strength. Analyzing the microstructural characteristics of finished sintered ores derived from ironsands is thus essential for a more intuitive and effective assessment of sintered ore quality.

Mineralogical analysis and microstructural observations were conducted on sintered ores prepared using different grain sizes and dosages of ironsands. The sintered ore samples were analyzed using SEM–EDS (Scanning Electron Microscopy with Energy Dispersive X-ray Spectroscopy) with a FEI Quanta 250 environmental scanning electron microscope to determine the mineral composition of various sintered ore samples. A total of 10 sets of sintered ore samples, including the reference samples, were selected for SEM analysis. The sample numbers for the micro-morphological experiments are presented in Table 5.11.

The experimental results, depicted in Figs. 5.17, 5.18, 5.19 and 5.20, show clear variations in the mineral types and their microstructures in the sintered ores derived from ironsands under different particle sizes and dosing amounts.

Micro-morphological analysis of sintered ore without ironsands

Figure 5.17 presents the results of scanning electron microscope (SEM) microanalysis of the benchmark sintered ore, which does not contain any added ironsands.

As depicted in Fig. 5.17, the primary mineral phase in the benchmark sintered ore is acicular calcium ferrate, with the majority of this phase being tetragonal calcium ferrate, which is both strong and reducible. Hematite is well-cemented within the calcium ferrate composite, and the calcium ferrate liquid phase is surrounded by numerous micro-zones. Notably, no skeletal crystalline rhombic hematite, which would be detrimental to the sintered ore, is observed. In Fig. 5.17a, the black areas (P) represent tiny holes in the sintered ore. The presence of micropores does not significantly impact the strength of the sintered ore, though thin-walled holes can weaken its structural integrity. Figure 5.17d illustrates the presence of black silicate embedded within the calcium ferrate composite, where the calcium ferrate phases are tightly cemented together.

Overall, the mineral composition of the benchmark sample is predominantly composed of calcium ferrate composites, which are effectively cemented with hematite and silicate. This represents an ideal sintered mineral structure.

Table 5.11 Experimental numbering for micro-morphology analysis of ironsands sintered ore

Number	Sintered ore sample
1	Reference frame
2	Primary particle size -2%-1
3	Primary particle size -5%-1
4	Primary particle size -8%-1
5	Intermediate particle size -2%-1
6	Intermediate particle size -5%-1
7	Intermediate particle size -8%-1
8	Fine particle size -2%-2
9	Fine particle size -5%-1
10	Fine particle size -8%-2

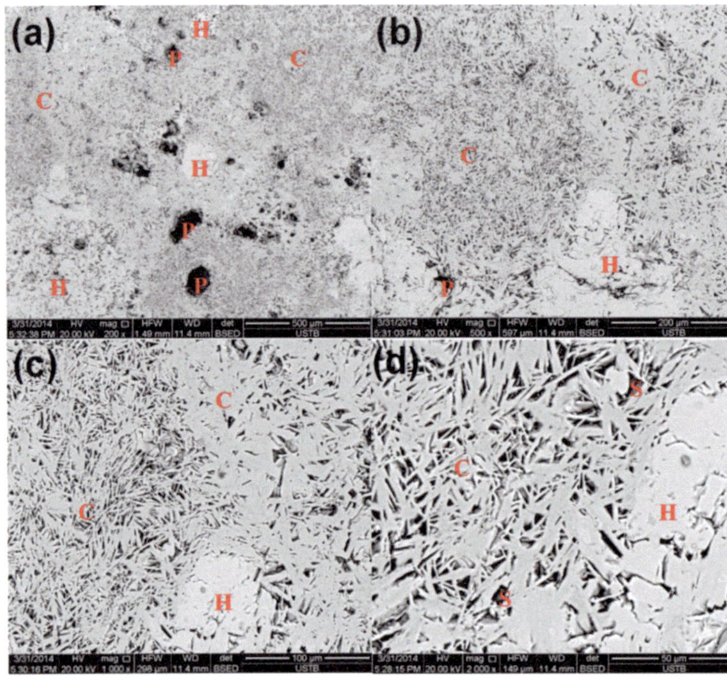

Fig. 5.17 Scanning electron microscope (SEM) microanalysis of the benchmark sintered ore sample at different magnifications. H: Fe_2O_3; M: Fe_3O_4; CF: complex calcium ferrate; S: silicate phase; P: pore

Micro-morphological analysis of sintered ore with original grain size of ironsands

Figure 5.18 shows the results of microscopic morphological analysis for sintered ore containing 2 to 8 wt% of original grain size ironsands.

From Fig. 5.18, it can be observed that the sintered ore containing 2–8 wt% original grain size ironsands has a mineral composition that is largely similar to the benchmark sample. The bonding phase is primarily calcium ferrate, while the silicate phase is embedded within the voids of the calcium ferrate and hematite phases. These phases are well cemented with one another.

Figure 5.18a shows sintered ore with 2 wt% original grain size ironsands, where the hematite phase is surrounded by a magnetite phase. This phenomenon is due to an insufficient oxidizing atmosphere in localized regions during the sintering process, which prevents complete oxidation and mineralization of magnetite, resulting in hematite forming around an inner core of magnetite. In Fig. 5.18b, which is an enlarged view of the red circle in Fig. 5.18a, EDS spectroscopy reveals that the area corresponds to a composite calcium ferrite. The structure of this composite calcium ferrite is lamellar, with the individual layers intricately interwoven in a three-dimensional arrangement. This structure contributes significantly to the strength of

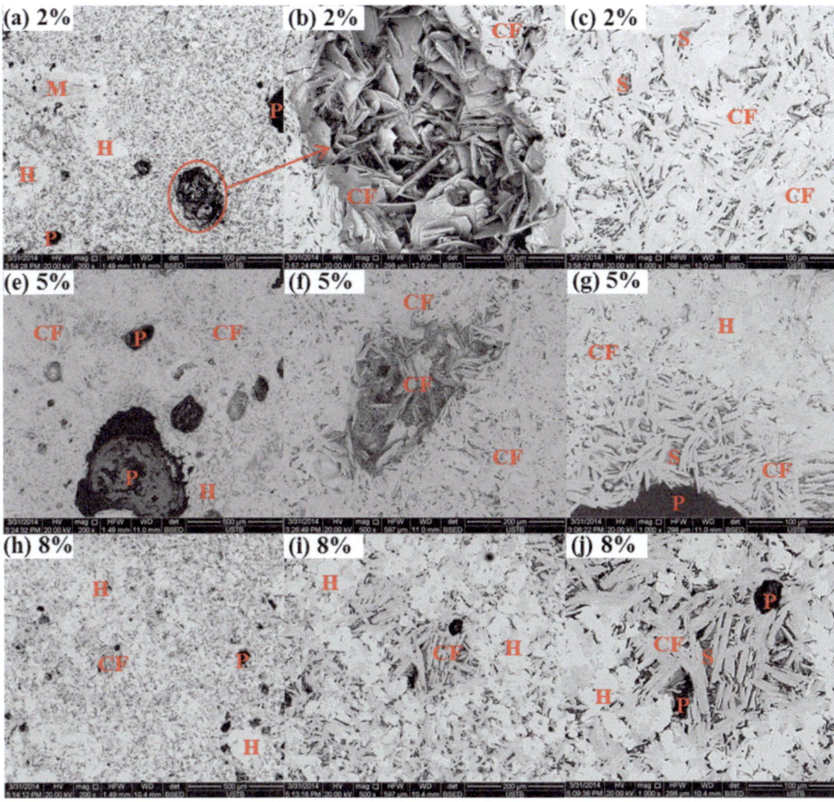

Fig. 5.18 SEM microanalysis of sintered ore samples with varying proportions of original grain size ironsands. H: Fe_2O_3; M: Fe_3O_4; CF: complex calcium ferrate; S: silicate phase; P: pore

the sintered ore, as the lamellar and intertwined configuration increases its mechanical strength.

Additionally, the laminated three-dimensional structure increases the specific surface area of the sintered ore, enhancing the area exposed to reducing gases during reduction. This increased surface area improves the reducibility of the sintered ore. When the dosing of original grain size ironsands increases to 8 wt% (Fig. 5.18h–j), a significant increase in the hematite phase is observed. This hematite phase predominantly takes the form of anhedral crystals or semi-autocrysts, which are tightly cemented with the calcium ferrite phase, forming the primary matrix of the sintered ore. The calcium ferrite phase is well-developed, appearing as laths, stripes, or an intertwined fused structure.

Overall, compared to the benchmark sample without ironsands, the addition of original grain size ironsands does not significantly alter the microstructure of the sintered ore. However, when the proportion of ironsands reaches 8 wt%, the hematite

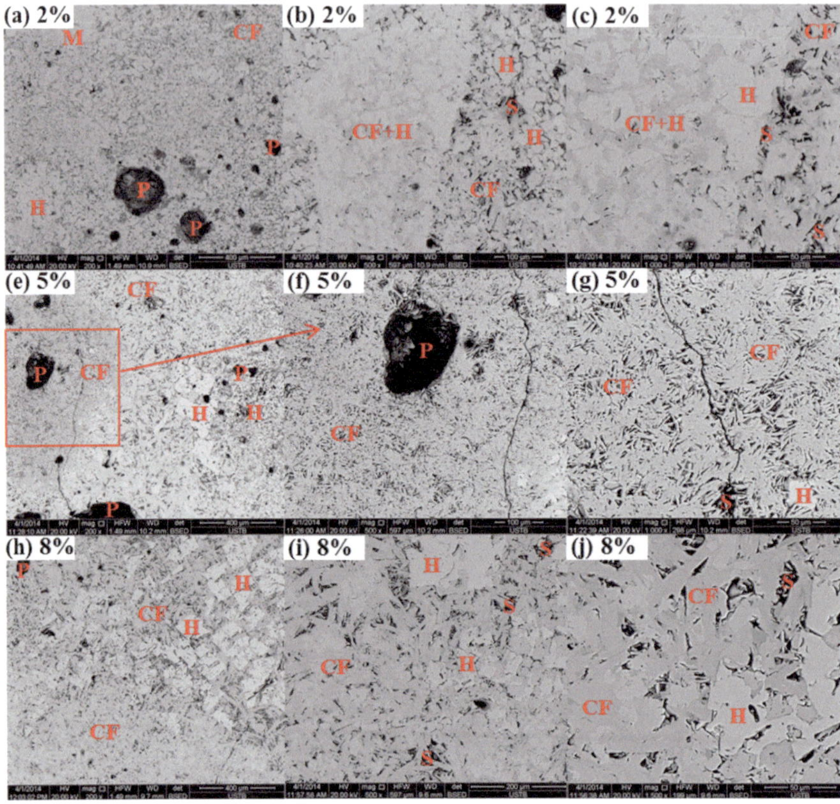

Fig. 5.19 SEM microanalysis of sintered ore samples with varying proportions of intermediate-grained ironsands. H: Fe_2O_3, M: Fe_3O_4; CF: complex calcium ferrate; S: silicate phase; P: pore

phase increases, which negatively impacts the low-temperature reduction and pulverization performance of the sintered ore. This is due to the volumetric expansion of hematite during its reduction to magnetite.

Micro-morphological analysis of sintered ore with intermediate-grained ironsands

Figure 5.19 presents the results of SEM microanalysis of sintered ore containing 2–8 wt% of intermediate-sized ironsands.

As shown in Fig. 5.19a, the mineral composition of the sintered ore primarily consists of calcium ferrite, magnetite, and hematite, with a relatively minor presence of silicate liquid phase. The hematite phase is predominantly distributed in granular form and is cemented with calcium ferrite. Microstructural examination reveals holes within the sintered ore, and the hematite grains surrounding these holes are well-developed. This phenomenon is related to the stronger oxidizing atmosphere around the holes, which is favorable for hematite formation.

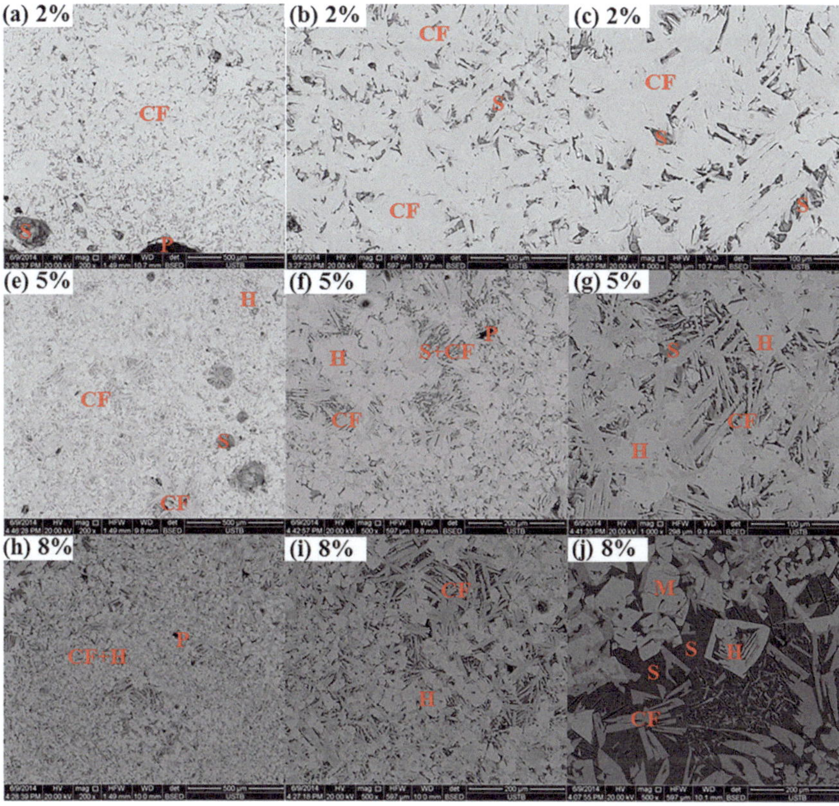

Fig. 5.20 SEM microanalysis of sintered ore samples with varying proportions of fine-grained ironsands

In Fig. 5.19b, c, large areas of fused structures consisting of calcium ferrite and hematite are observed, where the phases are combined in an anhedral-crystal form. These fused structures effectively prevent the low-temperature reduction pulverization of hematite, which enhances the overall strength of the sintered ore. In Fig. 5.19b, a clear boundary is visible between the fused hematite-calcium ferrite microzone and the adjacent calcium ferrite microzone. This boundary is attributed to the formation of the hematite-calcium ferrite melt microzone, which occurs due to larger iron ore particles that are mineralized by quicklime and retain their original morphology, while the surrounding calcium ferrite bonding phase cements it in place.

Figure 5.19e reveals a needle-like interwoven calcium ferrite phase surrounding the holes and granular hematite. A significant amount of regenerated hematite is formed around the holes, which negatively impacts the metallurgical properties of the sintered ore, increasing the low-temperature reduction chalking during the reduction process. The red box in Fig. 5.19e is magnified in Fig. 5.19f, showing a concentrated distribution of the needle-like calcium ferrite phase.

Figure 5.19h displays an increase in the rhombic hematite phase within the sintered ore, encapsulated by fused and acicular calcium ferrite. This structure leads to volume expansion during the reduction of the sintered ore, which adversely affects its metallurgical properties. However, in Fig. 5.19i, j, hematite appears in the form of anhedral crystals or semi-autocrysts, which do not negatively impact the metallurgical properties of the sintered ore.

In general, for sintered ore containing intermediate-grained ironsands, the microstructure is similar to that of the reference sample when the ironsands proportion is between 2 and 5 wt%. However, when the proportion reaches 8 wt%, the mineral composition and microstructure of the sintered ore begin to deteriorate, leading to potential challenges in the ore's performance.

Micro-morphological analysis of sintered ore with fine-grained ironsands

Figure 5.20 presents the results of scanning electron microscope (SEM) analysis for sintered ore containing 2–8 wt% fine-grained ironsands.

From Fig. 5.20a–c, it is clear that when the proportion of fine-grained ironsands is 2 wt%, the mineral composition of the sintered ore is primarily composed of calcium ferrite and silicate phases. The silicate phase is mostly distributed within the voids of the calcium ferrite phase, filling these voids. The calcium ferrite phase is mainly fused and lath-like in structure, resulting in a compact structure with high strength and good reducibility—characteristics of an ideal mineral structure for sintered ore. Thus, the microstructure of sintered ore with 2 wt% fine-grained ironsands is comparable to the reference sample.

As the proportion of fine-grained ironsands increases to 5 wt% (Fig. 5.20e–g), the hematite phase begins to emerge. The calcium ferrite phase alternates with the silicate phase, which significantly reduces the reducibility of the sintered ore. In Fig. 5.20h–j, when the fine-grained ironsands proportion reaches 8 wt%, the sintered ore exhibits an intertwined structure of calcium ferrite and hematite phases. The hematite phase predominantly appears in granular form, and the relative content of calcium ferrite decreases. The calcium ferrite phase is cemented to the silicate phase, and hematite appears loosely interspersed between the hematite and calcium ferrite phases. This structure weakens the strength of the sintered ore, which is detrimental to its performance. In Fig. 5.20j, skeletal crystalline rhombic hematite is evident, encapsulated by the silicate phase. Since both silicate and hematite phases are brittle, this structure compromises the blast furnace metallurgical performance of the sintered ore. Furthermore, EDS analysis revealed the presence of Ti in the silicate glass phase, which exacerbates the deterioration of the low-temperature reduction pulverization index of the sintered ore.

Overall Summary

(1) The benchmark samples are primarily composed of a calcium ferrate complex, which is well-cemented with hematite and silicate. This mineral composition represents an ideal structure for sintered ore.
(2) The addition of original grain size ironsands has little noticeable effect on the primary mineral types or the microstructure of the sintered ore.

(3) When the proportion of intermediate-grained ironsands is between 2 and 5 wt%, the mineral types and microstructure of the sintered ore closely resemble those of the reference sample. However, when the proportion increases to 8 wt%, the microstructure begins to deteriorate.
(4) At an 8 wt% proportion of fine-grained ironsands, the Ti element is detected in the silicate glass phase. This presence of Ti negatively impacts the low-temperature reduction pulverization performance of the sintered ore.

5.4.3 Particle Size Distribution

The particle size distribution of the finished sintered ore with the addition of ironsands was analyzed through particle size screening. The results, presented in Table 5.12, provide insight into the particle size composition of the finished sintered ore under different particle sizes and dosing levels of ironsands. The corresponding average particle sizes are depicted in Figs. 5.21 and 5.22.

From the perspective of varying dosing ratios, the following trends were observed. Original-Grained Ironsands: The average particle size shows minimal variation across different dosing ratios. Specifically, the particle size is slightly smaller when 5 wt% is added, while the differences between 2 and 8 wt% additions are negligible. Intermediate-Grained Ironsands: For 2 and 5 wt% additions, the average particle size does not show significant changes, but for the 8 wt% addition, the average particle size is somewhat improved. Fine-Grained Ironsands: As the addition of fine-grained ironsands increases, the average particle size of the sintered ore decreases correspondingly.

From the perspective of different ironsands particle sizes, the average particle size of the finished sintered ore containing original and fine-grained ironsands fluctuates around that of the baseline sample. In contrast, the average particle size of the sintered ore with intermediate-grained ironsands is less favorable.

This variation is attributed to the fact that both original and fine-grained ironsands can act as nucleating particles and adhere as fine powder within the sintered material. However, intermediate-sized ironsands tend to disrupt the permeability of the sintered material layer, resulting in a reduced average particle size. This reduction in particle size leads to an increase in the amount of powder, thereby raising the volume of return ore and ore returning to the blast furnace. Such an effect compromises the permeability of the blast furnace and disrupts the smooth flow within the furnace.

The particle size composition of the finished ore is a critical indicator of sintered ore quality. The results of the particle size distribution of the finished sintered ore are presented in Figs. 5.23, 5.24 and 5.25.

As shown in Fig. 5.23, when sintering with original-grain ironsands, increasing the proportion of ironsands in the sintering process leads to a reduction in the proportion of both large sintered ore (>40 mm) and returned ore (<5 mm), with these reductions occurring to varying extents. Notably, the smallest proportion of returned ore (smaller than 5 mm) is observed when the proportion of ironsands is increased to 8 wt%.

Table 5.12 Particle size distribution of finished sintered ore

Number	> 40 mm	40–25 mm	25–16 mm	16–10 mm	10–5 mm	5–0 mm	Average particle size mm
Reference frame	16.84	21.04	13.28	11.71	15.53	21.59	21.21
Primary particle size -2%-1	18.00	22.49	11.78	12.67	14.66	20.40	21.98
Primary particle size -2%-2	16.26	23.57	13.89	11.57	14.99	19.71	21.76
Primary particle size-5%-1	10.24	31.52	10.63	12.17	14.78	20.65	20.75
Primary particle size-5%-2	9.02	25.12	16.73	11.71	18.55	18.88	19.49
Primary particle size -8%-1	14.57	25.76	13.99	12.64	15.38	17.66	21.76
Primary particle size -8%-2	9.92	27.27	16.73	13.01	14.76	18.33	20.50
Intermediate particle size -2%-1	14.17	22.49	13.15	12.99	19.52	17.67	20.69
Intermediate particle size -2%-2	8.64	25.28	13.16	14.79	19.75	18.38	19.10
Intermediate particle size -5%-1	11.00	27.84	13.14	12.70	17.10	18.23	20.63
Intermediate particle size -5%-2	7.69	22.17	15.25	15.71	20.20	18.98	18.21
Intermediate particle size -8%-1	22.59	22.54	11.68	10.46	14.97	17.77	23.94
Intermediate particle size -8%-2	32.37	15.45	10.00	9.75	14.50	17.93	26.06
Fine particle size -2%-1	29.23	16.89	12.39	10.48	14.83	16.17	25.52
Fine particle size -2%-2	20.26	23.69	13.62	12.14	15.02	15.27	23.71

(continued)

Table 5.12 (continued)

Number	> 40 mm	40–25 mm	25–16 mm	16–10 mm	10–5 mm	5–0 mm	Average particle size mm
Fine particle size -5%-1	9.80	26.72	16.59	12.41	18.29	16.18	20.37
Fine particle size -5%-2	10.83	31.87	12.88	12.32	16.16	15.95	21.62
Fine particle size -8%-1	9.94	27.06	12.48	13.20	18.57	18.75	19.90
Fine particle size -8%-2	11.77	30.75	10.23	11.48	17.40	18.37	21.23

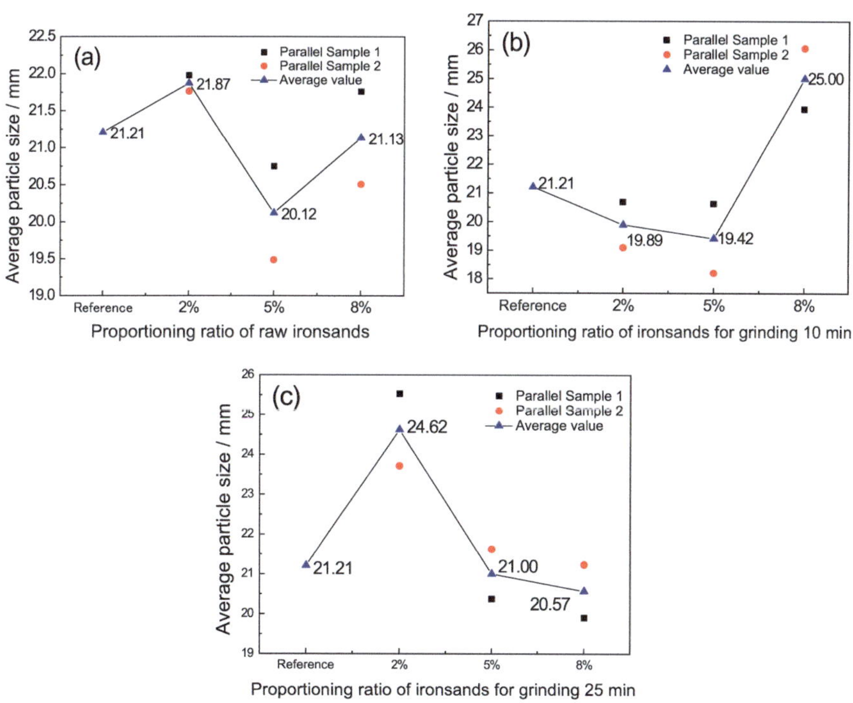

Fig. 5.21 Average particle size of finished sintered ore corresponding to different proportions of ironsands

Additionally, the proportion of sintered ore in the 25–40 mm size range is significantly improved, indicating a notable enhancement in the particle size composition of the finished sintered ore.

By comparing the particle size distribution of sintered ore with different proportions of original-grain ironsands, it becomes evident that the addition of 8 wt%

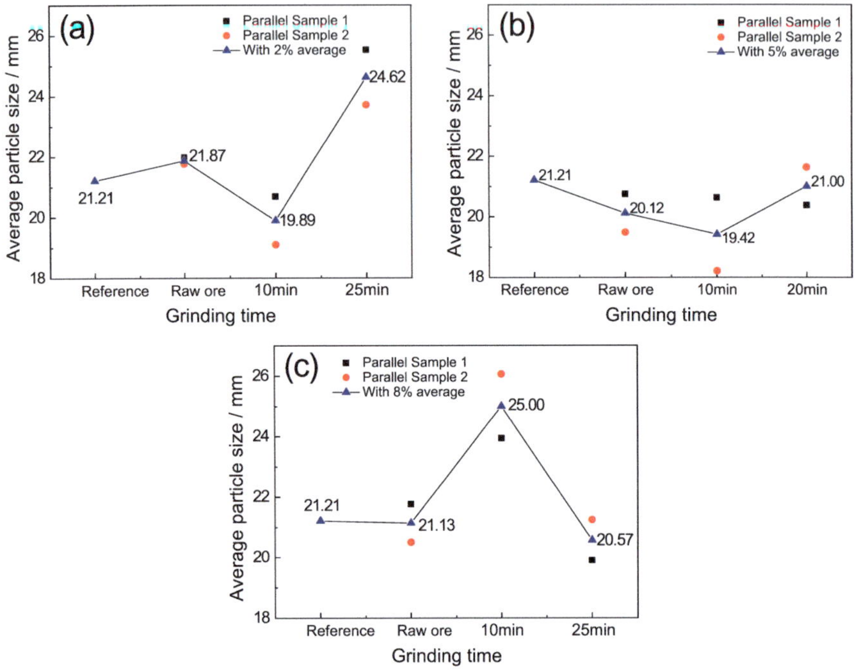

Fig. 5.22 Relationship between ironsands particle size and average particle size of the finished sintered ore

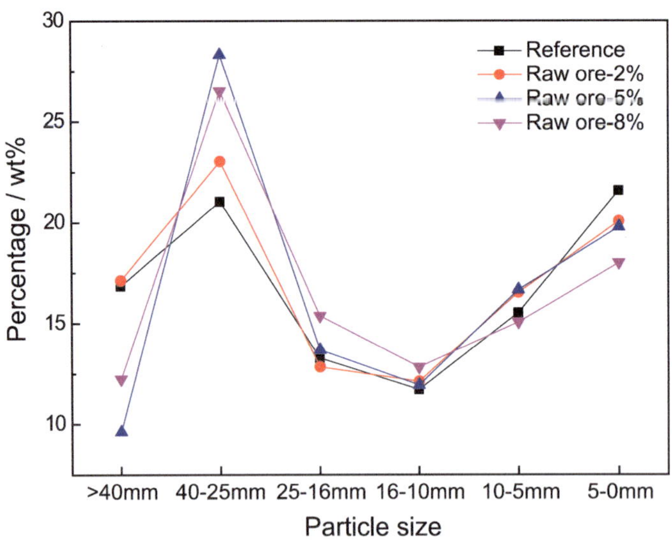

Fig. 5.23 Particle size distribution of finished sintered ore with varying proportions of original-sized ironsands

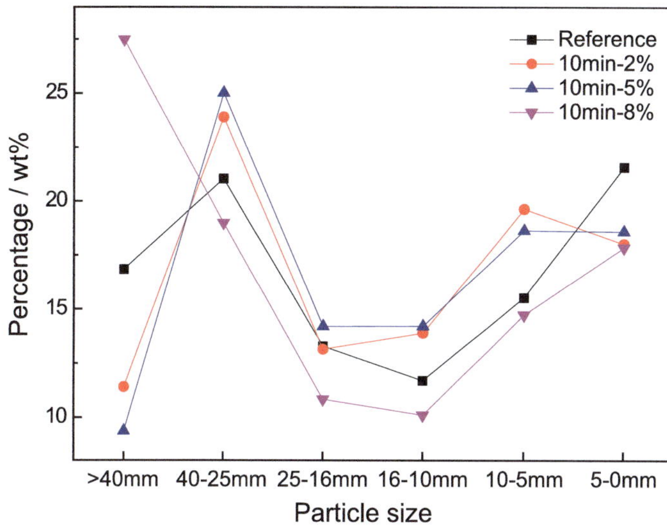

Fig. 5.24 Particle size distribution of finished sintered ore with varying proportions of intermediate-sized ironsands

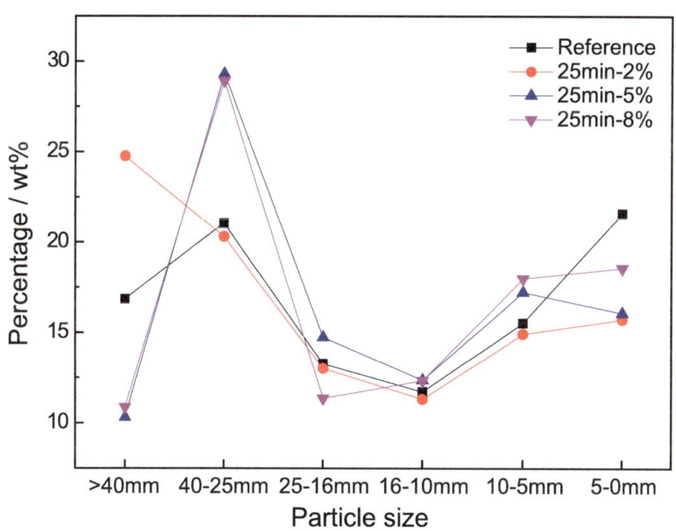

Fig. 5.25 Particle size distribution of finished sintered ore using different proportions of fine-grained ironsands

original-grain ironsands results in a superior particle size composition in the finished sintered ore.

As shown in Fig. 5.24, when sintering is conducted with intermediate-grain ironsands, the particle size composition of the sintered ore improves significantly compared to the baseline experiment when the proportion of ironsands is less than 5 wt%. In this case, the proportion of returned ore (smaller than 5 mm) decreases substantially, while the proportion of sintered ore in the 25–40 mm size range increases to around 25 wt%. However, when the proportion of ironsands exceeds 5 wt%, the proportion of large sintered ore (> 40 mm) increases significantly, and the proportion of intermediate-sized sintered ore decreases. Additionally, the amount of returned ore (smaller than 5 mm) further decreases, and the sintered cake becomes more difficult to break.

Based on the particle size distribution results for sintered ore with intermediate-grade ironsands, it is not recommended to add these ironsands in large quantities. If they must be used, adding 2–5 wt% is preferable, as the particle size distribution of the finished sintered ore deteriorates significantly when the proportion of intermediate-grade ironsands exceeds 8 wt%.

As shown in Fig. 5.25, when fine-grained ironsands are added to the sintering process, the proportion of large sintered ore (larger than 40 mm) increases and the proportion of returned ore (smaller than 5 mm) decreases when the proportion of ironsands is 2 wt%. However, when the proportion of ironsands exceeds 5 wt%, the proportion of sintered ore in the 25–40 mm size range becomes close to 30%, while the proportions of both large sintered ore (> 40 mm) and returned ore (less than 5 mm) significantly decrease. This results in a more rational distribution of sintered ore sizes when the ironsands proportion is higher than 5 wt%.

Based on the sintered ore particle size distribution results for fine-grained ironsands, it is evident that adding 5–8 wt% of ironsands leads to a more reasonable particle size distribution of the finished sintered ore. The comparison of particle size composition under different ironsands proportioning conditions, shown in Figs. 5.26, 5.27 and 5.28, further supports this finding.

According to Fig. 5.26, when the proportion of ironsands is 2 wt%, increasing the grinding time of the ironsands significantly alters the particle size composition of the sintered ore. For fine-grained ironsands, the proportion of returned ore is smaller, and the proportion of large sintered ore is larger. With intermediate-grained ironsands, although the proportion of sintered ore in the 25–40 mm range is higher, there is also a notable increase in the proportion of small grains (5–10 mm). In contrast, when sintering with original-grained ironsands, the proportion of returned ore is relatively high but still lower than in the baseline experiment. While the proportion of intermediate-grade (25–40 mm) sintered ore is higher, the proportion of small-sized sintered ore (5–10 mm) is also increased, which could negatively affect the permeability of the blast furnace.

Therefore, when the proportion of ironsands is controlled at 2 wt%, using intermediate-grained ironsands yields a better size distribution of finished sintered ore, improving its suitability for blast furnace use.

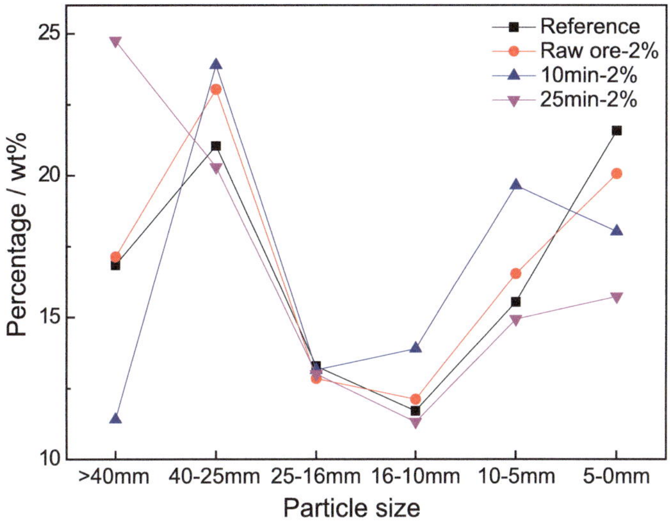

Fig. 5.26 Particle size distribution of sintered ore with 2 wt% ironsands under different grinding durations

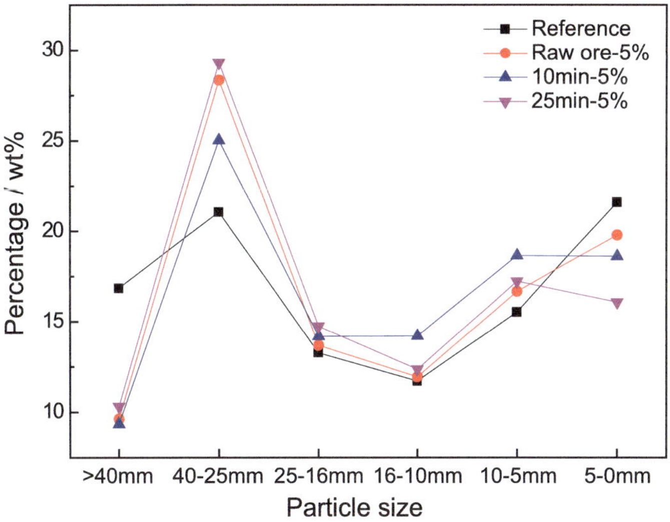

Fig. 5.27 Particle size distribution of sintered ore with 5 wt% ironsands under different grinding durations

According to Fig. 5.27, when the proportion of ironsands is 5 wt%, sintering with fine-grained ironsands results in a smaller proportion of large sintered ore, a larger proportion of sintered ore in the 25–40 mm grain size range, and a lower proportion of both small-grained finished ore and returned sintered ore. This leads to a more

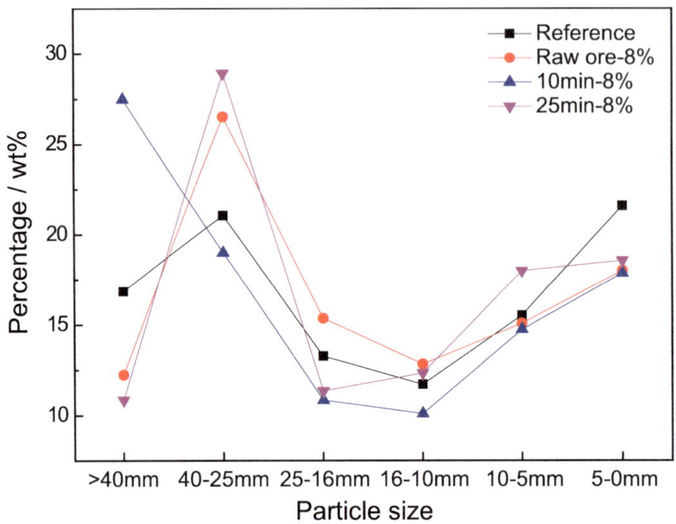

Fig. 5.28 Particle size distribution of sintered ore with 8 wt% ironsands under different grinding durations

desirable particle size distribution of the sintered ore. When using the original grain size of ironsands, the sintered ore size distribution is more similar to that of the fine-grained ironsands. Therefore, when adding 5 wt% of ironsands, it is more appropriate to use either raw or fine-grained ironsands to achieve a better sintered ore particle size distribution.

As shown in Fig. 5.28, when the proportion of ironsands is 8 wt%, there is a noticeable difference in the particle size composition of the sintered ore under different grinding time conditions. The sintered ore made with the original grain size of ironsands has a much higher proportion of the 25–16 mm particle size range compared to the sintered ore made with fine-grained ironsands, despite the lower proportion of the 25–40 mm range.

In conclusion, when 8 wt% ironsands is used, the finished sintered ore exhibits a more reasonable grain size composition when the original grain size of ironsands is used.

5.4.4 Drum Strength

The theoretical drum quantity of finished sinter ore was calculated based on the mass percentages of three grain size fractions: greater than 40 mm, 40–25 mm, and 25–16 mm. A 7.5 kg sample of each grain size was processed in the drum equipment. After drumming and sieving, the mass percentages of the finished sintered ore were recorded and presented in Table 5.13.

Table 5.13 Drum strength of finished sintered ore/wt%

Number	Drum set			Post-sieve volume	
	> 40 mm	40–25 mm	25–16 mm	+ 6.3 mm	− 6.3 mm
Reference frame	3.61	1.97	1.92	5.03	2.36
Primary particle size -2%-1	3.59	1.88	2.02	5.07	2.34
Primary particle size -2%-2	3.61	2.13	1.77	5.00	2.40
Primary particle size -5%-1	4.35	1.47	1.68	5.29	2.13
Primary particle size -5%-2	3.52	2.34	1.64	5.01	2.38
Primary particle size -8%-1	3.69	2.00	1.81	5.29	2.11
Primary particle size -8%-2	3.59	2.20	1.71	5.19	2.22
Intermediate particle size -2%-1	3.47	2.03	2.00	5.29	2.10
Intermediate particle size -2%-2	3.56	1.85	2.08	5.11	2.29
Intermediate particle size -5%-1	3.89	1.84	1.77	5.32	2.09
Intermediate particle size -5%-2	3.13	2.15	2.22	5.02	2.40
Intermediate particle size -8%-1	3.78	1.96	1.76	5.15	2.24
Intermediate particle size -8%-2	3.29	2.13	2.08	4.98	2.43
Fine particle size -2%-1	3.19	2.34	1.98	4.94	2.47
Fine particle size -2%-2	3.59	2.07	1.84	5.31	2.10
Fine particle size -5%-1	3.60	2.23	1.67	4.88	2.52
Fine particle size -5%-2	4.19	1.69	1.62	5.07	2.32
Fine particle size -8%-1	3.85	1.77	1.88	5.23	2.18
Fine particle size -8%-2	4.40	1.46	1.64	5.03	2.37

The drum strength was determined by the mass percentage of the sieve fraction greater than 6.3 mm, and the results were compared with the benchmark sample drum strength of 65.47%, marked with a solid line for reference. The comparison of drum strengths is shown in Fig. 5.29.

From Table 5.13 and Fig. 5.29, it can be observed that the drum strength remained unchanged when the original grain size was dosed up to 8 wt%. However, the drum strength of sintered ore with intermediate or fine-grained ironsands showed a slight decline. This is primarily due to the strong cementation of hematite and calcium ferrite in the sintered ore when the original grain size is used, forming a stable matrix with well-developed lath-like or intertwined melt structures, which enhance the ore's strength.

On the other hand, when intermediate or fine-grained ironsands were added at 8 wt%, more rhombic hematite phases appeared, often in the form of anhedral-crystals or semi-autocrysts, surrounded by fused and acicular calcium ferrite. This microstructure has a detrimental effect on the cold strength of the sintered ore.

In conclusion, there is potential to further increase the proportion of original grain size ironsands without significantly affecting drum strength. However, for sintered ore containing intermediate or fine-grained ironsands, the drum strength deteriorates

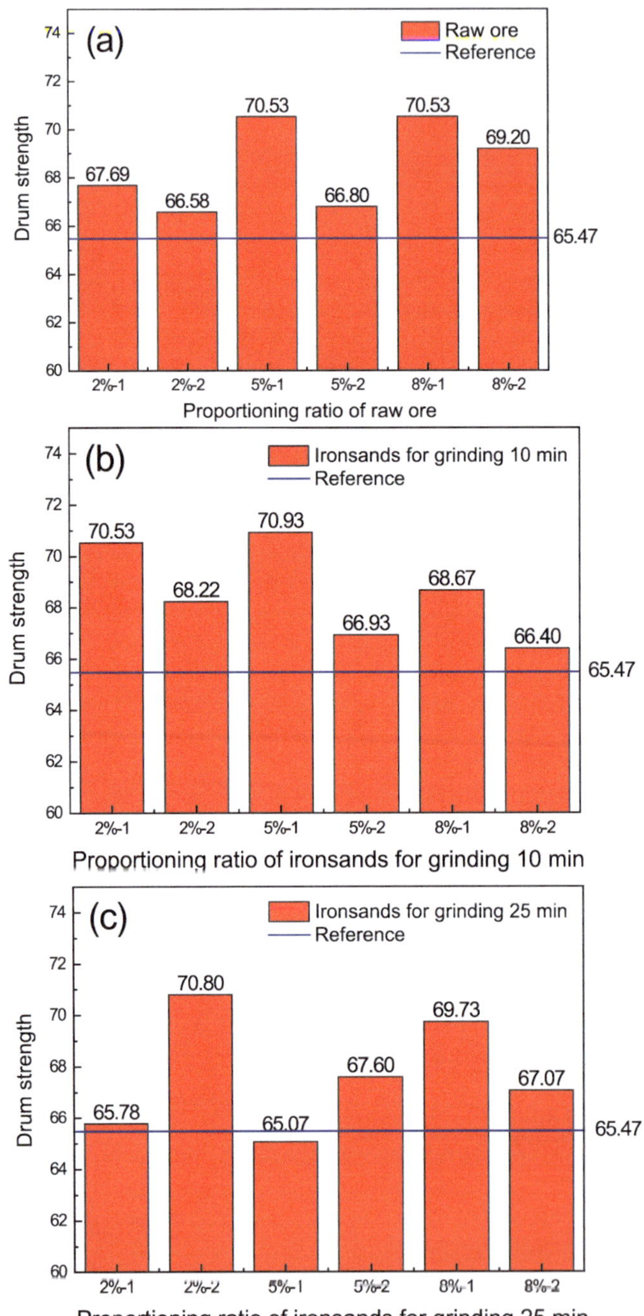

Fig. 5.29 Drum strength of sintered ore corresponding to different particle sizes and proportions of ironsands

significantly when the proportion exceeds 8 wt%, due to the adverse effects on sintering permeability.

5.5 High-Temperature Metallurgical Characteristics of Sintered Ore Containing Ironsands

5.5.1 *Low Temperature Reduction Pulverization Properties*

When iron-containing materials enter the upper part of the blast furnace at temperatures between 500 and 600 °C, they undergo reduction reactions, such as $Fe_2O_3 \rightarrow Fe_3O_4 \rightarrow FeO$, driven by the airflow and the reduction process. These changes induce crystallographic transformations in the iron-containing minerals, leading to pulverization of the ore in the lumpy zone. This increase in powder directly affects the internal airflow distribution within the furnace, as well as the downstream charge behavior. This section investigates the low-temperature reduction pulverization properties of sinter minerals containing ironsands.

Influence of chemical composition on low-temperature reduction pulverisation of sintered ore

The chemical composition of sintered ore, including elements such as CaO, MgO, Al_2O_3, FeO, and TiO_2, has a significant impact on the low-temperature reduction pulverization rate of the sintered ore. An increase in CaO, MgO, and FeO content in sintered ore helps improve its reduction pulverization performance. Conversely, higher contents of Al_2O_3 and TiO_2 tend to deteriorate its reduction pulverization performance.

Influence of FeO on low-temperature reductive pulverisation of sintered ores The FeO content in sintered ore is closely related to the amount of carbon present. In high-carbon sintering conditions, the sintered ore typically contains higher FeO, which reduces the amounts of Fe_2O_3 and $CaO{\cdot}Fe_2O_3$, thereby leading to hematite crystal transformations during the reduction process. This results in a lower pulverization rate, improving the low-temperature reduction pulverization index. For FeO contents below 8 wt%, the reduction in pulverization rate is most pronounced. When FeO content is between 8 and 10 wt%, the trend stabilizes, and beyond 10 wt%, the change in pulverization rate becomes negligible. Moreover, increasing FeO content enhances the room-temperature strength of sintered ore.

Effect of SiO_2 on low-temperature reductive pulverisation of sintered ore As SiO_2 content in sintered ore increases, the reduction of iron oxides slows down due to the increased formation of iron olivine, which impedes the low-temperature reduction pulverization. Additionally, SiO_2 contributes to the cementation of the sintered ore by forming silicate phases with Fe and Ca. A decrease in SiO_2 content disrupts the uniformity of the microstructure, shifting it from an interwoven fused structure to a mottled one, leading to increased crack formation and ultimately weakening the low-temperature reduction pulverization performance.

Effect of MgO on low-temperature reductive pulverisation of sintered ore Divalent Mg^{2+} ions can substitute for Fe^{2+} ions in magnetite, thereby reducing lattice defects and stabilizing the magnetite structure. This decreases the likelihood of forming secondary hematite, resulting in reduced low-temperature reduction pulverization. However, excessive MgO content can lead to the formation of complex compounds during sintering, which, due to variability in crystallization, induces internal stresses and microcracks in the sintered ore, ultimately exacerbating pulverization.

Experimental method for low-temperature reduction pulverisation rate of sintered ore containing ironsands

The low-temperature reduction pulverization rate of ironsands sinter ore was determined according to the method outlined in the Chinese national standard (GB/T 13242-91). The experimental parameters were as follows:

Specimen particle size: 10–12.5 mm.
Specimen mass: 500 g.
Reduction temperature: 500 $\pm$ 10 °C.
Reduction gas composition: $CO = 20\%$, $CO_2 = 20\%$, $N_2 = 60\%$
Gas flow: 15 L/min.
Reduction time: 60 min.

The specimens were first dried at 105 $\pm$ 5 °C for 2 h, then cooled to room temperature and stored in a desiccator. The weighed specimens were placed in a reduction tube, and nitrogen (or other inert gases) was passed through the tube at a standard flow rate of 5 L/min. After sealing the tube, it was heated to 500 °C at a rate of less than 10 °C/min. Once the temperature reached 500 °C, the flow rate of the protective gas was increased to 15 L/min. The reduction gas was then introduced at a flow rate of 15 L/min for 1 h. After reduction, inert gas was passed through the tube at 5 L/min, and the tube was removed from the furnace and cooled to below 100 °C.

Experimental results of low-temperature reduction pulverisation of finished sintered ore containing ironsands

To explore the influence of ironsands particle size and proportion on the low-temperature reduction pulverization index of finished sintered ore, the pulverization indices for different particle sizes and iron sand ratios were systematically analyzed. The results are presented in Table 5.14.

The low-temperature reduction pulverization index of sintered ore is primarily characterized by $RDI_{+3.15}$, with $RDI_{+6.3}$ and $RDI_{-0.5}$ serving as reference indices. For sintered ore, larger values of $RDI_{+3.15}$ and $RDI_{+6.3}$ and smaller values of $RDI_{-0.5}$ indicate better low-temperature reduction pulverization performance. Therefore, the average low-temperature reduction pulverization performance of sintered ore under

Table 5.14 Experimental results of low-temperature reduction pulverization performance of ironsands sinter ore

Experiment number	Pre-reduction weight/g	Reduced weight/g	Cryogenic reduction pulverization sieve size/g				Low temperature reduction pulverization index/%		
			> 6.3 mm	6.3–3.15 mm	3.15–0.5 mm	< 0.5 mm	$RDI_{+6.3}$	$RDI_{+3.15}$	$RDI_{-0.5}$
Reference frame	500.2	498.6	202.1	144.0	103.9	48.6	41	69	10
Primary particle size -2%-1	500.8	498.7	146.6	181.5	124.2	46.4	29	66	9
Primary particle size -2%-2	500.3	499.2	228.9	140.8	93.3	36.2	46	74	7
Primary particle size -5%-1	499.8	498.7	213.1	158.0	87.1	40.5	43	74	8
Primary particle size -5%-2	500.6	499.7	307.3	105.4	58.2	28.8	61	83	6
Primary particle size -8%-1	499.9	496.3	170.6	165.2	102.1	58.4	34	68	12
Primary particle size -8%-2	499.8	498.5	148.3	184.4	106.2	59.6	30	67	12
Intermediate particle size -2%-1	500.2	496.9	156.6	171.0	116.2	49.8	32	66	10
Intermediate particle size -2%-2	500.6	499.4	266.3	117.7	80.0	35.4	53	77	7
Intermediate particle size -5%-1	500.7	499.2	221.4	127.7	98.4	50.4	44	70	10
Intermediate particle size -5%-2	500.0	498.2	192.4	144.8	106.0	52.7	39	68	11
Intermediate particle size -8%-1	500.6	499.7	344.0	61.0	52.0	36.5	69	81	7

(continued)

Table 5.14 (continued)

Experiment number	Pre-reduction weight/g	Reduced weight/g	Cryogenic reduction pulverization sieve size/g				Low temperature reduction pulverization index/%		
			> 6.3 mm	6.3–3.15 mm	3.15–0.5 mm	< 0.5 mm	$RDI_{+6.3}$	$RDI_{+3.15}$	$RDI_{-0.5}$
Intermediate particle size -8%-2	499.5	497.9	80.8	146.7	173.5	95.1	16	46	19
Fine particle size -2%-1	499.6	496.6	113.0	153.7	156.8	71.0	23	54	14
Fine particle size -2%-2	500.5	496.7	175.3	165.5	105.8	50.1	35	69	10
Fine particle size -5%-1	500.0	492.1	139.5	182.3	118.9	51.4	28	65	10
Fine particle size -5%-2	500.4	497.6	223.3	128.8	96.1	47.4	45	71	10
Fine particle size -8%-1	500.5	498.2	123.6	187.4	123.3	63.9	25	62	13
Fine particle size -8%-2	500.9	498.7	113.6	182.5	131.3	66.0	23	59	13

different ironsands particle sizes (with consistent grinding times) and varying proportions of ironsands addition was analyzed, and the results are illustrated in Figs. 5.30, 5.31 and 5.32.

From Fig. 5.30, as the proportion of original grain size ironsands increases from 0 to 8 wt%, the low-temperature reduction pulverization index ($RDI_{+3.15}$) of the sintered

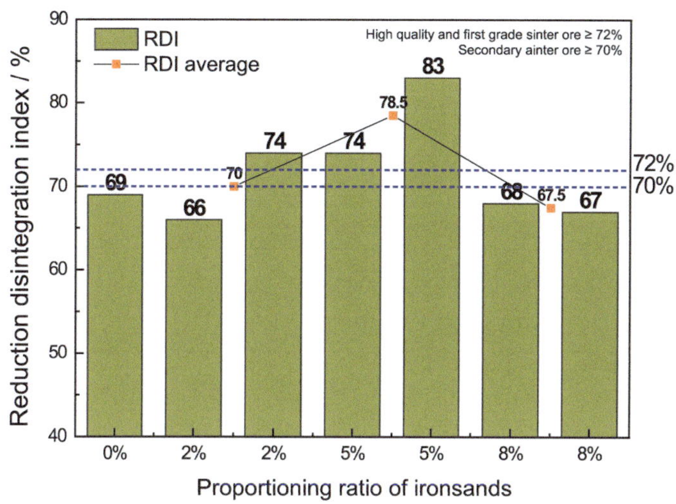

Fig. 5.30 Low-temperature reduction pulverization index of sintered ore with varying proportions of raw-grained ironsands

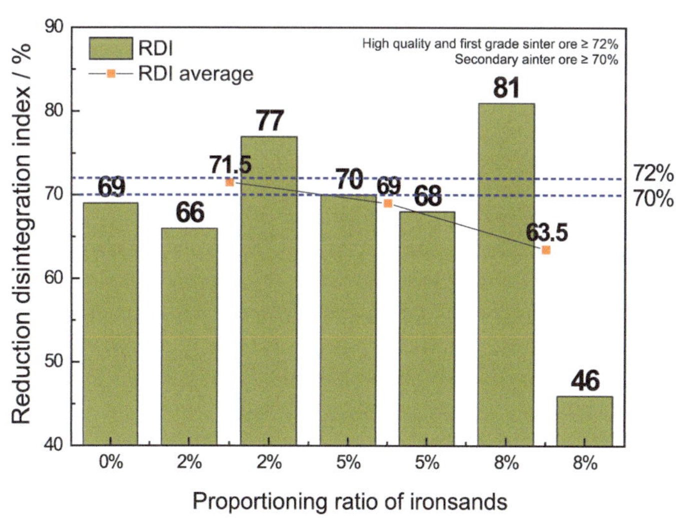

Fig. 5.31 Low-temperature reduction pulverization index of sintered ore with different proportions of intermediate-grained ironsands

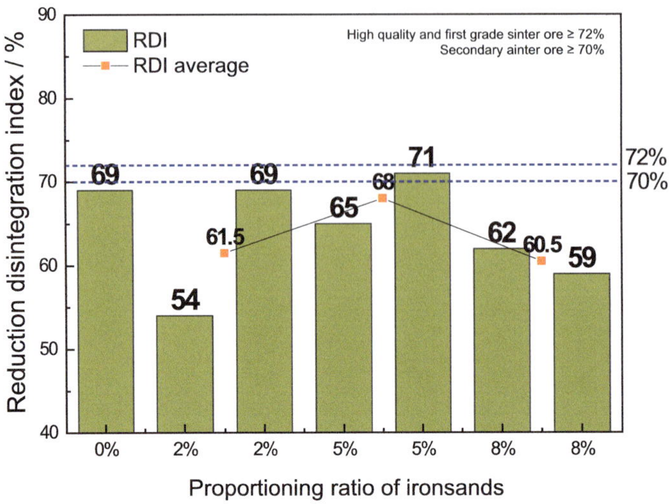

Fig. 5.32 Low-temperature reduction pulverization index of sintered ore with varying additions of fine-grained ironsands

ore remains relatively stable, fluctuating between 67 and 78%. However, when the proportion of original grain size ironsands reaches 8 wt%, the low-temperature reduction pulverization index of the sintered ore decreases below that of the baseline sample (69%), which is also lower than the standard for second-grade sintered ore.

This deterioration in pulverization performance is primarily attributed to the increase in the hematite phase in the sintered ore, as observed in the mineral composition and microstructural analysis in the previous section. The presence of a higher hematite content plays a key role in reducing the low-temperature reduction pulverization index.

In Fig. 5.31, the low-temperature reduction pulverization index ($RDI_{+3.15}$) of the sintered ore gradually decreases as the proportion of intermediate-grained ironsands increases from 2 to 8 wt%. When the ironsands proportion reaches 8 wt%, the $RDI_{+3.15}$ drops to just 63%. However, when the addition is 5 wt%, the $RDI_{+3.15}$ of the sintered ore with ironsands matches that of the benchmark sample, maintaining a value of 69%.

The deterioration in low-temperature reduction pulverization is primarily linked to the particle size falling within the intermediate range, which negatively affects the permeability of the sinter layer. Consequently, the sintered ore contains more rhombic hematite phases, which are encapsulated by fused and acicular calcium ferrite. This structural change causes volume expansion during sinter reduction, ultimately reducing the ore's low-temperature reduction pulverization performance.

As shown in Fig. 5.32, the low-temperature reduction pulverization index ($RDI_{+3.15}$) of the sintered ore fluctuates within a low range (60–68%) as the proportion of fine-grained ironsands increases from 2 to 8 wt%. The microstructure analysis of fine-grained ironsands reveals that the hematite phase in this sintered ore primarily exists as granular formations, while the relative content of the calcium ferrite phase decreases. The calcium ferrite phase, in turn, is cemented with the silicate phase. As a result, the hematite and calcium ferrite phases become more loosely connected, with the appearance of typical skeletal crystalline rhodochrosite encapsulated by the silicate phase. This structure significantly impairs the low-temperature reduction pulverization performance of the sintered ore, as both the silicate and hematite phases are brittle and contribute to the ore's tendency to powder. Additionally, the presence of titanium (Ti) in the silicate glass phase exacerbates the deterioration of the low-temperature reduction pulverization index.

Mechanism analysis of the influence of ironsands on the low-temperature reduction pulverisation of sintered ore

Based on the microstructure of ironsands sintered ore, it has been observed that when the proportion of ironsands increases and the particle size decreases, the hematite phase in the sintered ore becomes more prevalent, mainly existing in granular form. This structure is more porous, and when the sintered ore undergoes reduction at low temperatures, it experiences volumetric expansion. This expansion concentrates stress within the sintered ore, leading to the formation of cracks, which ultimately results in pulverization.

Moreover, as the proportion of ironsands increases, the TiO_2 content in the sintered ore rises. TiO_2 reacts with CaO to form $CaTiO_3$ in sintered ore with high basicity. This compound is primarily distributed within the glass phase of the sintered ore, contributing to a brittle texture. This brittleness weakens the sintered ore, promoting low-temperature reduction pulverization. When the proportion of ironsands reaches 8 wt%, the increase in both the hematite phase and TiO_2 content is the key factor behind the deterioration of the low-temperature reduction pulverization index of the ironsands sintered ore.

In summary, from the perspective of grain size, the addition of the original grain size ironsands has little impact on the low-temperature reduction pulverization performance of the sintered ore. This is because the increase in TiO_2 content and the hematite phase does not significantly affect the pulverization rate. However, when the ironsands particle size decreases, it exerts a stronger negative effect on the low-temperature reduction pulverization rate. Regarding the amount of ironsands added, all three grain sizes negatively affect the low-temperature reduction pulverization rate once the proportion reaches 8 wt%. Microscopic analysis indicates that the increased hematite phase and chalcocite content limit the maximum amount of ironsands that can be added to the sinter mix. Therefore, adding 8 wt% of original grain size ironsands results in a pulverization rate of 68.5%, which is close to the industry standard of 70% for secondary sintered ores, offering a valuable reference for determining the optimal grain size and proportion for addition.

5.5.2 Reduction Behavior

The reducibility of iron ore refers to the ease with which iron oxides in the ore react with gaseous reductants, such as CO and H_2. Reducibility is a critical parameter for evaluating iron-containing materials, as improved reducibility enhances the reduction rate, optimizes the utilization of blast furnace gas, and improves the technical and economic performance of blast furnace smelting. Ore with good reducibility allows for more metallic iron to be reduced in the medium temperature zone, facilitating indirect reduction. This reduces heat consumption in the high-temperature zone, helps decrease coke consumption, and improves the slagging process, promoting stable and efficient blast furnace operation. Ultimately, good reducibility leads to high-yield, high-efficiency, and high-quality blast furnace smelting.

Influence mechanism of chemical composition on reducibility The reduction of iron ore in the blast furnace is influenced by both the ore's intrinsic reducibility and the operational conditions within the blast furnace. Reducibility, in turn, is determined by the ore's mineral composition, physical properties, and structure. Key physical properties include particle size, porosity, and pore size. For sintered ores, the mineral composition and structure primarily involve the levels of SiO_2, FeO, basicity, as well as macro- and microstructural characteristics [9, 10].

Effect of basicity on reducibility of sintered ore The basicity of sintered ore has a significant impact on its reducibility. The reducibility of sintered ore varies with basicity, which is largely influenced by the ore's mineral composition. When the basicity is 1.2, the bonding phase of the sintered ore is dominated by $CaO{\cdot}SiO_2$ (CS), which is predominantly glassy in nature. This phase inhibits the reduction of the ore. As the basicity increases to 1.4, the formation of $CaO{\cdot}Fe_2O_3$ (CF) occurs, leading to the coexistence of both CS and CF phases, which in turn enhances the reducibility. With a further increase in basicity, the amount of calcium ferrate (CF) in the sintered ore gradually increases, thereby improving the reducibility. However, beyond a basicity of 2.5, the formation of dicalcium ferrite ($2\ CaO{\cdot}Fe_2O_3$), which is more difficult to reduce, begins to increase, thereby reducing the reducibility of the sintered ore.

Effect of SiO_2 on the reducibility of sintered ore The reducibility of sintered ore decreases progressively as the SiO_2 content increases. Higher SiO_2 levels increase the viscosity and surface tension of the sintered liquid phase, which negatively impacts the sintering permeability. This leads to reduced porosity in the sintered ore and the formation of a denser structure. While SiO_2 enhances the cold strength of the sintered ore, it has an adverse effect on its reducibility. Furthermore, SiO_2 can react with FeO to form iron olivine ($FeO{\cdot}SiO_2$), which is difficult to reduce and further hinders the reducibility of the sintered ore.

Effect of FeO on the reducibility of sintered ore FeO in sintered ore readily reacts with SiO_2 to form iron olivine. Reducing the FeO content not only increases the proportion of Fe_2O_3—an easily reducible phase—but also decreases the amount of iron olivine, which is difficult to reduce. Additionally, lowering FeO content can increase the porosity of the sintered ore, thereby enhancing the contact between the ore and the reducing gas, and thus improving the ore's reducibility.

The reducibility of various mineral phases in sintered ore follows this general order: Hematite > $CaO{\cdot}Fe_2O_3$ > $2CaO{\cdot}Fe_2O_3$ > Magnetite > Olivine. To optimize sintering, it is crucial to prevent the decomposition of primary hematite while minimizing the formation of secondary hematite. Moreover, it is essential to promote the formation of calcium ferrate in large quantities and to encourage the development of micropores. Acicular calcium ferrate is particularly easily reduced due to the presence of its abundant micropores [11].

Reductibility of sintered ore containing ironsands

Experimental method for the reducibility The experimental conditions for the reducibility of sintered ironsands are as follows:

Reduction gas composition: CO: 30 ± 0.5 vol% (V/V), N_2: 70 ± 0.5 vol% (V/V);
Flow rate of reducing gas: 15 ± 1 L/min;
Experimental temperature: 900 ± 10 °C.

Upon completion of the reduction process, the reduction degree of the sintered ore is calculated using the reduction degree formula, as shown in Eq. 5.7:

Reduction degree formula:

$$R_t = \left[\frac{0.11w_1}{0.43w_2} + \frac{m_1 - m_t}{m_0 \times 0.43w_2} \times 100\right] \times 100 \tag{5.7}$$

where

R_t degree of reduction of the pellet at time t, %;
m_0 mass of the specimen at the beginning of reduction, g;
m_1 mass of the specimen when heated to 900°C before reduction, g;
m_t mass of the specimen after reduction at time t, g;
w_1 content of FeO in the specimen before reduction, wt%;
w_2 content of total iron in the specimen before reduction, wt%.

Results of the sintered ore containing ironsands reduction experiment The sintering cup experiments, conducted with the reference sample sintered ore and different ironsands dosages, yielded the results as presented in Table 5.15.

To explore the effects of different particle sizes and proportions of ironsands on the reduction behavior of sintered ore, the results of the reduction experiments were analyzed. The reduction trends associated with the addition of ironsands to sintered ore are discussed in detail. For ease of analysis, the experimental results from two groups of parallel samples were averaged, as shown in Table 5.16.

The reduction index (RI) of sintered ore under identical particle size compositions and varying ironsands ratios is analyzed in Figs. 5.33, 5.34, 5.35 and 5.36.

Table 5.15 Experimental results on the reducibility of sintered ores from ironsands mines

Experiment number	Pre-reduction weight/g	Reduced weight/g	TFe/wt%	FeO/wt%	RI/%
Reference frame	500.4	393.7	55.68	4.56	91.17
Primary particle size -2%-1	500.5	398.3	55.41	5.21	88.13
Primary particle size -2%-2	500.6	395.4	55.44	5.03	90.49
Primary particle size -5%-1	500.4	400.9	55.44	5.96	86.18
Primary particle size -5%-2	500.0	398.2	55.32	6.03	88.40
Primary particle size -8%-1	500.1	399.3	55.18	6.68	88.07
Primary particle size -8%-2	500.9	402.7	55.42	6.03	85.08
Intermediate particle size -2%-1	500.3	405.9	55.73	6.90	81.93
Intermediate particle size -2%-2	500.4	411.6	55.97	7.33	77.12
Intermediate particle size -5%-1	500.3	399.4	55.42	6.68	87.74
Intermediate particle size -5%-2	499.6	402.2	55.40	6.97	85.09
Intermediate particle size -8%-1	500.1	402.2	55.79	4.96	83.90
Intermediate particle size -8%-2	500.3	400.6	55.81	4.45	85.10
Fine particle size -2%-1	500.5	396.4	55.30	4.88	89.75
Fine particle size -2%-2	500.5	395.8	55.67	4.89	89.66
Fine particle size -5%-1	500.5	402.8	55.83	8.48	85.23
Fine particle size -5%-2	499.8	399.7	55.63	6.68	86.83
Fine particle size -8%-1	500.7	412.4	55.52	7.47	77.34
Fine particle size -8%-2	499.3	398.0	55.69	7.54	88.22

Table 5.16 Average results of reduction experiments on ironsands sintered ore

Sinter ore number	RI/%
Reference frame	91.17
Primary particle size -2%	89.31
Intermediate particle size -2%	79.52
Fine particle size -2%	89.70
Primary particle size -5%	87.29
Intermediate particle size -5%	86.41
Fine particle size -5%	86.03
Primary particle size -8%	86.57
Intermediate particle size -8%	84.50
Fine particle size -8%	82.78

The reduction index of the benchmark sample is included for comparison, with the point where the proportion of ironsands added is 0 wt% representing the benchmark sample.

As shown in Fig. 5.33, with an increase in the proportion of the original granular ironsands, the reduction index (RI) of the ironsands-sintered ore exhibits a slight decreasing trend. However, the overall difference is not significant. When the proportion of original granular ironsands reaches 8 wt%, the reduction index (RI) of the sintered ore is 86.57%, which is 4.6% lower than that of the benchmark sample. It is important to note that the metallurgical industry standard for high-quality sintered ore requires a reduction index (RI) of $\geq$ 78%. For first- and second-grade sintered

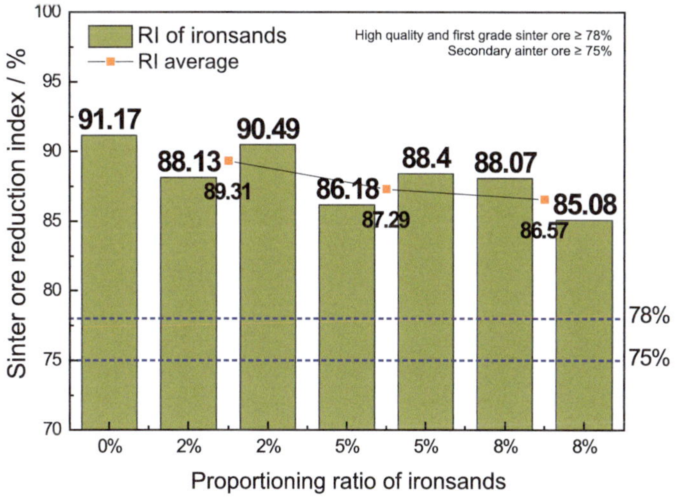

Fig. 5.33 Reduction index (RI) of sintered ore under varying allocation ratios of raw-grained ironsands

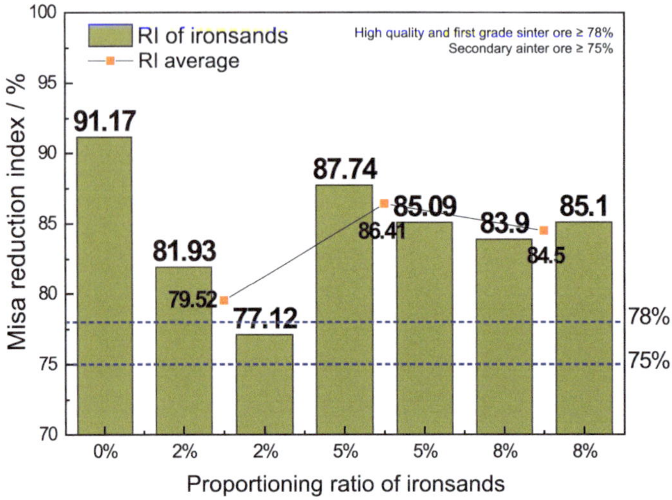

Fig. 5.34 Reduction index (RI) of sintered ore with different dosing ratios of intermediate-grained ironsands

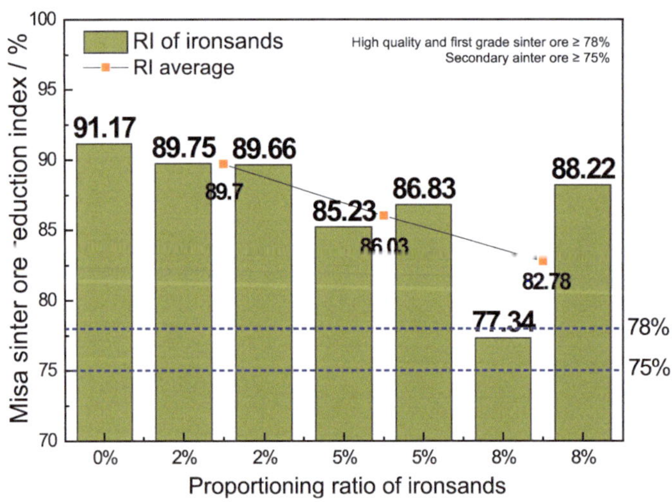

Fig. 5.35 Reduction index (RI) of sintered ore with varying proportions of fine-grained ironsands

ore with a basicity of 1.5–2.5, the reduction indices must meet the standards of RI $\geq$ 78% and RI $\geq$ 75%, respectively. Therefore, the reduction index of the sintered ore with 8 wt% original granular ironsands still exceeds the industry standard for high-quality sintered ore.

Furthermore, compared to the reference sample, the reduction of sintered ore with ironsands is slightly lower due to the primary mineral composition of ironsands being

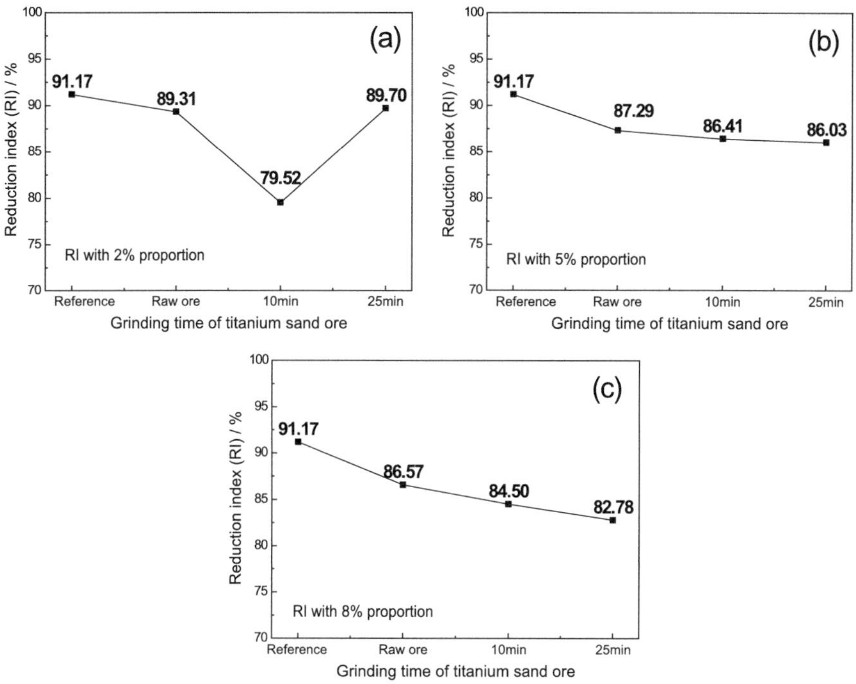

Fig. 5.36 Influence of dosing ratios of different ironsands particle sizes on the reduction index (RI) of sintered ore. **a** 2 wt% ironsands; **b** 5 wt% ironsands; **c** 8 wt% ironsands

titanomagnetite. The titanium element in titanomagnetite is strongly dissolved in the magnetite phase, resulting in a reduction potential of the ironsands itself that is not as high as that of ordinary iron ore powder. Consequently, the addition of ironsands inevitably leads to a slight reduction in the overall reduction performance of the sintered ore.

Figure 5.34 illustrates the trend of the reduction index of sintered ore as the proportion of intermediate-grained ironsands increases from 2 to 8 wt%. As seen in the figure, compared to the reduction index of the benchmark sintered ore, the reduction index of the sintered ore with intermediate-grained ironsands is relatively lower. Specifically, the reduction index for sintered ore with 2 wt% intermediate-grained ironsands is 79.52%, which is 11.65% lower than that of the benchmark. However, the reduction index still exceeds the metallurgical industry standard of 78%, categorizing it as high-quality sintered ore.

As shown in Fig. 5.35, the reduction index (RI) of sintered ore decreases gradually with the increase in the proportion of fine-grained ironsands. The trend mirrors that observed with original-grain ironsands. When the proportion of fine-grained ironsands reaches 8 wt%, the reduction index (RI) of the sintered ore is 82.78%, which is 8.4% lower than the benchmark value.

The influence of ironsands with different grain sizes on the reduction degree of sintered ore is further summarized in Fig. 5.36.

From Fig. 5.36a, it can be observed that when the proportion of ironsands is 2 wt%, the sintered ore containing intermediate-grained ironsands exhibits the lowest reduction index of 79.52%. The reduction index of sintered ore with original-grained and fine-grained ironsands are nearly identical, both slightly lower than the reduction index of the reference sample. In Fig. 5.36b, the reduction index of sintered ore with 5 wt% ironsands shows a decreasing trend as the grinding time is extended. In Fig. 5.36c, it is evident that, with an 8 wt% ironsands proportion, the reduction index of sintered ore also decreases as the grinding time increases.

The results indicate that as grinding time increases, the reduction index of sintered ore decreases with finer ironsands particle sizes. Specifically, the reduction index of sintered ore with fine-grained ironsands is approximately 3.8% lower than that of sintered ore with original-grained ironsands. This phenomenon becomes more pronounced with higher proportions of ironsands. Moreover, as grinding increases production costs, it is recommended to use original-grained ironsands for sintering. Even with an 8 wt% addition, the reducibility of sintered ore remains within the high-quality range defined by industry standards.

Mechanism analysis of the influence of ironsands on the reducibility of sinter ore

In blast furnace smelting, improved reducibility of sintered ore enhances the utilization rate of blast furnace gas and reduces the coke ratio. Based on the reduction index (RI) of sintered ores with varying proportions of ironsands, it is evident that the reducibility of sintered ore decreases to different extents with an increase in the proportion of ironsands. This phenomenon can be attributed to the fact that ironsands consist of vanadium-titanium magnetite, which is a more refractory ore type. As the proportion of ironsands increases, the amount of magnetite, which is more difficult to reduce, also increases in the sintered ore. This results in a decrease in the overall reducibility of the sintered ore, leading to a reduction in the reduction index (RI).

Despite this, the experimental data show that the reduction index (RI) of sintered ore still exceeds the metallurgical industry standard requirement of 78%. Even when the proportion of ironsands reaches 8 wt%, the reduction index remains above 80%. Therefore, there is still potential for further optimization of the ironsands proportion from the perspective of the reduction index.

Additionally, as the particle size of ironsands decreases, the reduction index of sintered ore tends to decrease, with the effect being more pronounced when the proportion of ironsands is higher. Since grinding increases the cost of sintering production, it is recommended to use the original grain size of ironsands directly in sintering production. This approach ensures that the reducibility of the sintered ore remains within the high-quality range, despite the reduction in reducibility with higher proportions of ironsands.

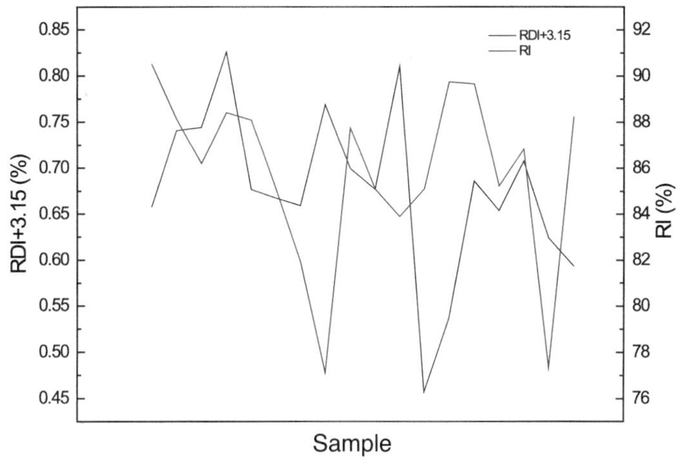

Fig. 5.37 Relationship between low-temperature reduction pulverization characteristics and reducibility of sintered ores

This study also examines the relationship between the reducibility of ironsands sintered ore and the low-temperature reduction pulverization index. A comparative analysis between the low-temperature reduction sub-index ($RDI_{+3.15}$) and the reduction index (RI) after the addition of ironsands is presented in Fig. 5.37.

In Fig. 5.37, each data point represents a sintered ore. Most sintered ores exhibit an inverse relationship between the reduction index (RI) and the low-temperature reduction pulverization index ($RDI_{+3.15}$). Specifically, when the low-temperature reduction pulverization index ($RDI_{+3.15}$) of a sintered ore is poor, its reduction index (RI) tends to be relatively higher.

This relationship can be explained as follows: when the low-temperature reduction pulverization of sintered ore is more pronounced, the ore is pulverized into smaller particles before entering the medium-temperature reduction zone in the blast furnace. These smaller particles increase the contact area between the sintered ore and the reducing gases, thereby accelerating the reduction rate. Consequently, the reduction index (RI) increases. In contrast, if the pulverization is less severe, the reduction index tends to be lower. This is why both the low-temperature reduction pulverization index and the reduction index are important indicators in metallurgical work. During the production of sintered ore, both the pulverization performance and the reduction properties must be optimized to ensure high metallurgical performance.

In summary, the reduction performance of sintered ore containing ironsands can be outlined as follows:

(1) The reduction index (RI) of all sintered ores containing ironsands is lower than that of the benchmark. This is because ironsands primarily consist of vanadium-titanium magnetite, which is more difficult to reduce. As the proportion of ironsands increases, the amount of the difficult-to-reduce magnetite phase in the sintered ore also increases, thus diminishing the reducibility of the sintered

ore and resulting in a lower reduction index (RI). However, the reduction index (RI) of all sintered ores with ironsands still exceeds the metallurgical industry standard requirement of 78% for high-quality sintered ores.

(2) A decrease in the particle size of ironsands does not improve the reducibility of sintered ore. This effect becomes more pronounced as the proportion of ironsands increases. Fine-grained ironsands tend to be more dispersed and distributed within the sintered ore, which significantly inhibits the reduction of iron oxides. Consequently, the presence of titanomagnetite, the main phase in ironsands, has a more pronounced impact on the reducibility of the sintered ore. Therefore, it is recommended that original-grain ironsands be directly used in sintering production, especially when rich ore powder serves as the primary iron-containing material.

(3) When the low-temperature reduction pulverization of sintered ore is more severe, the ore is broken into smaller particles before entering the medium-temperature reduction zone in the blast furnace. This increases the contact area between the ore and reducing gases, accelerating the reduction rate and thereby increasing the reduction index (RI) within the same period. Therefore, there is an inverse relationship between the low-temperature reduction pulverization index (RDI + 3.15) and the reduction index (RI) of the sintered ore.

5.5.3 Softening–Melting and Dropping Characteristics

Influencing factors of the blast furnace soft melting zone

The soft melting process of sinter and other iron ores in the blast furnace can be divided into four distinct stages:

The first stage is that the high-level oxides of iron are reduced to wustite, and metallic iron shells are generated at the periphery of the iron oxides. Before the softening of the iron ore, a certain degree of sintering of the wustite and metallic iron occurs, resulting in shrinkage of the ore stock.

The second stage is softening, during which there is a large degree of deformation of the layer, however, as there is still a certain amount of void space in the layer, the differential pressure does not increase to a large extent, while each particle may have a large deformation.

The third stage is the exudation of the FeO-containing melt, which fills the voids in the layer, while the melt is gradually reduced to a higher melting temperature, resulting in a rapid increase in the differential pressure of the layer.

The fourth stage is melting, as the temperature increases, the ore gradually melts, and finally forms liquid drops, causing the height of the material layer to further decrease.

Stage 1 In this stage, the higher oxides of iron are reduced to wüstite, and metallic iron shells form around the periphery of the iron oxides. Prior to the softening of the ore, some degree of sintering occurs between the wüstite and the metallic iron, resulting in the shrinkage of the ore bed.

Stage 2 During softening, there is significant deformation within the layer. However, as void spaces remain in the layer, the differential pressure does not increase substantially, allowing each particle to undergo considerable deformation.

Stage 3 At this stage, FeO-containing melts begin to exude, filling the voids within the layer. As the melt is progressively reduced to a higher melting temperature, the differential pressure within the layer increases rapidly.

Stage 4 As the temperature rises, the ore gradually melts, ultimately forming liquid droplets. This leads to a further decrease in the height of the material layer.

The shape, thickness, and position of the soft melt zone play a crucial role in determining the distribution of blast furnace gas in the middle and lower sections of the furnace. This directly affects the permeability of the blast furnace, as the soft melt zone dictates the distribution of the temperature field and gas flow. Therefore, its shape and position significantly influence the overall smelting process.

In the daily operation of the blast furnace, ensuring the uniform and stable descent of the charge and controlling the proper distribution of the blast furnace gas flow are key factors for maintaining stable and efficient furnace operation. Several solid–liquid reactions occur in the soft melt zone, and as such, the characteristics of this zone directly impact the secondary distribution of the blast furnace gas flow. This, in turn, plays a critical role in the stable and smooth operation of the furnace.

Analytical study of the softening melt-drop properties of titanium-containing sinter ores

Experimental method for softening melt-drop properties of ironsands sinter ores
The experimental method employed in this study aims to simulate the actual conditions of the blast furnace, including unit load weight, temperature control system, and atmospheric composition, with a maximum temperature reaching 1600 °C. The experimental temperature and layer shrinkage are automatically recorded by a computer system.

The experimental conditions for the softening and melt-drop properties of iron ore are summarized in Table 5.17. A displacement sensor was used to measure the vertical displacement (in millimeters) of the material surface. The softening temperature was identified as the point at which the material surface reached a shrinkage rate of 10%. The temperature corresponding to a 40% shrinkage rate was considered the endpoint of the softening process. The difference between these two temperatures was defined as the softening temperature range.

Additionally, the temperature at which the first iron drop occurred was recorded as the "drop temperature." The temperature corresponding to the 40% material surface shrinkage was designated as the "melting temperature." The temperature interval between the drop temperature and the melting temperature defined the "melting range." (Table 5.18).

A total of 10 groups of sintered ore samples, including the reference samples, were selected to conduct the softening drop performance experiments. The experimental sample numbers are provided in Table 5.19.

Table 5.17 Experimental parameters for softening–melting performance evaluation

Number	Item	Experimental condition
1	Sample	The particle size is 10–12.5 mm, and the sample height is about 60 mm
2	Crucible	Graphite crucible, inner diameter 75 mm
3	Reduce gas composition	30 vol%CO + 70 vol%N_2
4	Experimental load and reducing gas flow rate	1 kg/cm^2, 10 L/min

Table 5.18 Standard parameters of molten drop experiment and their significance

Parameter	Significance
T_{10}	Softening starting temperature (°C), the temperature at which the sample softens and shrinks by 10%
T_{40}	Softening end temperature (°C), the temperature at which the sample softens and shrinks by 40%
ΔT_A	Softening interval (T_{40}–T_{10}) (°C)
T_d	Dropping temperature
ΔT_d	Melting zone (T_{40}–T_{10}) (°C)

Table 5.19 Experimental numbers for softening–melting and dropping tests of ironsands sintered ore

Number	Sintered ore sample	Number	Sintered ore sample
1	Reference frame	1	Reference frame
2-1	Primary particle size -2%-1	2-2	Primary particle size -2%-2
3-1	Primary particle size -5%-1	3-2	Primary particle size -5%-2
4-1	Primary particle size -8%-1	4-2	Primary particle size -8%-2
5-1	Intermediate particle size -2%-1	5-2	Intermediate particle size -2%-2
6-1	Intermediate particle size -5%-1	6-2	Intermediate particle size -5%-2
7-1	Intermediate particle size -8%-1	7-2	Intermediate particle size -8%-2
8-1	Fine particle size -2%-1	8-2	Fine particle size -2%-2
9-1	Fine particle size -5%-1	9-2	Fine particle size -5%-2
10-1	Fine particle size -8%-1	10-2	Fine particle size -8%-2

Softening melt drop experiment results for ironsands sintered ore The sintered cup experiments, conducted using the reference sample sintered ore and varying particle size ranges and ironsands dosages, yielded typical differential pressure and displacement curves for the softening melt drop experiment, as shown in Fig. 5.38. The experimental results of the softening melt drop are summarized in Table 5.20.

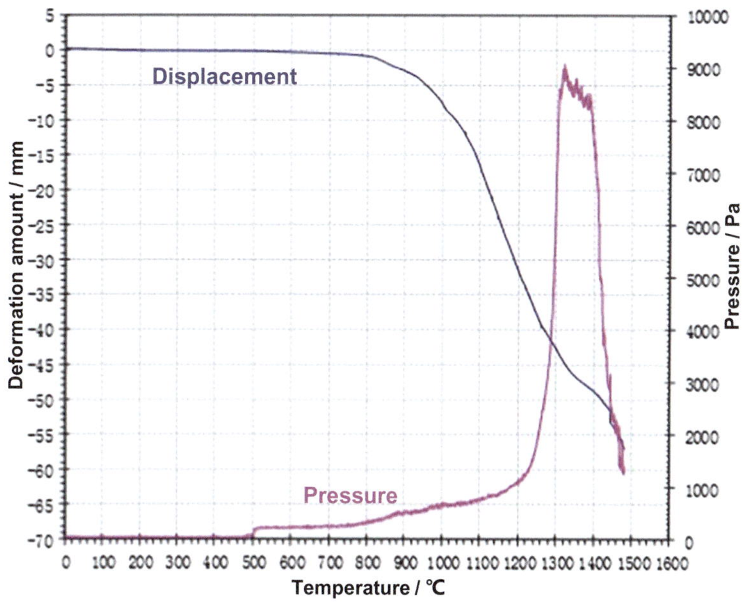

Fig. 5.38 Differential pressure and displacement curves from a typical softening–melting experiment of ironsands-based sintered ore

Mechanism analysis of the influence of ironsands on the softening melt-drop properties of sintered ores

Research on the influence of ironsands on the softening interval of sintered ore The size of the softening zone is a critical factor influencing permeability in the blast furnace. In general, a narrower softening interval of sintered ore leads to improved gas permeability in the blast furnace. Additionally, the softening behavior of sintered ore primarily occurs during the indirect reduction stage in the blast furnace. A wider softening interval suggests insufficient development of indirect reduction, which can indicate poor reducibility. Therefore, a narrower softening interval not only enhances blast furnace permeability but also reflects the good metallurgical performance of the sintered ore.

The softening temperature range ($\triangle T_A$) of each group of ironsands sintered ore was analyzed, with the results shown in Fig. 5.39. When original particle size ironsands were added, the softening range of the sintered ore tended to widen with increasing addition ratios. For example, when the proportion of original particle size ironsands reached 8 wt%, the softening temperature range of the sintered ore expanded to 304 °C, which is 53 °C higher than that of the reference sample. Consequently, increasing the proportion of original particle size ironsands during sintering is not conducive to improving the softening range of sintered ore.

When intermediate-grade ironsands were added during the sintering process, the results, as shown in Fig. 5.40, indicated that as the proportion of ironsands increased,

Table 5.20 Softening–melting and dropping experimental results of ironsands sintered ore/°C

Number	Corresponding sinter	T_{10}	T_{40}	T_d	ΔT_A	ΔT_d
1	Reference frame	1049	1300	1475	251	175
2-1	Primary particle size -2%-1	1049	1316	1480	267	164
2-2	Primary particle size -2%-2	1043	1338	1488	295	150
2-Average	2-Average	1046	1327	1484	281	157
3-1	Primary particle size -5%-1	1054	1321	1474	267	153
3-2	Primary particle size -5%-2	1074	1337	1504	263	167
3-Average	3-Average	1065	1329	1489	264	160
4-1	Primary particle size -8%-1	1049	1349	1500	300	151
4-2	Primary particle size -8%-2	1057	1365	1484	308	119
4-Average	4-Average	1053	1357	1492	304	135
5-1	Intermediate particle size -2%-1	1008	1213	1476	205	263
5-2	Intermediate particle size -2%-2	994	1193	1482	199	289
5-Average	5-Average	1001	1203	1479	202	276
6-1	Intermediate particle size -5%-1	1041	1287	1503	246	216
6-2	Intermediate particle size -5%-2	1033	1269	1493	236	224
6-Average	6-Average	1037	1278	1498	241	220
7-1	Intermediate particle size -8%-1	1001	1269	1491	268	222
7-2	Intermediate particle size -8%-2	987	1245	1465	258	220
7-Average	7-Average	994	1257	1473	263	216
8-1	Fine particle size -2%-1	1021	1250	1479	229	229
8-2	Fine particle size -2%-2	1045	1260	1493	215	233
8-Average	8-Average	1033	1255	1486	222	231
9-1	Fine particle size -5%-1	1012	1275	1488	263	213
9-2	Fine particle size -5%-2	1004	1271	1474	267	203
9-Average	9-Average	1008	1273	1481	265	208
10-1	Fine particle size -8%-1	998	1261	1479	263	218
10-2	Fine particle size -8%-2	1012	1277	1469	265	192
10-Average	10-Average	1005	1269	1474	264	205

the softening interval of the sintered ore gradually expanded. At a proportion of 2 wt% ironsands, the softening interval significantly decreased to only 202 °C. However, when the proportion of ironsands increased to 8 wt%, the softening interval of the sintered ore was just 8 °C larger than that of the benchmark sintered ore. This suggests that while adding intermediate-grade ironsands improves the softening interval of the sintered ore to some extent, the proportion of intermediate-grade ironsands should not exceed 8 wt%.

As shown in Fig. 5.41, when fine-grained ironsands were added to the sintering mix, the softening interval of the sintered ore initially increased and then stabilized

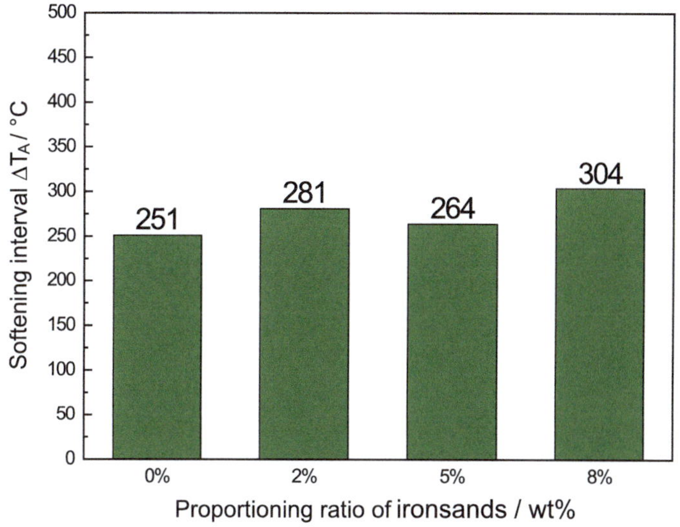

Fig. 5.39 Softening temperature interval (ΔT_A) of sintered ore under various dosing ratios of raw-grained ironsands

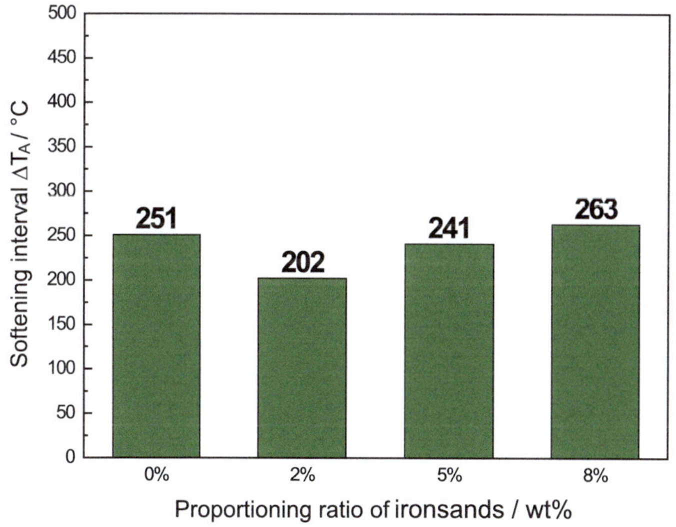

Fig. 5.40 Softening temperature interval (ΔT_A) of sintered ore under different dosing ratios of intermediate-grained ironsands

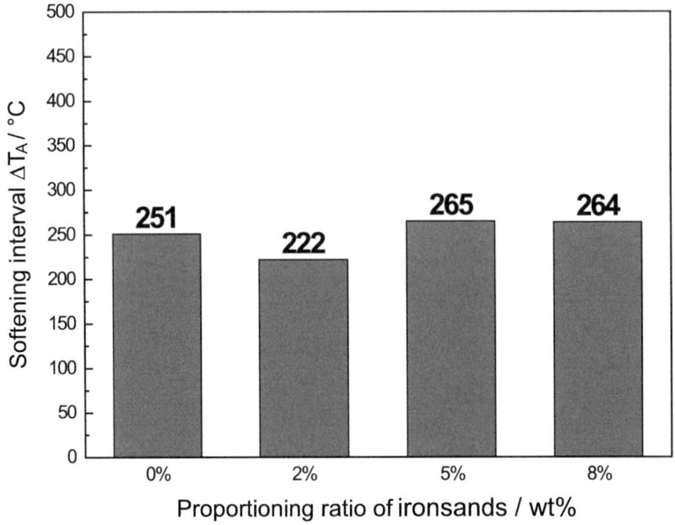

Fig. 5.41 Softening temperature interval (ΔT_A) of sintered ore under various dosing ratios of fine-grained ironsands

as the proportion of ironsands increased from 2 to 8 wt%. At 2 wt%, the softening temperature interval of the sintered ore was 222 °C, approximately 30 °C lower than that of the reference sample sintered ore. However, when the proportion of ironsands ranged from 5 to 8 wt%, the softening interval remained stable, approximately 10 °C wider than that of the reference sample sintered ore.

In conclusion, when fine-grained ironsands were used, the softening temperature interval of the sintered ore decreased compared to the increase observed with original-grade ironsands. The softening interval of the sintered ore with fine-grained ironsands increased by about 10 °C compared to the reference sample, while the softening interval of sintered ore with original-grade ironsands increased by about 50 °C.

Effect of ironsands on the drop temperature of sintered ore The experimental results for the drop temperature of sintered ore with the same particle size composition but varying proportions of ironsands are analyzed to facilitate comparison. For reference, a 0 wt% proportion of ironsands is considered the baseline sample. The results are presented in Fig. 5.42.

As shown in Fig. 5.42, with the increasing proportion of original-grain ironsands, the drop temperature of the sintered ore gradually increases. The drop temperature rises from 1,475 °C in the reference sample to 1,492 °C when the proportion of original-grain ironsands reaches 8 wt%. This increase in sinter drop temperature can reduce the height ratio of the soft-melt zone in the blast furnace, which facilitates the indirect reduction of sintered ore and enhances the effective use of heat and gas in the furnace.

Figure 5.43 illustrates that with the increase in the proportion of intermediate-grain ironsands, the drop temperature initially increases, then decreases. The highest

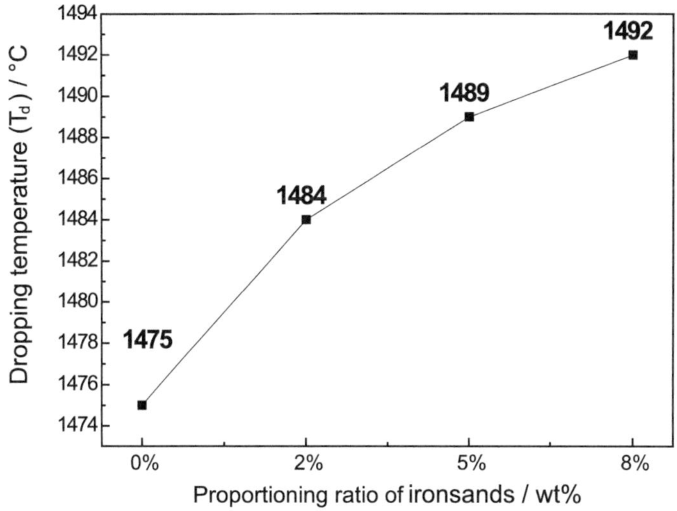

Fig. 5.42 Sintered ore dripping temperature (T_d) at different raw-grained ironsands dosing ratios

drop temperature, 1498 °C, is observed when the proportion of ironsands is 5 wt%. However, at 8 wt%, the drop temperature of sintered ore decreases slightly to 1473 °C, which is slightly lower than the baseline sample.

Figure 5.44 reveals that with the addition of fine-grained ironsands, the drop temperature of the sintered ore decreases gradually. At a 2 wt% proportion, the drop

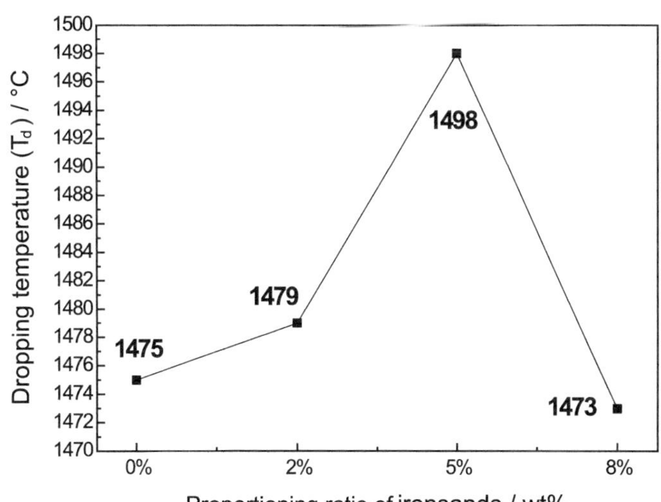

Fig. 5.43 Sintered ore dripping temperature (T_d) under different dosing ratios of intermediate-grained ironsands

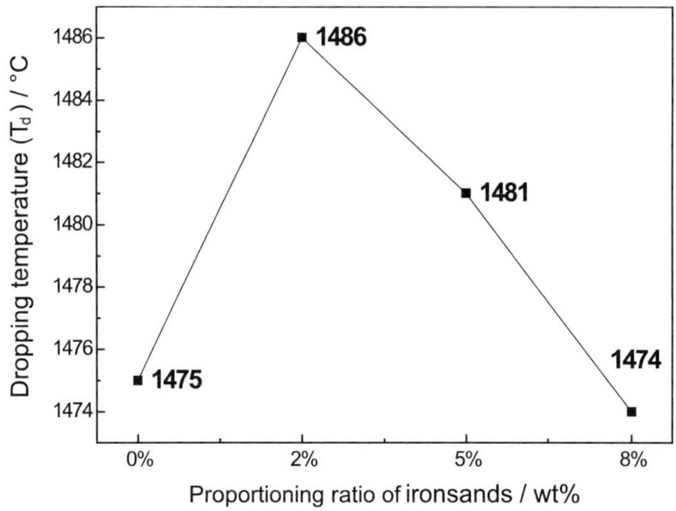

Fig. 5.44 Sintered ore dripping temperature (T_d) under varying dosing ratios of fine-grained ironsands

temperature is the highest at 1486 °C. When the proportion increases to 8 wt%, the drop temperature decreases to 1474 °C. Overall, the addition of fine-grained ironsands results in the least increase in drop temperature. The highest drop temperature recorded is 1486 °C, which is only 11 °C higher than that of the reference sample. When the proportion of fine-grained ironsands reaches 8 wt%, the drop temperature is nearly the same as the reference sample.

A comparison between Figs. 5.42 and 5.44 indicates that the drop temperature of sintered ore is least affected by the addition of fine-grained ironsands. This is because the small particle size of fine-grained ironsands facilitates their melting within other liquid phases, thereby exerting a less significant impact on the overall drop temperature.

Influence of ironsands proportion on the melting temperature range of sintered ore The melting interval in the molten drop experiment refers to the temperature range between the point where the sintered ore reaches 40% deformation and the onset of slag drop, marking the transition to the liquid phase. Sintered ore with superior metallurgical performance requires a narrower melting interval, which helps to enhance the permeability of the material layer in the blast furnace, increase gas utilization, and reduce the coke ratio. The melting temperature intervals (ΔT_d) for various groups of ironsands sintered ore were analyzed, and the results are shown in Figs. 5.45, 5.46 and 5.47.

From Fig. 5.45, it can be seen that with the increase in the proportion of original-grain ironsands during sintering, the melting interval of the sintered ore exhibits a narrowing trend. When the proportion of original-grain ironsands reaches 5 wt%, the melting interval is 160 °C, which is 15 °C less than that of the reference sample.

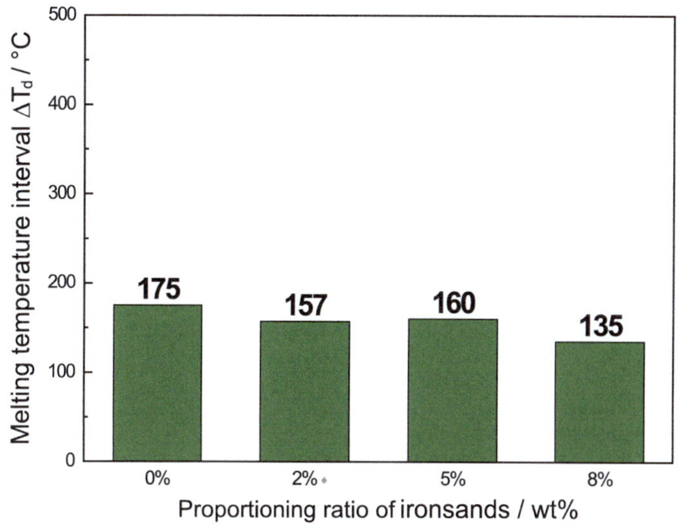

Fig. 5.45 Melting temperature interval (ΔT_d) of sintered ore with different proportions of raw-grained ironsands

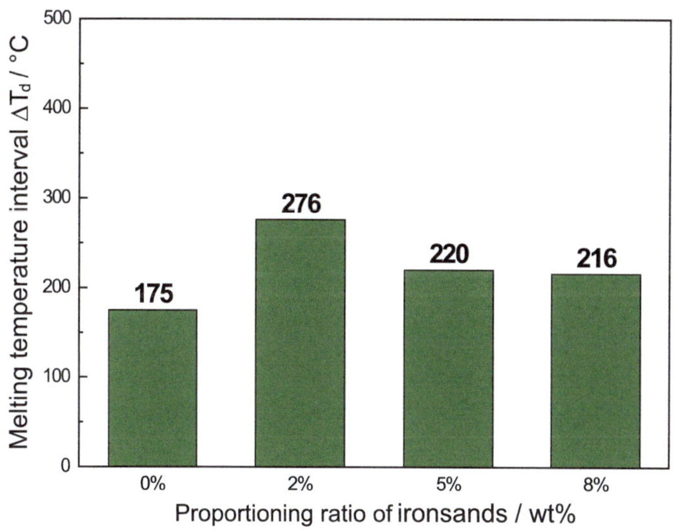

Fig. 5.46 Melting temperature interval (ΔT_d) of sintered ore under different dosing ratios of intermediate-grained ironsands

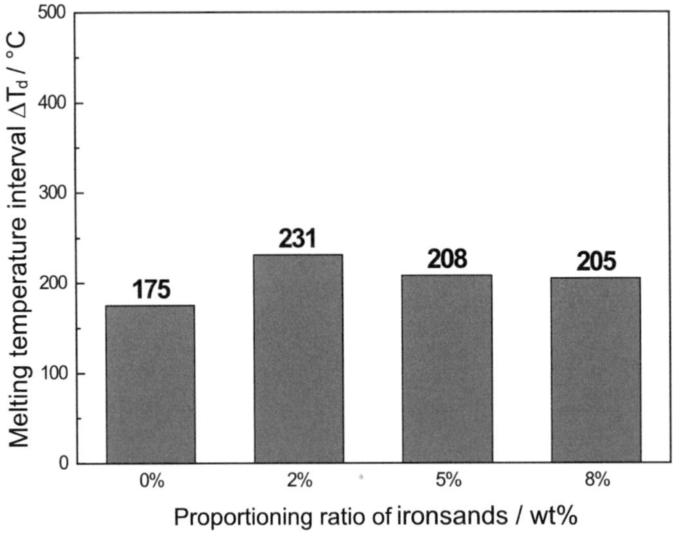

Fig. 5.47 Melting temperature interval (ΔT_d) of sintered ore under varying dosing ratios of fine-grained ironsands

At 8 wt%, the melting interval further narrows to 135 °C, which is 40 °C less than that of the reference sample.

This indicates that increasing the proportion of original-grain ironsands improves the melting interval of sintered ore, making the slag-iron liquid phase less viscous and easier to drop, thus benefiting the quality of the sintered ore.

In Fig. 5.45, when intermediate-grained ironsands are used in sintering, the melting interval of sintered ore initially decreases with increasing ironsands proportion. At 2 wt%, the melting interval reaches 276 °C, which is about 100 °C higher than that of the reference sample. When the proportion of ironsands is between 5 and 8 wt%, the melting interval decreases but remains 40–45 °C higher than the reference sample.

Thus, while the melting interval tends to decrease with increasing intermediate-grained ironsands, it remains higher than that of the reference sample. This is mainly due to the higher reduction potential of sintered ore with intermediate-grained ironsands and the earlier formation of low-melting-point substances.

As seen in Fig. 5.46, when fine-grained ironsands are added, the melting interval of the sintered ore increases compared to the reference sample. However, as the proportion of sintered ore increases, the melting interval gradually decreases, similar to the trend observed with intermediate-grained ironsands. At 2 wt%, the melting interval reaches 231 °C, which is 56 °C higher than that of the reference sample. The influence of fine-grained ironsands on the melting interval is less pronounced than that of intermediate-grained ironsands. When the proportion reaches 5–8 wt%, the melting interval stabilizes at around 30 °C higher than that of the reference sample.

When the proportion of sintered ore with original-grain ironsands increases, the softening interval exhibits a distinct widening trend. At 8 wt%, the softening interval reaches 304 °C, which is 53 °C higher than the reference sample. Conversely, when sintered ore contains intermediate and fine-grained ironsands, the softening interval tends to decrease compared to the original-grain ironsands, with a slight increase of around 10 °C when fine-grained ironsands reach 8 wt%. Similarly, the drop temperature of sintered ore increases with higher proportions of ironsands, showing the greatest increase when using original-grain ironsands, where the maximum drop temperature reaches 1498 °C.

In summary, as the proportion of original-grain ironsands increases, the melting interval of sintered ore narrows, while the addition of intermediate- and fine-grained ironsands tends to widen the melting interval. The most significant effect on the melting interval is observed with intermediate-grained ironsands. For the softening temperature interval, the trend is an increase with the rise in ironsands dosage, while for the melting temperature interval, an increase in dosage results in a decrease in the melting interval.

Study on the influence of different ironsands grain sizes on the softening temperature interval of sintered ore The softening temperature intervals (ΔT_A) were analyzed for each group of softening melt drop experiments.

As shown in Fig. 5.48a, when the proportion of ironsands is 2 wt%, the sintered ore with the addition of original-grain ironsands exhibits the largest softening temperature range, reaching 281 °C, which is 30 °C higher than that of the baseline sample. Conversely, the addition of intermediate- and fine-grained ironsands results in a softening temperature range that is lower than the baseline sample by approximately 50 °C and 30 °C, respectively.

From Fig. 5.48b, when the proportion of ironsands is 5 wt%, the softening temperature interval of sintered ore is about 10 °C higher than the reference sample when either original-grain or fine-grained ironsands are added. However, when intermediate-grained ironsands are added, the softening temperature interval of sintered ore is 10 °C lower than that of the reference sample.

As seen in Fig. 5.48c, when the proportion of ironsands is 8 wt%, the softening interval of sintered ore shows a decreasing and stabilizing trend as the particle size of the ironsands becomes finer. When original-grain ironsands are added, the softening temperature interval of sintered ore is about 50 °C higher than the reference sample. In contrast, the addition of intermediate- and fine-grained ironsands results in a softening temperature interval that is about 12 °C higher than the reference sample.

In summary, it can be concluded that, within the addition range of 2–8 wt%, refining the particle size of ironsands reduces the softening temperature interval of ironsands sintered ore.

Study on the effect of different ironsands grain sizes on the drip temperature of sintered ore As shown in Fig. 5.49, when the proportion of ironsands is 2 wt%, the variation in the drip temperature of ironsands sintered ore is minimal as the particle size of ironsands becomes finer, fluctuating between 1479 and 1488 °C. Similarly, when the proportion of ironsands is 5 wt%, the drip temperature shows only slight

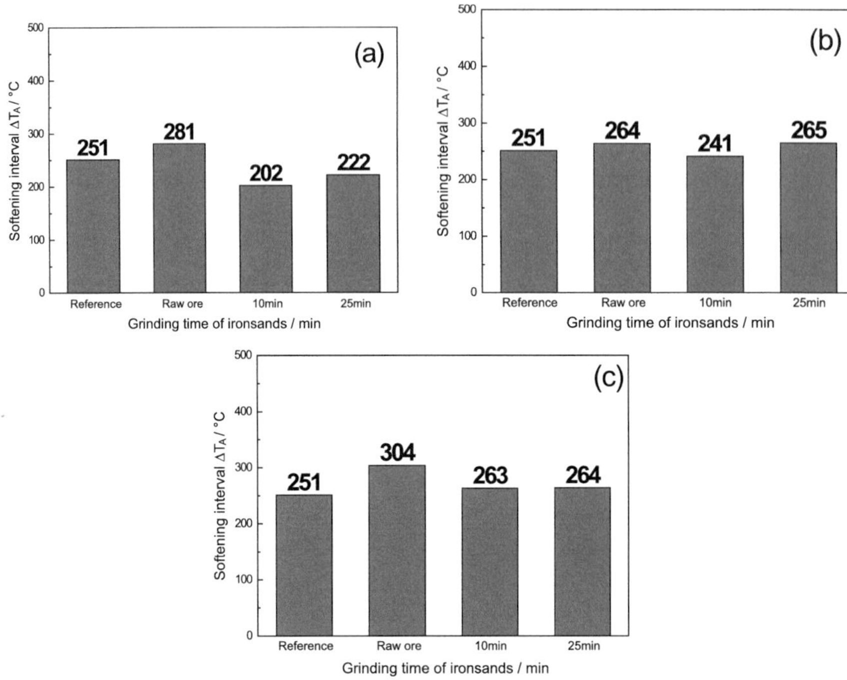

Fig. 5.48 Softening temperature interval (ΔT_A) of sintered ore with 2 wt%, 5 wt%, and 8 wt% ironsands for different particle sizes. **a** 2 wt% ironsands; **b** 5 wt% ironsands; **c** 8 wt% ironsands

variation, as depicted in Fig. 5.49b. When the proportion of ironsands is 5 wt%, Fig. 5.49b shows that the droplet temperature of sintered ore generally increases as the particle size of the ironsands becomes finer. However, as seen in Fig. 5.49c, when the proportion of ironsands increases to 8 wt%, the droplet temperature decreases as the ironsands particle size becomes finer. When fine-grained ironsands are added, the droplet temperature of the sintered ore is nearly the same as that of the reference sample. In conclusion, it can be inferred that the impact of changes in the particle size of ironsands on the drip temperature of ironsands sintered ore is minimal.

Study on the effect of different grain sizes of ironsands on the drip temperature of sintered ore As shown in Fig. 5.50a, when the proportion of ironsands is 2 wt%, the melting interval of sintered ore increases as the particle size of the ironsands becomes finer. Specifically, the melting temperature interval of sintered ore with the addition of intermediate and fine-grained ironsands is 55–100 °C larger than that of the reference sample. In Fig. 5.50b, c, when the proportion of ironsands increases to 5 wt% and 8 wt%, respectively, the melting interval of sintered ore shows a similar trend to that observed with the 2 wt% ironsands addition. That is, refining the particle size of ironsands results in an increase in the melting interval.

Overall, regarding the softening interval of sintered ore, refining the particle size of ironsands reduces the softening temperature interval for sintered ores with a 2–8 wt%

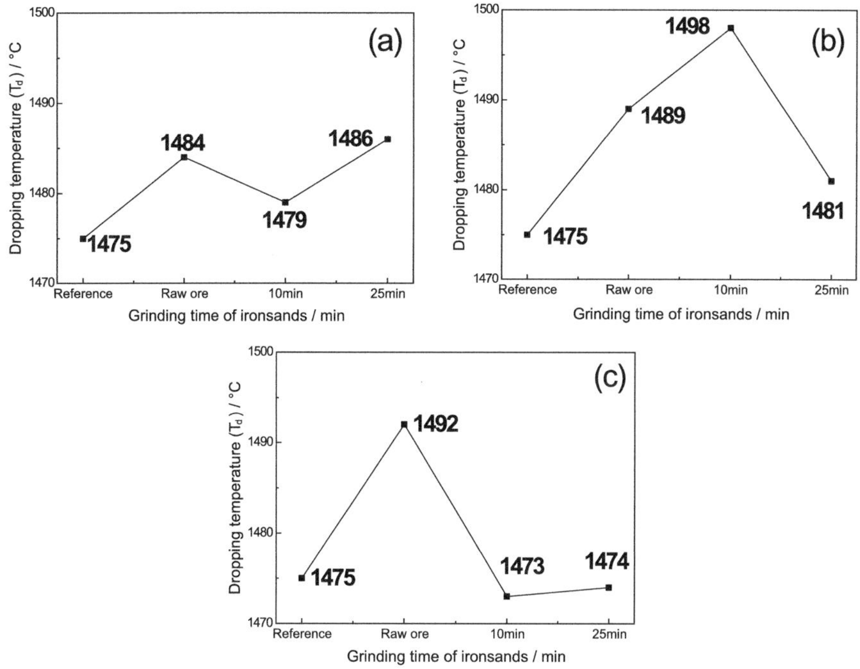

Fig. 5.49 Dropping temperature (T_d) of sintered ore with 2–8 wt% dosing ratios of different particle sizes of ironsands. **a** 2 wt% ironsands; **b** 5 wt% ironsands; **c** 8 wt% ironsands

dosing ratio. However, concerning the drip temperature of sintered ore, the change in ironsands particle size has minimal effect. In terms of the drip interval, refining the ironsands particles results in an increase in the interval when the proportion of ironsands is within the 2–8 wt% dosing range.

Integrated permeability study of ironsands sinter ore In order to better measure the melt droplet performance of the charge, the concept of the total eigenvalue S of the melt droplet performance is introduced. The S value characterises the integral of the differential pressure over the temperature in a certain temperature range. Therefore, the smaller the value of S is, the lower the differential pressure is within a certain temperature range, which means the better the permeability of the charge, and the better the performance of the charge droplet. The calculation method is shown in Eq. 5.8.

Integrated permeability study of ironsands sinter ore To better assess the melt droplet performance of the charge, the concept of the total eigenvalue, S, is introduced. The S-value characterizes the integral of the differential pressure over the temperature within a specified temperature range. Therefore, the smaller the S-value, the lower the differential pressure within that range, indicating better permeability of the charge and superior droplet performance. The calculation method is presented in Eq. 5.8:

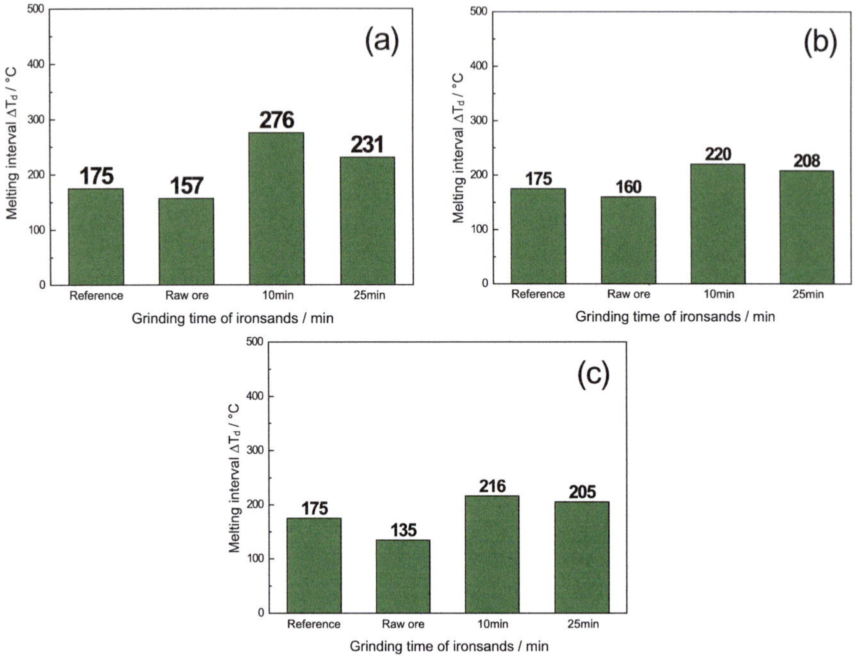

Fig. 5.50 Melting temperature interval (ΔTd) of sintered ore with 2–8 wt% dosing ratios of different ironsands particle sizes. **a** With 2 wt% ironsands; **b** with 5 wt% ironsands; **c** with 8 wt% ironsands

$$S = \int_{Ts}^{Td} (\Delta P - \Delta Ps)\mathrm{d}T \tag{5.8}$$

where

T_d sinter drop temperature, °C;
T_s temperature at which the differential pressure sharply rises, °C;
ΔP_s differential pressure at the steep rise moment, Pa;
ΔP differential pressure at time t, Pa.

This yields the total eigenvalue S of the ironsands sinter ore, as shown in Fig. 5.51.

As shown in Fig. 5.51, with the increase in ironsands dosing from 2 to 8 wt%, the total eigenvalue SSS of ironsands sintered ore exhibits a gradual upward trend across all three grain size conditions. This indicates that the permeability of ironsands sintered ore deteriorates with increasing ironsands content. This can primarily be attributed to the rise in titanium oxides as the ironsands content increases. Additionally, the formation of titanium carbon-nitride particles under the reducing atmosphere (which involves exposure to reducing gases and direct contact with coke) significantly raises the viscosity of the molten slag iron. This increase in viscosity enhances the resistance to gas penetration through the molten layer, leading to a

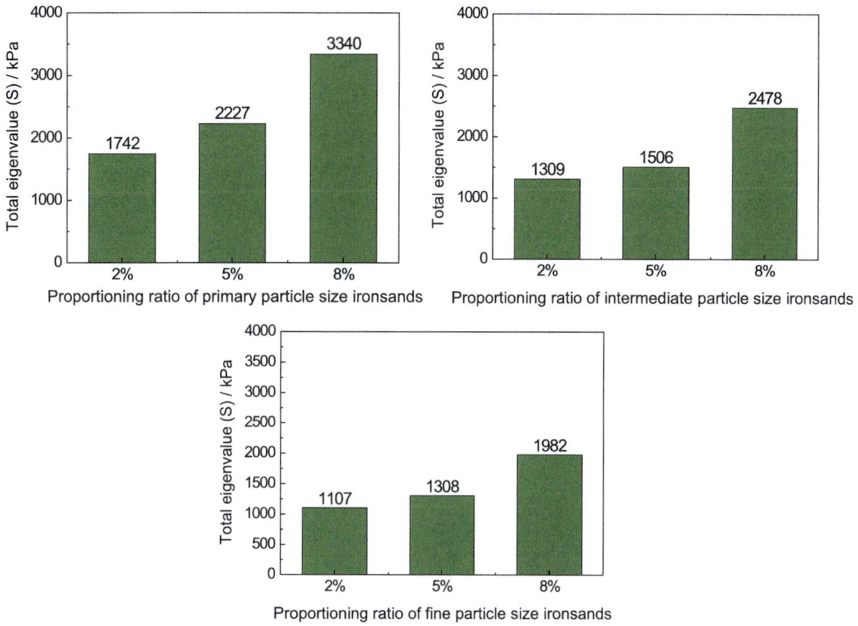

Fig. 5.51 Total eigenvalues of molten droplets of ironsands sinter at different ironsands dosage

higher pressure differential. Consequently, the gas permeability of the ironsands sintered ore decreases as the ironsands dosing increases.

5.6 Effects of Ironsands Addition on Sintering Process Parameters

Influence of ironsands addition on the sintering utilisation coefficient

In sintering production, the utilisation coefficient is a key indicator of overall sintering efficiency. By examining this coefficient, a more comprehensive understanding can be gained regarding the impact of different ironsands proportions on sintering performance. The sintering utilisation coefficients for varying proportions of ironsands were analyzed, and the results are shown in Fig. 5.52.

As observed in Fig. 5.52a, the sintering utilisation coefficients significantly improved with the addition of ironsands, with values consistently higher than those of the baseline samples. The highest sintering utilisation coefficient of 1.9 $t{\cdot}m^2{\cdot}h^{-1}$ was achieved when the proportion of raw granular ironsands was 5 wt%.

According to Fig. 5.52b, the sintering utilisation coefficient exhibited a trend of initially increasing and then decreasing as the proportion of intermediate-grade ironsands increased during the sintering process. The coefficients for 2 and 5 wt%

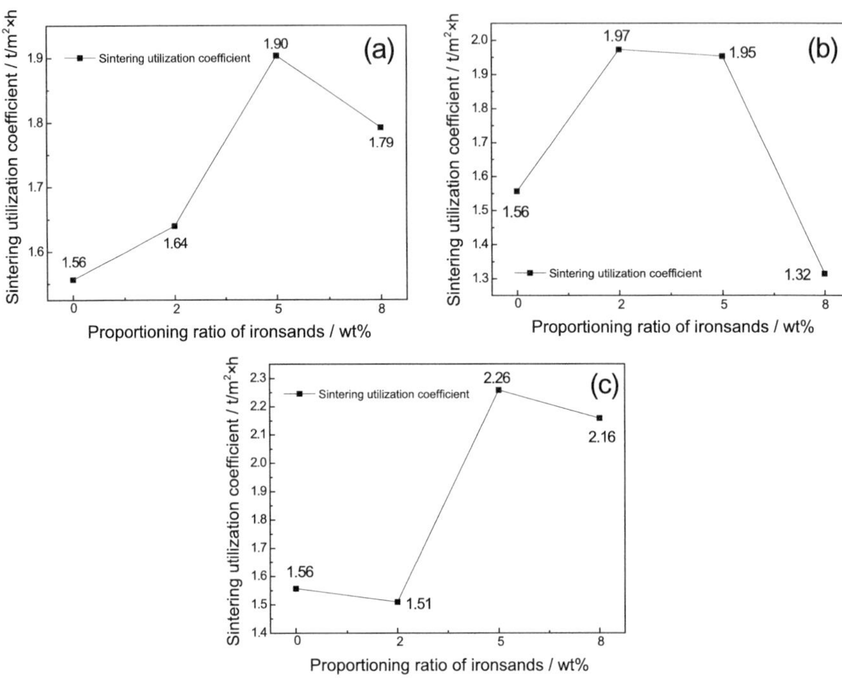

Fig. 5.52 Sintering utilization factor at various dosing ratios of ironsands. **a** Primary-grained ironsands; **b** intermediate-grained ironsands; **c** fine-grained ironsands

intermediate-grade ironsands were both high, exceeding 1.95 t·m^2·h^{-1}, while the coefficient for 8 wt% was 1.32 t·m^2·h^{-1}, still higher than the baseline sample.

From Fig. 5.52c, when 2 wt% fine-grained ironsands was added, the sintering utilisation coefficient showed a slight decrease but remained at the same level as the benchmark experiment, approximately 1.5 t·m^2·h^{-1}. As the proportion of fine-grained ironsands increased, the sintering utilisation coefficient rose substantially, reaching a peak of 2.26 t·m^2·h^{-1}. Thus, the addition of fine-grained ironsands notably improved the sintering utilisation coefficient.

In conclusion, with the exception of the intermediate-grained ironsands at 8 wt%, which resulted in decreased sintering permeability and a reduction in the sintering utilization factor, ironsands generally enhanced the sintering utilization factor under the other conditions. Therefore, to optimize the sintering utilization factor, when the proportion of ironsands exceeds 5 wt%, it is recommended to use either original-grade ironsands or fine-grained ironsands for sintering to achieve optimal results. Additionally, if ore grinding is not feasible due to time constraints or cost considerations, utilizing original-grade ironsands for sintering can still result in a satisfactory sintering utilization factor.

Influence of ironsand addition on vertical sintering speed

Similar to the sintering utilization factor, vertical sintering speed is a crucial index that reflects the efficiency of the sintering process. If the vertical sintering speed is too slow, the sintering utilization factor will decrease, thereby negatively impacting sintering production efficiency. On the other hand, if the vertical sintering speed is too fast, it can deteriorate the mineral composition of the sintered ore, reduce the quality of the sintered ore, lower the strength of the rotary drum, and ultimately decrease the yield of sintered ore. Therefore, maintaining an optimal vertical sintering speed is essential for efficient sintering production.

The impact of different ratios of ironsands on vertical sintering speed was analyzed based on sintering experiments, as shown in Fig. 5.53.

As shown in Fig. 5.53a, for original-grade ironsands, the vertical sintering speed first increases and then decreases with the rising proportion of ironsands added during the sintering process. When the proportion of original-grade ironsands reaches 5 wt%, the vertical sintering speed is maximized at 24.67 mm/min, representing a 16.2% improvement over the baseline experiment. Although the vertical sintering speed decreases when the proportion of ironsands reaches 8 wt%, it still remains higher than that of the baseline experiment. On average, the vertical sintering speed with

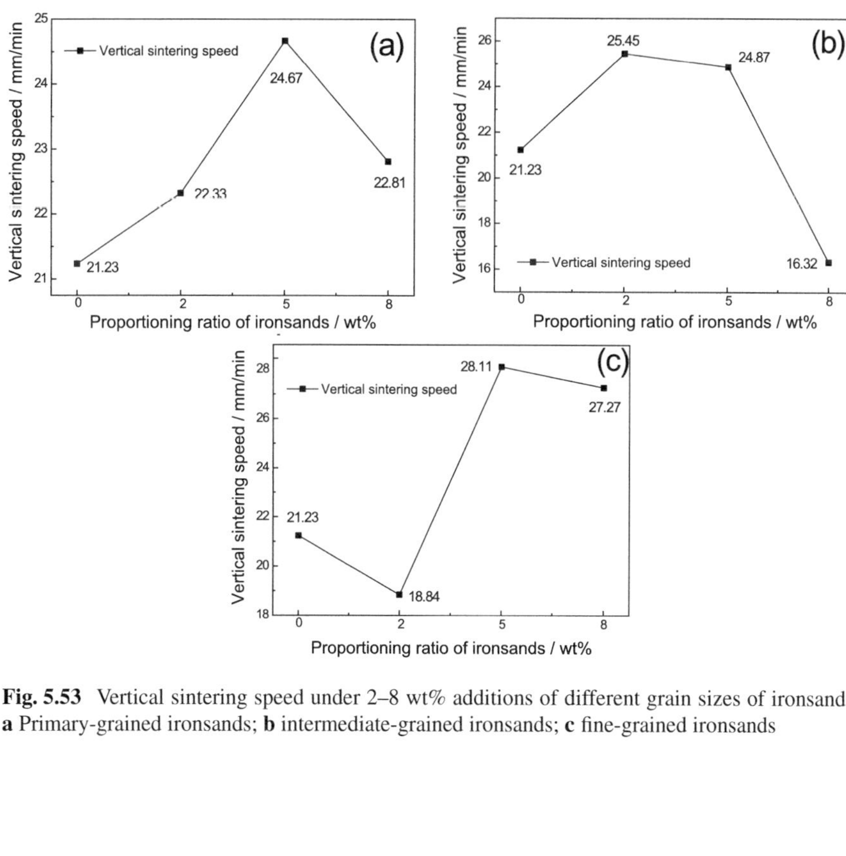

Fig. 5.53 Vertical sintering speed under 2–8 wt% additions of different grain sizes of ironsands. **a** Primary-grained ironsands; **b** intermediate-grained ironsands; **c** fine-grained ironsands

original-grade ironsands increased by 9.6% compared to the benchmark sample, which is highly beneficial for enhancing sintering production efficiency.

From Fig. 5.53b, it is observed that with the increase in the proportion of intermediate-grained ironsands, the vertical sintering speed follows a trend of first increasing and then decreasing. Notably, when the proportion of ironsands reaches 8 wt%, the vertical sintering speed decreases to 16.32 mm/min. This reduction is primarily attributed to the poor granulation effect caused by the intermediate-grained ironsands, which prevented the formation of spherical particles during the granulation process. This, in turn, negatively affected sintering air permeability, as favorable spherical particles enhance permeability. Furthermore, the finer grain size of intermediate-grained ironsands introduces more powder into the mixture, affecting the permeability of the material layer during sintering, thereby contributing to the decline in vertical sintering speed for this set of experiments.

In Fig. 5.53c, when the proportion of fine-grained ironsands is 2 wt%, there is little change in the vertical sintering speed compared to the baseline sample. However, when the proportion exceeds 5 wt%, the vertical sintering speed increases. Overall, the addition of fine-grained ironsands enhances the vertical sintering speed and improves sintering productivity.

Through the above analyses, it can be concluded that the vertical sintering speed varies under different conditions of ironsand proportions and grain sizes, yet a discernible pattern emerges. Overall, reducing the particle size of ironsands tends to improve the vertical sintering speed, although the effect is not particularly pronounced. Therefore, it is recommended to use original-grade ironsands for sintering production, as this approach not only saves on milling costs but also yields a satisfactory vertical sintering speed.

Impact of ironsand addition on sintering yield

Sintered ore yield is a key technical index calculated based on the composition and particle size of sintered ore, serving as an important measure of sintering production efficiency. Both high and low sintered ore yields directly affect sintering efficiency. The results for sintered ore yield under different ratios and grain sizes of ironsands are presented in Fig. 5.54.

As shown in Fig. 5.54a, with an increasing proportion of sintered ironsands, the sintered ore yield exhibits a significant increase. When the proportion of ironsands reaches 8 wt%, the sintered ore yield peaks at 80.9%, a 4.7% increase compared to the baseline experiment.

From Fig. 5.54b, it can be observed that as the proportion of intermediate-grade ironsands increases in the sintering process, the sintered ore yield initially increases significantly and then fluctuates at a high level, maintaining a yield above 80%. On average, this represents a 4.4% increase in sintered ore yield compared to the benchmark experiment.

According to Fig. 5.54c, when the proportion of fine-grained ironsands increases in the sintering process, the sintered ore yield first increases and then decreases. The maximum yield of 83.52% occurs at 2 wt% fine-grained ironsands. Even at 8 wt%

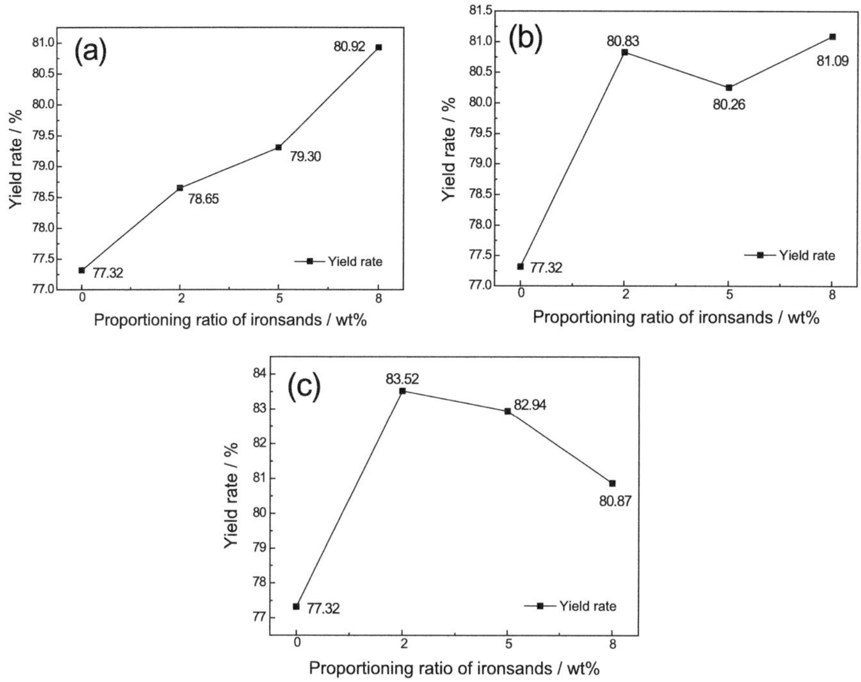

Fig. 5.54 Sintered ore yield under 2–8 wt% additions of different ironsands grain sizes. **a** Primary-grained ironsands; **b** intermediate-grained ironsands; **c** fine-grained ironsands

fine-grained ironsands, the yield remains at 80.87%, still higher than the baseline yield.

In summary, the addition of ironsands in the sintering process leads to different trends in sintered ore yield depending on the grain size of the ironsands. Overall, the addition of ironsands improves sintered ore yield to varying degrees, thereby positively influencing the efficiency of sintering production.

Influence of ironsand addition on sinter return ratio

The amount of sintered ore return is a critical factor influencing sintering yield. A lower return ratio indicates better quality sintered ore and reduced fuel consumption, as less ore needs to be reprocessed in the second sintering stage. The return ratio for the benchmark sample is 21.59%, as shown in Fig. 5.55, with a solid line for easy comparison.

When comparing the return ratio under different dosing conditions, the return ratio for all types of ironsands (primary, intermediate, and fine-grained) is lower than that of the benchmark sample. For the primary-grained ironsands, as the dosing amount increases, the return ratio gradually decreases from 20.40% and 19.71% to 17.66% and 18.33%. For sintered ore with intermediate-grained ironsands, the return ratio remains relatively stable across the 2–8 wt% range, but it is still lower than the

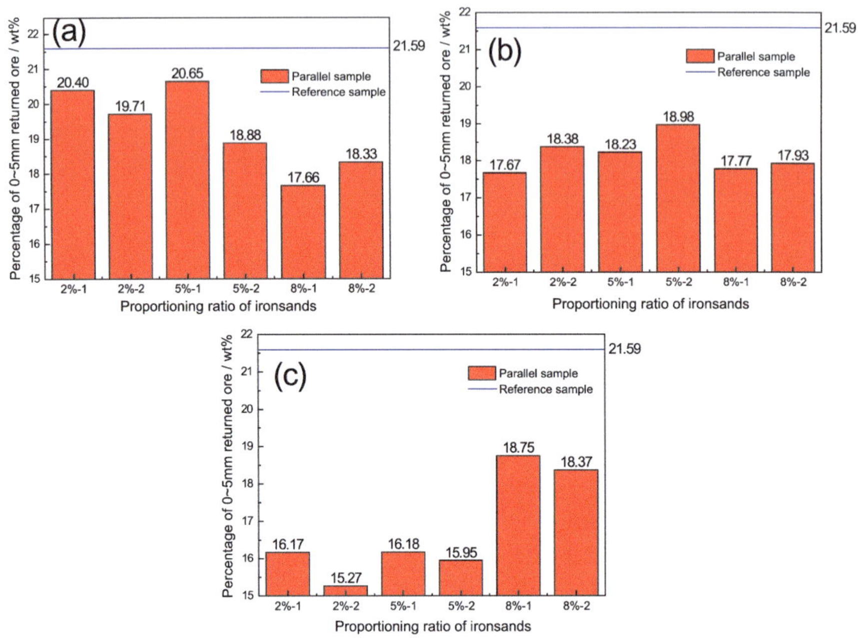

Fig. 5.55 Return ratio of ironsands under varying dosages of the same particle size. **a** Primary-grained ironsands; **b** intermediate-grained ironsands; **c** fine-grained ironsands

benchmark sample. In the case of fine-grained ironsands, the return ratio is better at 2 wt% and 5 wt%, but it increases by 2% when the dosing amount reaches 8 wt%.

The comparison of the effects of different ironsand fineness on the sintering return ratio under the same proportion of ironsands is illustrated in Fig. 5.56.

As shown in Fig. 5.56a, b, comparing the return ratios corresponding to different grinding times of ironsands, it can be seen that at 2 and 5 wt% ironsands, a finer particle size of ironsands results in a lower sintering return ratio. For example, when 2 wt% is added, the return ratio for original-grade ironsands is 20.40% and 19.71%, while for fine-grained ironsands, the return ratio decreases to 16.17% and 15.27%, lower than that of the benchmark sample. At 8 wt%, the fineness of the ironsands has little impact on the return ratio, which ranges between 17.6 and 18.8%.

Effect of ironsand addition on sintered solid combustion consumption

Solid combustion consumption is a key index for evaluating energy utilization efficiency in the sintering process. It has direct practical significance for energy conservation and consumption reduction. The results of sintering experiments with different ironsand ratios, as shown in Figs. 5.57 and 5.58, were analyzed to assess the impact on solid combustion consumption.

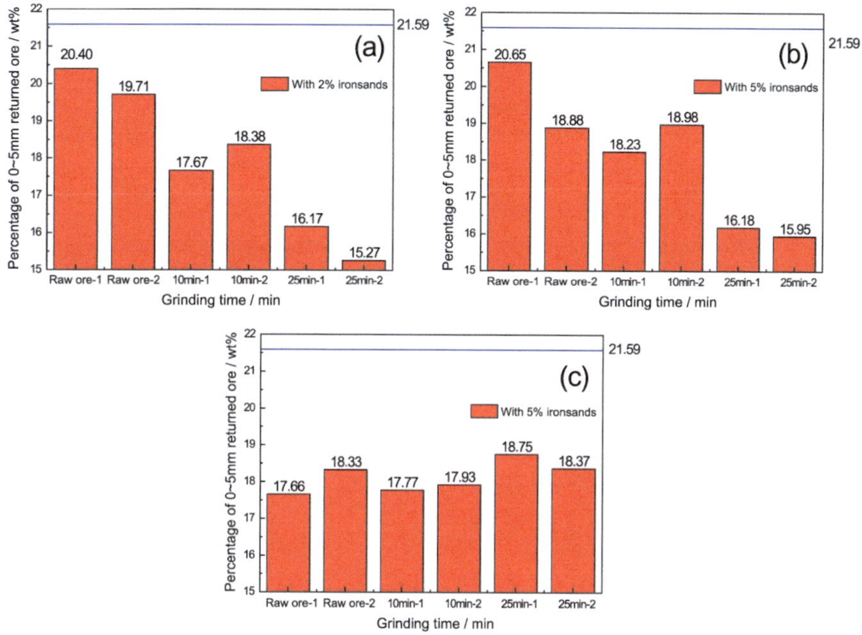

Fig. 5.56 Influence of ironsands particle size on the sintering return rate under identical dosing levels. **a** Primary-grained ironsands; **b** intermediate-grained ironsands; **c** fine-grained ironsands

As observed in Fig. 5.57, the addition of ironsands to the sintering process leads to an overall reduction in solid fuel consumption. Specifically, with increasing proportions of primary and intermediate-grained ironsands, solid combustion consumption tends to decrease. When fine-grained ironsands are added, the solid combustion consumption shows only a slight increase when the proportion reaches 8 wt%. This suggests that the inclusion of ironsands in the sintering process can significantly reduce solid fuel consumption, which is advantageous for energy saving and consumption reduction in sintering.

Further analysis of solid fuel consumption under different grinding time conditions is presented in Fig. 5.58.

From Fig. 5.58, when the proportion of ironsands is 2 wt%, decreasing the particle size of the ironsands leads to a sharp decline in solid fuel consumption. When the ironsands reach the fine-grained stage, the solid fuel consumption drops to 49.2 kg/t, which is 8.6% lower than the baseline experiment. When the proportion of ironsands increases to 5 wt% (see Fig. 5.58b), the solid fuel consumption continues to decrease with longer milling times, and the consumption at each milling time condition remains significantly lower than that of the benchmark experiment. As shown in Fig. 5.58c, when the proportion of ironsands reaches 8 wt%, an increase in milling time results in a general decrease in solid fuel consumption, with a final value of 53.83 kg/t, lower than the baseline sample. Therefore, adding ironsands effectively

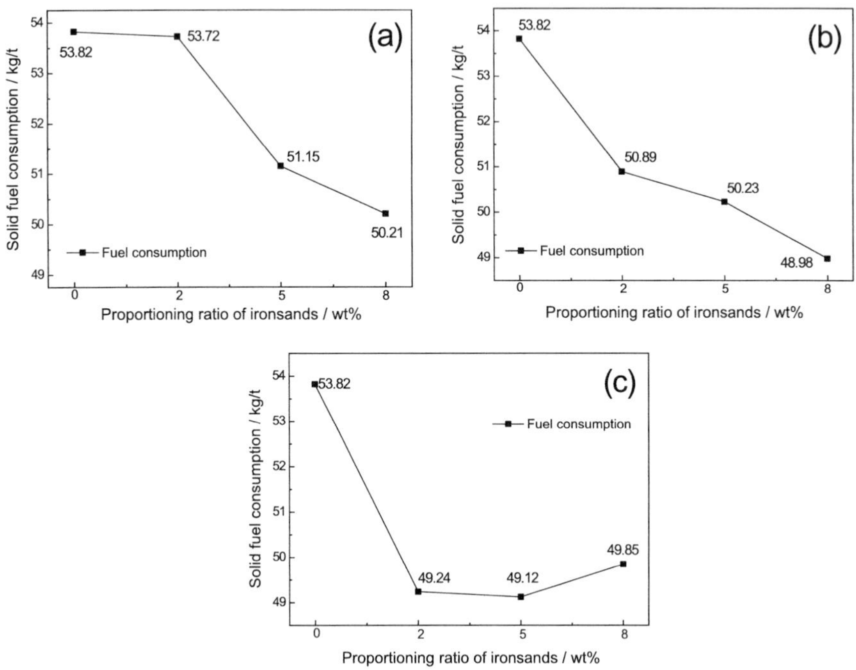

Fig. 5.57 Effect of ironsands on solid fuel combustion consumption at different dosing levels for the same grain size. **a** Primary-grained ironsands; **b** intermediate-grained ironsands; **c** fine-grained ironsands

reduces solid combustion consumption, with the most significant reduction observed with fine-grained ironsands milled for 25 min.

In summary, it can be concluded that, at lower dosages, reducing the particle size of ironsands significantly reduces solid fuel consumption during sintering. However, beyond an 8 wt% dosage, the effect of particle size reduction on solid fuel consumption becomes less pronounced.

Effect of ironsand addition on sintering burn rate

The sintering loss rate refers to the percentage of water, gas, and powder in the sintered mix that is lost during the sintering process due to solid fuel consumption, moisture volatilization, decomposition of iron ore and flux, dust blow-off, and other factors. This rate directly impacts the yield of sintered ore, and ideally, the sintering loss should be minimized.

The experimental results for sintering burn rate under different ironsand grain sizes and ratios are analyzed, with the results shown in Figs. 5.59 and 5.60.

As shown in Fig. 5.59a, as the proportion of original-grained ironsands added to the sintering process increases, the sintered ore burn rate shows an overall decreasing trend. When the proportion of original-grained ironsands reaches 8 wt%, the burn rate is minimized at 14.1%, which is 8.3% lower than the baseline experiment.

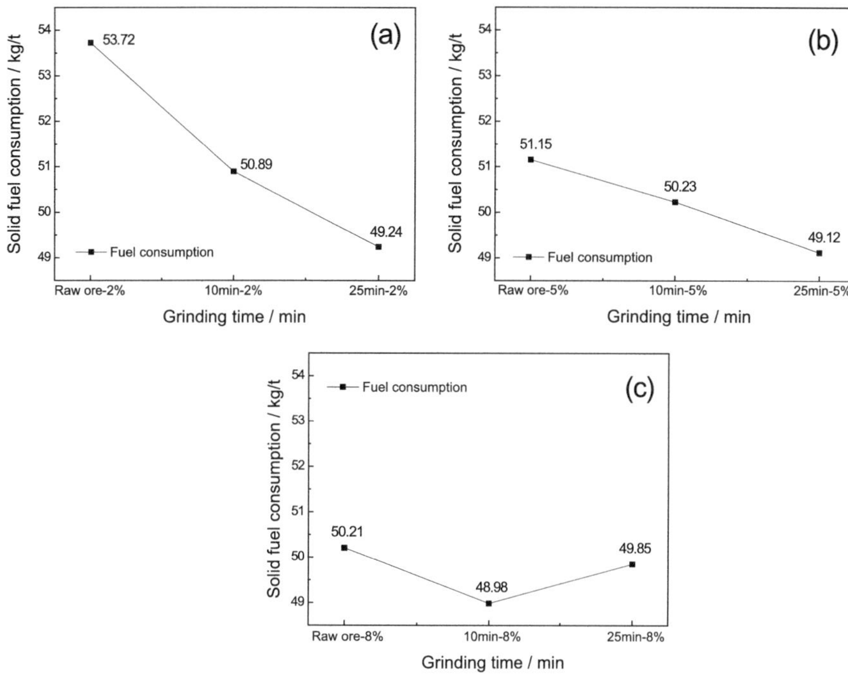

Fig. 5.58 Effect of ironsands grain size on solid fuel combustion consumption at constant dosing levels. **a** 2 wt% ironsands; **b** 5 wt% ironsands; **c** 8 wt% ironsands

Figure 5.59b demonstrates that as the proportion of intermediate-grained ironsands increases in the sintering process, the burn rate continues to decrease. At 8 wt%, the burn rate reaches a minimum of 14.3%, a reduction of 7.3% compared to the baseline burn rate. Figure 5.59c shows that with increasing proportions of fine-grained ironsands, the burn rate also exhibits a decreasing trend.

Overall, the addition of ironsands to the sintering process can reduce the burn rate to varying degrees. This is primarily due to the dense mineral phase of ironsands and their extremely low crystalline water content, which reduces the burn rate as the proportion of ironsands increases. The reduction in sintering loss rate enhances energy savings and consumption reduction, contributing positively to the improvement of energy utilization in the sintering process.

A comparison of the burn rate under different ironsand grain size conditions is shown in Fig. 5.60.

From Fig. 5.60a, when the proportion of ironsands is 2 wt%, a decrease in ironsand grain size leads to a gradual reduction in the burn rate. In Fig. 5.60b, when the proportion of ironsands is 5 wt%, the burn rate initially decreases and then increases with prolonged milling time. The lowest burn rate is observed with the finest grade of ironsands, at 14.2%. Figure 5.60c shows that when the proportion of ironsands reaches 8 wt%, the burn rate increases slightly as the particle size decreases, but the overall increase is minimal. On average, the sintering rate under different grinding

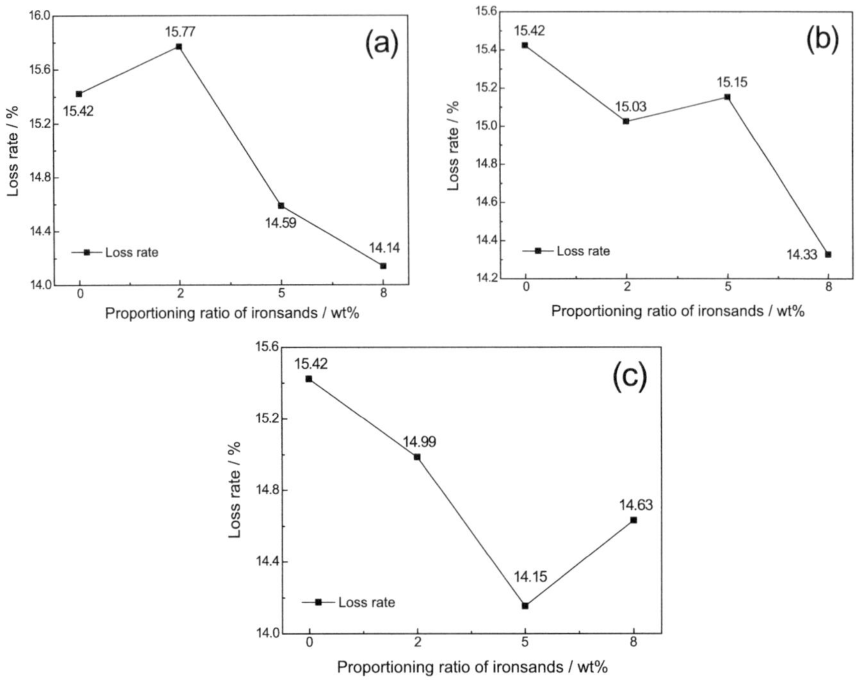

Fig. 5.59 Burnout rate of ironsands under different dosing levels for the same grain size. **a** Primary-grained ironsands; **b** intermediate-grained ironsands; **c** fine-grained ironsands

times is 6.8% lower than that of the baseline experiment. When the proportion of ironsands reaches 8 wt%, the sintering process yields a relatively low sintering rate, which can be attributed to the mineral composition of the ironsands.

In summary, to minimize the sintering loss rate, it is recommended to use fine-grained ironsands for sintering when the proportion of ironsands is less than 5 wt%. For proportions exceeding 5 wt%, original-grade ironsands should be used for sintering production.

5.7 Industrial Testing of Ironsands Sintering

The first industrial sintering test of Indonesian ironsands

Indonesian ironsands have been subjected to industrial testing in a vanadium and titanium ore smelting enterprise in China to assess their feasibility for sintering processes. This test aimed to explore the potential of incorporating Indonesian ironsands into the current raw material mix, replacing 5–10 wt% of vanadium-containing concentrates. The composition of the Indonesian ironsands used in the test is presented in Table 5.21, while the test program is outlined in Table 5.22.

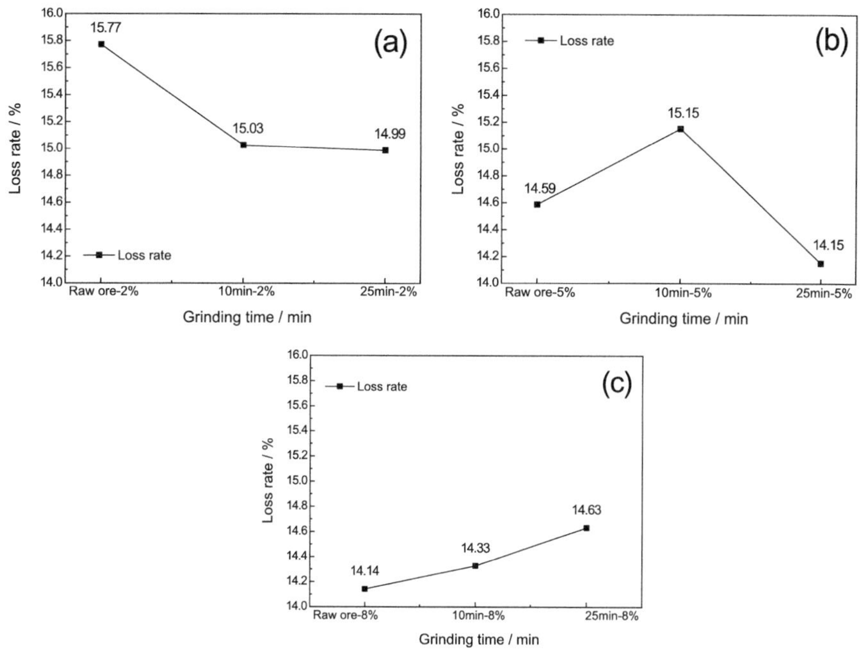

Fig. 5.60 Influence of ironsands particle size on burnout rate at constant dosing levels. **a** 2 wt% ironsands; **b** 5 wt% ironsands; **c** 8 wt% ironsands

Table 5.21 Composition/wt% of Indonesian ironsands for industrial tests

CaO	MgO	SiO_2	TFe	V_2O_5	TiO_2	Al_2O_3	S	FeO	P
0.57	1.85	2.99	55.97	0.613	12.05	2.56	0.012	21.77	0.033

The test results are shown in Tables 5.23, 5.24, 5.25 and 5.26. When 5–10 wt% of Indonesian ironsands was added to replace the vanadium-containing concentrate, the sintered ore drum strength ranged from 56.5% to 64%, the vertical sintering speed ranged from 14.5 to 16.1 mm/min, the average particle size ranged from 14.3 to 16.6 mm, and the low-temperature reducing pulverization index ranged from 19.7 to 38.3%. These results indicate that Indonesian ironsands can be successfully incorporated into the sintering process.

In the case of Option 2, which involved adding 10 wt% Indonesian ironsands, the sintered ore exhibited the best overall performance in terms of drum strength, vertical sintering speed, average particle size, and low-temperature reduction pulverization index. Therefore, Option 2 is the preferred choice.

Conversely, Option 3, which involved adding 5 wt% Indonesian ironsands, 15 wt% PB, 10 wt% Mac, and 5 wt% Miaogou Pu powder, resulted in poor performance across all indices. As a result, Option 3 is not recommended for use.

Table 5.22 Ironsands sintering industrial test proportioning scheme/wt%

Programme	Mixing ratio					Sintering machine ratio							
	Xiaoying vanadium powder	Miaogou ordinary powder	PB powder	MAC powder	Indonesian ironsands	Impurity	Mixture	Impurity	Calcium ash	Magnesium ash	Fuel	High return	Self-return
1	60.0		25	10	5		74.9	8	8.4	3.2	5.5	10	15
2	55.0		25	10	10		75.0	8	8.3	3.2	5.5	10	15
3	57.5	5	15	10	5	7.5	83.3		8	3.2	5.5	10	15

Table 5.23 Results of industrial test production parameters for sintering of ironsands mines in Indonesia

Programme	Sintering time min	Burn rate %	Vertical sintering speed mm/min	Yield rate %	Drum index %	Utilization factor $t/m^2 \cdot h$	Fuel consumption kg/t
1	38.00	7.45	15.00	71.62	61.20	1.01	66.38
2	35.45	8.21	16.08	78.92	64.00	1.21	60.74
3	39.33	6.04	14.49	70.29	56.53	0.97	66.62

Table 5.24 Sintered ore particle size composition of sintered industrial test of ironsands in Indonesia

Programme	> 40/ wt%	40–25/ wt%	25–15/ wt%	15–10/ wt%	10–5/ wt%	< 5/wt%	Average particle size/mm
1	4.86	14.89	18.04	19.37	15.68	27.16	14.30
2	7.28	19.74	18.98	19.01	14.89	20.09	16.58
3	8.72	13.27	18.37	18.31	13.64	27.7	14.82

Table 5.25 Chemical composition of sintered ore from the industrial sintering test of ironsands in Indonesia/wt%

Programme	CaO	MgO	SiO_2	TFe	V_2O_5	TiO_2	Al_2O_3	S	FeO	P
1	10.83	2.33	4.68	54.63	0.28	2.10	2.32	0.112	14.15	0.051
2	10.95	2.31	5.18	55.18	0.23	1.73	2.15	0.054	11.57	0.042
3	8.02	3.36	3.66	54.94	0.37	3.76	2.30	0.032	5.39	0.038

Indonesia ironsands second sintering industrial test

The second industrial test was carried out on the raw material structure of common ore and vanadium-titanium ore with the addition of this kind of Indonesian ironsands, and the proportioning scheme is shown in Table 5.26.

The second industrial sintering test of Indonesian ironsands

A second industrial sintering test was conducted using a raw material mix comprising common ores and vanadium–titanium ores, incorporating Indonesian ironsands. The proportioning scheme is presented in Table 5.26.

The results of the test are shown in Tables 5.27 and 5.28. Schemes 1 and 2 involved the addition of 5 wt% and 10 wt% Indonesian ironsands, respectively, to a general ore base. In terms of sinter quality, both schemes produced drum indices of 70.53% and 68.27%, and average particle sizes of 16.58 mm and 15.07 mm, respectively, all of which meet production requirements. However, the vertical sintering speed in both cases was relatively low. Although Scheme 1 achieved a slightly better rate at

Table 5.26 Proportioning scheme for the second industrial sintering test of ironsands in Indonesia

Programme	Mixing ratio							Sintering machine ratio					
	Vanadium powder	Ordinary powder	PB powder	MAC powder	North African powder	Indonesian ironsands	Impurity	Mixture	Impurity	Calcium ash	Magnesium ash	Coke powder	Return ore
1		40	35	10	10	5		74.5	8	9.5	3	5	20
2		40	30	10	10	10		74.4	8	9.6	3	5	20
3	57.5		30			5	7.5	83.5		8	3	5.5	30

Table 5.27 Process parameters of the second industrial sintering test of Indonesian ironsands

Programme	Vertical sintering speed/mm/min	Yield rate/%	Drum index/%	Utilization factor	Fuel consumption/kg/t	Average particle size/mm
1	13.58	75.17	70.53	0.92	60.25	16.58
2	11.01	72.91	68.27	0.77	57.75	15.07
3	9.79	60.62	63.33	0.57	70.79	14.09
3	11.52	78	68.53	0.88	54.82	17.54
Average of ordinary ore	18.8	79.6	69.6	1.4	58.8	17.9
Average of vanadium ore	16.5	76.9	67.5	1.2	57	17.4

13.58 mm/min, it remains insufficient to enhance the output of the sintering machine, thereby resulting in a low utilization factor.

In terms of melting and dripping performance, Scheme 1 exhibited better behavior, while Scheme 2 presented excessive melting, with a melting interval of 476 °C and a melting zone thickness of 51.19 mm, both of which are too high. Consequently, Scheme 2 is not considered suitable for production.

Regarding Scheme 3, results from both trials indicate that this raw material configuration is unsuitable. In the second trial, the maximum vertical sintering speed reached only 11.52 mm/min, which is below the commonly accepted threshold of 13 mm/min. Although the drum index reached 68.53%, and the yield and finished particle size distribution met basic requirements, the metallurgical properties were suboptimal. In the first test, the softening interval of 137 °C was acceptable, but the melting interval of 425 °C and melting zone thickness of 49.2 mm were excessive. In the second test, although the softening interval increased to 176 °C, the melting interval remained high at 398 °C, and the melting zone thickness was still too large at 44.87 mm.

In general, for optimal sintering performance, the softening interval should be less than 170 °C, the melting interval should be below 300 °C, and the melting zone thickness should not exceed 30 mm. Based on these criteria, none of the configurations in Scheme 3 meet the necessary conditions for efficient industrial sintering.

5.8 Summary

In this chapter, the fundamental mineralogical properties of ironsands were examined in detail, including chemical composition, microstructure, particle size distribution, and abrasive characteristics. Through sintering cup experiments, the effects of three particle size distributions and three different dosing levels of ironsands on the sintering process were systematically studied. These effects encompassed both

Table 5.28 Operating parameters of the second industrial sintering test using Indonesian ironsands

Programme	$RDI_{+3.15mm}$	Differential pressure steep rise temperature Ts/°C	Softening zone T40-T10/°C	Dropping temperature Td/°C	Melting zone Td-Ts/°C	Thickness of molten zone ΔH/mm	Characteristic value S
1	30.29	1209	126	1475	266	23.64	473
2	33.61	1009	109	1485	476	51.19	636
3	34.17	1037	137	1462	425	49.02	610
3	35.43	1047	176	1445	398	44.87	467
Ordinary ore	31.7	1273	69	1488	220	24.8	191
Vanadium ore	33.6	1208	140	1450	244	26.7	269

the physical properties of the resulting sintered ore and key sintering process parameters. The metallurgical performance of sintered ores—particularly low-temperature reduction degradation, reducibility, and softening–melting–dripping behaviors—was evaluated under different particle sizes and dosage conditions. The main findings are summarized as follows:

(1) The ironsands used in this study had a total iron grade of 55.63 wt% and a TiO_2 content of 11.41 wt%. Iron and titanium were primarily present in the form of titanomagnetite, indicating the potential of this material as a high-grade iron-bearing feedstock suitable for blast furnace operations.
(2) The addition of ironsands had minimal influence on the overall iron grade and basicity of the final sintered product. FeO and MgO contents increased but remained within the standards for high-quality sinter. Al_2O_3 content decreased, which is favorable for enhancing the sinter's metallurgical performance. TiO_2 exhibited a linear increase, while harmful elements such as P and S remained low, contributing to improved environmental and metallurgical properties.
(3) Raw ironsands particles were mainly spherical or ellipsoidal, with smooth surfaces and low specific surface area, resulting in relatively poor granulation and adherence properties compared to conventional fine ores. After 25 min of ball milling, over 65 wt% of the ironsands reached sizes below 200 mesh, meeting requirements for adherent fine materials in sintering. Microstructural analysis showed that the baseline sample mainly contained calcium ferrite well bonded with hematite and silicate. The addition of coarse-grained ironsands did not significantly alter the dominant mineral phases. However, at 8 wt% dosing of intermediate- and fine-grained ironsands, the presence of skeletal rhombohedral hematite increased, negatively impacting the physical and metallurgical properties of the sinter.
(4) When 2–8 wt% of coarse-grained ironsands was added, the average particle size of the resulting sinter remained stable. For intermediate and fine-grained ironsands, average particle size decreased with increasing dosage. Coarse-grained ironsands had negligible impact on drum strength across the tested range, while intermediate-grained ironsands caused a decline in drum strength at 8 wt%, likely due to their being outside the optimal size range for both nucleation and adherence during granulation.
(5) Solid fuel consumption during sintering decreased with increasing ironsands addition. For small additions (2–5 wt%), reduced particle size led to lower fuel consumption; however, at higher additions (8 wt%), further size reduction showed minimal impact. Ironsands addition improved sinter yield and utilization and reduced the burnout rate compared to the baseline. At 8 wt% addition of coarse-grained ironsands, all key performance indices met production requirements, with most surpassing those of the baseline sample. While finer particles promote yield, caution is advised due to their potential effects on sinter quality.

(6) At 2–5 wt% dosing, the low-temperature reduction degradation performance of ironsands sinter was comparable to that of the baseline. At 8 wt%, TiO_2 content increased and reacted with CaO to form perovskite, which concentrated in the glassy phase, adversely affecting the low-temperature reduction degradation. Fine-grained ironsands promoted skeletal rhombohedral hematite formation, also contributing to increased degradation. Coarse-grained ironsands did not significantly alter the LTRDI.
(7) Due to the formation of solid solutions involving titanium, the reducibility of sinter declined slightly with increasing ironsands content, particularly at 8 wt%. Nevertheless, the reducibility remained above the threshold for high-quality sinter. With decreasing particle size, reducibility showed a downward trend, likely because finer ironsands reduced the permeability of the sintering bed and led to a denser microstructure, inhibiting gas diffusion.
(8) Increasing the dosage of coarse-grained ironsands widened the softening interval of the sinter. In contrast, decreasing the particle size led to a narrowing of this interval. At 8 wt% addition of fine-grained ironsands, the softening interval increased by only ~ 10 °C. Higher ironsands content increased droplet temperature due to the high melting point of titanium compounds, with the maximum droplet temperature reaching 1498 °C. However, droplet temperature was not significantly affected by particle size. As ironsands content increased, the melting interval narrowed for coarse particles but widened for intermediate and fine particles, with intermediate-grained ironsands showing the greatest increase.
(9) With increased ironsands dosing, TiO_2 content rose correspondingly. Under reducing conditions (via reducing gas and coke contact), titanium carbonitride particles formed and substantially increased the viscosity of molten slag. This hindered gas penetration, increased differential pressure, and reduced sinter bed permeability. Consequently, the permeability of sinter deteriorated with both increasing ironsands dosage and finer particle sizes, as reflected by higher comprehensive gas resistance (S) values.
(10) Overall, the incorporation of ironsands in the sintering process led to varying degrees of improvement in key technical indices. Sinter yield increased with higher ironsands content and with reduced particle size. However, the sintering utilization factor decreased significantly when medium-grained ironsands were used, due to prolonged sintering times caused by reduced bed permeability. Both coarse and fine-grained ironsands improved sintering utilization. Solid fuel consumption dropped markedly across all dosing schemes, with only minor differences between fine- and coarse-grained ironsands. Although finer ironsands positively influenced parameters such as sinter return rate and utilization, their impact on sinter quality must be carefully considered.

References

1. Bosi F, Halenius U, Skogby H. Crystal chemistry of the magnetite-ulvospinel series. Am Mineral. 2009;94(1):181–9.
2. Wang YM, Yuan ZF, Guo ZC, et al. Reduction mechanism of natural ilmenite with graphite. Trans Nonferrous Metals Soc China. 2008;18(4):962–8.
3. Wang Z, Pinson D, Chew S, et al. Behavior of New Zealand ironsand during iron ore sintering. Metall Mater Trans B-Proc Metall Mater Proc Sci. 2016;47(1):330–43.
4. Yu ZW, Li GH, Liu C, et al. Comparison of reduction disintegration characteristics of TiO_2-rich burdens prepared with sintering process and composite agglomeration process. Int J Min Met Mater. 2016;23(4):389–98.
5. Wang ZY, Zhang JL, Xing XD, et al. Phase transitions and atomic-scale migration during the preoxidation of a titania/ferrous oxide solution. JOM. 2016;68(2):656–67.
6. Yu ZW, Li GH, Jiang T, et al. Effect of basicity on titanomagnetite concentrate sintering. ISIJ Int. 2015;55(4):907–9.
7. Jung SM. Effects of $CaO/CaCO_3$ on the carbothermic reduction of titanomagnetite ores. Metall Mater Trans B-Proc Metall Mater Proc Sci. 2015;46(3):1162–74.
8. Geng C, Sun TC, Yang HF, et al. Effect of Na_2SO_4 on the embedding direct reduction of beach titanomagnetite and the separation of titanium and iron by magnetic separation. ISIJ Int. 2015;55(12):2543–9.
9. Schwebel GL, Filippou D, Hudon G, et al. Experimental comparison of two different ilmenites in fluidized bed and fixed bed chemical-looping combustion. Appl Energy. 2014;113:1902–8.
10. Bernasowski M. Theoretical study of the hydrogen influence on iron oxides reduction at the blast furnace process. Steel Res Int. 2014;85(4):670–8.
11. Dang J, Hu X, Zhang G, et al. Kinetics of reduction of titano-magnetite powder by H_2. High Temp Mater Process (Lond). 2013;32(3):229–36.

Chapter 6
Pelletizing Processes for Ironsands

This chapter presents an experimental investigation into the application of ironsands in oxidized pellet production. Parameters such as raw material composition, oxidation roasting temperature, and physical properties of both green and fired pellets are optimized. Additionally, the high-temperature metallurgical behavior of ironsands pellets is analyzed to evaluate their feasibility for industrial pelletizing applications.

The structure of this chapter is as follows:

(1) Due to the coarse particle size and low specific surface area of the raw granular ironsands, a milling pretreatment is performed. The particle size distribution and specific surface area of the ironsands concentrate powder obtained after milling are measured to determine suitable milling process parameters.
(2) Milled ironsands powders at various proportions are mixed with conventional iron ore fines and a fixed amount of bentonite. The mixtures are then processed into green pellets. A portion of the qualified green pellets is selected for testing, including compressive strength, drop strength, and moisture content, to evaluate their mechanical integrity.
(3) The qualified green pellets are dried in a constant-temperature oven at 105 °C. The dried pellets are then subjected to preheating and roasting under different thermal regimes. The compressive strength of the roasted pellets is tested to study the influence of different preheating and roasting conditions on pellet strength.
(4) The roasted pellets are tested for key metallurgical properties, including reducibility, low-temperature reduction degradation, and reduction expansion. Based on these technical indices, changes in pellet quality due to the addition of ironsands are analyzed. The underlying mechanisms and influencing factors are examined to assess the feasibility of using ironsands in the pelletizing process.

The experimental methodology is illustrated in Fig. 6.1.

Z. Wang et al., *Smelting and Utilization Technology of Ironsands*,
https://doi.org/10.1007/978-981-95-2868-4_6

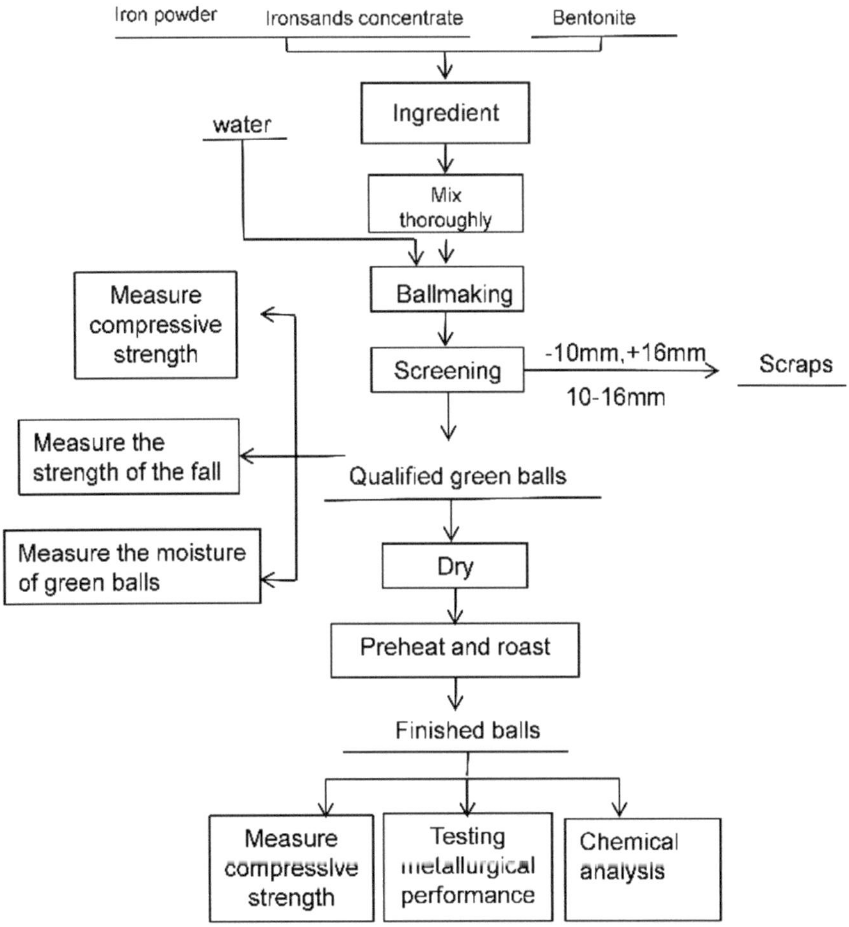

Fig. 6.1 Research process for pelletizing and lumping ironsands

6.1 Characteristics of Raw Materials for Pellet Formation

Composition and properties of ironsands-based raw materials

The chemical compositions of the mineral powders used in the ironsands pelletizing experiments are presented in Tables 6.1 and 6.2:

Table 6.1 Chemical composition of ironsands/wt%

TFe	FeO	SiO_2	CaO	Al_2O_3	MgO	TiO_2	V_2O_5	Burn
55.63	29.60	4.13	0.60	3.38	3.74	11.41	0.48	– 3.14

Table 6.2 Chemical composition of the other two mineral powders used in the experiment/wt%

Name	TFe	FeO	SiO_2	CaO	Al_2O_3	MgO	Burn	– 0.074 mm
Peruvian powder	69.27	29.28	2.11	0.51	0.35	0.53	– 2.35	88.51
Self-processing powder	67.57	29.71	3.19	0.77	0.61	1.02	– 2.06	80.30

As shown in Tables 6.1 and 6.2, the total iron content of the ironsands is 55.63 wt%, with an FeO content of 29.60 wt%. The TiO_2 and V_2O_5 contents are 11.41 wt% and 0.48 wt%, respectively. In comparison, the two types of ore powders—Peruvian powder and self-processed powder—exhibit higher total iron contents of 69.27 wt% and 67.57 wt%, respectively, with FeO contents exceeding 29 wt%, indicating that both are high-grade magnetite ores. The ironsands have relatively higher contents of SiO_2, Al_2O_3, and MgO compared to the Peruvian and self-processed powders, reaching 4.13 wt%, 3.38 wt%, and 3.74 wt%, respectively.

Binders are essential in the pelletizing process. The selection of an appropriate binder not only enhances product quality and reduces energy consumption, but also improves the surface characteristics of the material, thereby promoting pellet formation and enhancing the strength of the final pellets. Moreover, binders can significantly influence the metallurgical properties of the pellets. In this study, bentonite was selected as the binder [1]. The experimentally measured physical properties of the bentonite are listed in Table 6.3, and its chemical composition is shown in Table 6.4.

Ball milling characteristics of ironsands

The experimental ironsands exhibit a coarse particle size, with virtually no particles smaller than 0.074 mm. This is significantly coarser than what is required for the balling process, which demands finer particles for effective agglomeration. Moreover, the particles are predominantly spherical or ellipsoidal in shape and possess a low specific surface area, resulting in poor balling performance. Therefore, a

Table 6.3 Properties of bentonite clay

Cellulose price ml/15 g	Inflation index	Water absorption %/2 h	Amount of blue absorption g/100 g	Montmorillonite content/wt%	< 0.074 mm/wt%
14	9	158	35	81.45	91.5

Table 6.4 Chemical composition of bentonite/wt%

TFe	SiO_2	CaO	Al_2O_3	MgO	S	P	Na_2O	K_2O
2.32	57.1	2.55	17.17	4.45	0.023	0.028	1.92	0.71

grinding pretreatment is essential to refine the original granular ironsands to meet the requirements of pellet production.

A Bond work index ball mill was used for the grinding process, as shown in Fig. 6.2.

For each trial, 2 kg of ironsands was milled using the Bond ball mill. After grinding for a designated time, the milled material was removed and a representative portion was subjected to particle size analysis using a 0.074 mm standard sieve. The weight percentage of particles smaller than 0.074 mm was calculated. Additionally, specific surface area tests were conducted on selected samples. The proportion of particles below 0.074 mm and the corresponding specific surface area values are presented in Fig. 6.3 and Table 6.5.

As shown in Fig. 6.3, the proportion of particles smaller than 0.074 mm increases with longer grinding times. However, there is no sharp increase at any specific time

Fig. 6.2 Photograph of the ball mill apparatus

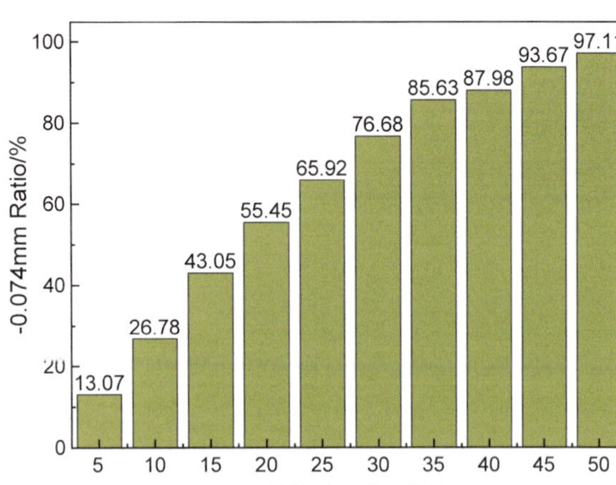

Fig. 6.3 Relationship between grinding time and particle size distribution

Table 6.5 Specific surface area of ironsands at grinding times of 25–35 min

Grinding time/min	25	30	35
Specific surface area/$cm^2\ g^{-1}$	1891	2276	2741

point, indicating that the ironsands are relatively resistant to grinding and refinement. After 5 min of grinding, only 13.07 wt% of the material is smaller than 0.074 mm. This proportion rises to 55.45 wt% at 20 min—an increase of 42.38 wt%. Extending the grinding time to 35 min further increases the proportion to 85.63 wt%, representing an additional 30.18 wt%. At 50 min, the proportion reaches 97.11 wt%, a 11.48 wt% gain over the 35-min result.

The rate of particle size reduction is more rapid in the early stages of grinding. Between 5 and 20 min, the proportion of sub-0.074 mm particles increases by approximately 12.2 wt% more than during the 20 to 35-min interval. This indicates that the initial phase of grinding is more efficient in particle size reduction. Between 20 and 35 min, the rate of increase in fine particles averages about 10 wt% every 5 min. In contrast, from 35 to 50 min, the increase slows to approximately 3 wt% for every 5-min increment, reflecting the diminishing returns in the later stages of grinding.

According to Table 6.5, when the grinding time reaches 25 min, the specific surface area of the ironsands concentrate is 1891 $cm^2\ g^{-1}$. Extending the grinding time to 35 min increases the specific surface area to 2741 $cm^2\ g^{-1}$, at which point the fraction of particles below 0.074 mm also reaches 85.63 wt%. This satisfies the particle size requirement for the balling process. Considering grinding costs, particle size distribution, and specific surface area, the ironsands concentrate ground for 35 min was selected as the raw material for the subsequent pelletizing experiments.

6.2 Preparation and Quality Assessment of Green Pellets

To investigate the effect of different ironsands concentrate dosages on pellet quality, varying proportions of ironsands concentrate powder were mixed with ordinary iron ore concentrate. The experimental scheme is detailed in Table 6.6.

The pelletizing mixture was prepared using 4 kg of iron-bearing raw materials. Bentonite, used as a binder, was added at 1.7 wt% of the total mass of iron-containing

Table 6.6 Experimental schemes/wt%

Serial number	Peruvian ore	Self-processing powder	Ironsands	Bentonite
1	85	5	10	1.7
2	70	0	30	1.7
3	50	0	50	1.7
4	30	0	70	1.7
5	0	0	100	1.7

minerals. The components were manually mixed, followed by the addition of water to bring the moisture content slightly below the optimal level required for green ball formation. Once the mixture was thoroughly homogenized, it was transferred to the pelletizing process.

Pellets were formed using a disc pelletizer, with each batch consisting of 4 kg of raw material, in accordance with the machine's rated capacity. Green pellets with a diameter of 10–16 mm were considered qualified. Ten qualified green pellets were selected to measure average compressive strength and drop strength. Once the quality standards were met, the remaining pellets were dried for subsequent preheating and roasting experiments.

The disc pelletizer used in this experiment had the following specifications:

Diameter: 800 mm;
Rotational speed: 38 r/min;
Side height: h = 150 mm;
Inclination: Q = 47°.

The disc pelletizer is shown in Fig. 6.4.

Drop Strength Testing of Green Pellets Ten green pellets were dropped freely from a height of 0.5 m onto a 10 mm-thick steel plate. The number of drops each pellet could withstand without breaking was recorded, and the average was taken as the drop strength. The testing apparatus is shown in Fig. 6.5.

Compressive Strength Testing of Green Pellets Ten green pellets were selected for compressive strength testing using a particle strength tester. The average value was taken as the compressive strength result. The compressive strength testing equipment is shown in Fig. 6.6.

For each test group, 4 kg of dry raw material was prepared according to the predetermined mixing ratios. Approximately 5 wt% water was added and manually

Fig. 6.4 Disc pelletizing machine used for pellet formation

Fig. 6.5 Device for measuring the drop strength of green pellets

Fig. 6.6 Equipment for measuring the compressive strength of green pellets

mixed, followed by pelletizing using the 800 mm disc pelletizer. Additional water was added as needed during pelletizing. The resulting green pellets (10–16 mm in diameter) were tested for drop strength and compressive strength. The pellet quality results are summarized in Table 6.7.

The results for the number of drops sustained by the raw balls are presented in Fig. 6.7. As shown in the figure, the highest drop resistance was recorded at 4.3 drops when the ironsand content was 30 wt%, while the lowest was 3.3 drops at 70 wt%

Table 6.7 Quality parameters of green pellets

Batch number	1	2	3	4	5
Raw ball drop intensity/times	4.0	4.3	4.1	3.3	3.6
Raw ball compressive strength/N	8.1	8.2	8.2	8.9	8.2

ironsand. Nevertheless, all samples exhibited drop resistances exceeding three drops, indicating that the mechanical integrity of the raw balls under varying ironsand ratios meets the production requirements.

The compressive strength results of the raw balls are shown in Fig. 6.8. According to the figure, the highest compressive strength, 8.9 N, was achieved at an ironsand content of 70 wt%. Even when 100% ironsand was used in pellet preparation, the compressive strength remained high at 8.5 N. Across all tested ironsand ratios, the compressive strength exceeded 8 N, thereby fulfilling the mechanical strength criteria for industrial production.

Based on the analysis of raw ball properties after incorporating ironsand into ordinary iron powder, the following conclusions can be drawn:

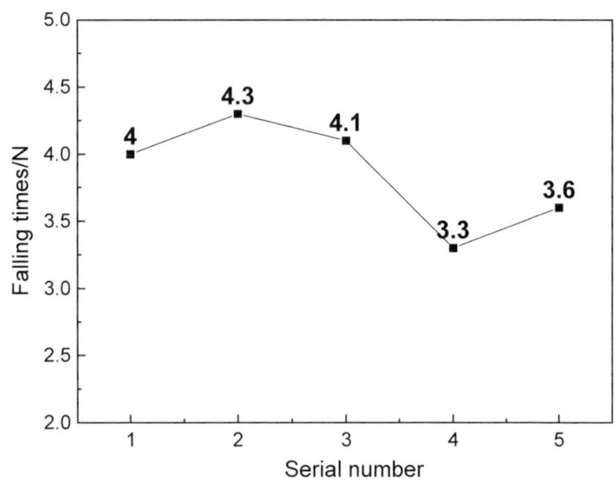

Fig. 6.7 Drop count of green pellets under standardized testing conditions

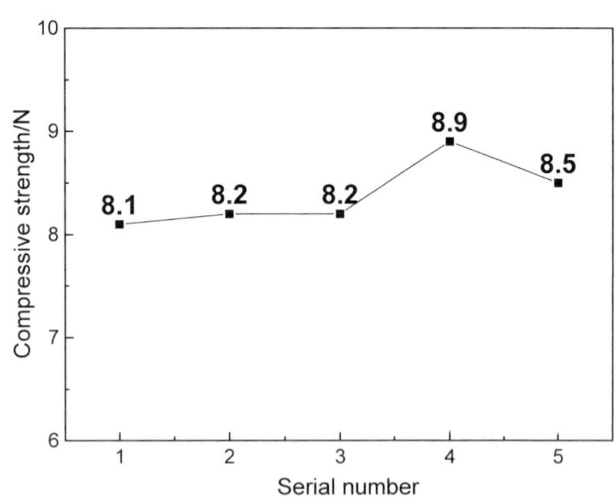

Fig. 6.8 Compressive strength of green pellets

(1) During the pretreatment of the original granular ironsand, a grinding time of 35 min resulted in 85.63 wt% of the ironsand concentrate achieving a particle size below 0.074 mm. The specific surface area reached 2741 cm^2/g, meeting the particle size and surface area requirements for the pelletization process.
(2) Regarding drop strength, the highest value was 4.3 drops at 30 wt% ironsand, and the lowest was 3.3 drops at 70 wt%. For compressive strength, all formulations with varying ironsand content exhibited strengths above 8 N. These results suggest that the proportion of ironsand has limited impact on the overall pelletizing process. Even with 100% ironsand substitution, the resulting raw balls satisfy the mechanical requirements for production.

6.3 Preparation and Quality Assessment of Finished Pellets

Raw pellets must undergo three sequential thermal treatment stages—drying, preheating, and roasting—to be transformed into finished pellets.

Drying Stage Conducted at temperatures between 200–400 °C, this stage primarily facilitates the evaporation and removal of free moisture. Additionally, a portion of the crystalline water may decompose and be released during this phase.

Preheating Stage Typically carried out at temperatures ranging from 900–1000 °C, this stage expels any residual crystalline moisture not eliminated during drying. Key reactions during preheating include the oxidation of magnetite (Fe_3O_4) to hematite (Fe_2O_3), the decomposition and oxidation of sulfides, the breakdown of carbonate minerals, and the initiation of certain solid-phase reactions.

Roasting Stage: Performed at temperatures generally between 1200–1300 °C, this stage continues the decomposition, oxidation, and desulfurization reactions initiated during preheating. It is characterized by the crystallization and recrystallization of iron oxides, grain growth, further solid-phase reactions, partial melting of low-melting-point compounds, the formation of a liquid phase, volumetric shrinkage of the pellet, and densification of its internal structure.

The preheating and roasting experiments were conducted using two horizontal tube furnaces. The selected experimental conditions included a preheating temperature of 950 °C for 15 min, followed by a roasting temperature of 1260 °C, also maintained for 15 min. The experimental setup is illustrated in Fig. 6.9.

The chemical composition of the roasted finished pellets was analyzed, and the results are presented in Table 6.8. With the gradual increase in the proportion of ironsand, the overall iron grade in the finished pellets decreases, while the contents of SiO_2 and TiO_2 increase. This trend is primarily attributed to the inherent chemical composition of the ironsand. Additionally, since ironsand is predominantly in the magnetite phase, the FeO content in the finished pellets shows an upward trend—even after oxidative roasting—due to incomplete oxidation.

Fig. 6.9 Horizontal tube furnace used for pellet heat treatment

Table 6.8 Chemical composition of finished pellets/wt%

Serial number	TFe	FeO	SiO_2	TiO_2
1	64.9	0.32	2.80	1.60
2	62.9	0.35	3.31	3.42
3	60.4	0.62	3.85	6.89
4	58.0	1.12	4.18	8.87
5	53.8	1.72	4.44	11.2

The compressive strength of the pellets was tested using a maximum applied pressure of 10^4 N and a loading rate of 15 ± 5 mm/min. The tested pellet samples had particle sizes ranging from 10 to 16 mm. For each composition, ten pellets were tested, and the average compressive strength was recorded as the representative value. The experimental setup is shown in Fig. 6.10. The compressive strength results for the five pellet compositions are shown in Fig. 6.11.

As illustrated in Fig. 6.11, the highest compressive strength was recorded at 2674 N when the ironsand content was 10 wt%. When the ironsand proportion reached 100 wt%—i.e., when the pellets were prepared entirely from ironsand—the compressive strength decreased to 2260 N, which is 414 N lower than that of the 10 wt% sample. Despite this reduction, the compressive strength still exceeds the blast furnace requirement of 2000 N for qualified pellet feed.

The decline in compressive strength with increasing ironsand content is primarily due to a decrease in pellet grade and a reduced proportion of Fe_3O_4. This results in less magnetite being oxidized to hematite during roasting, which in turn leads to a reduction in the formation of microcrystalline bonds between hematite particles—an

Fig. 6.10 Equipment for determining the compressive strength of sintered pellets

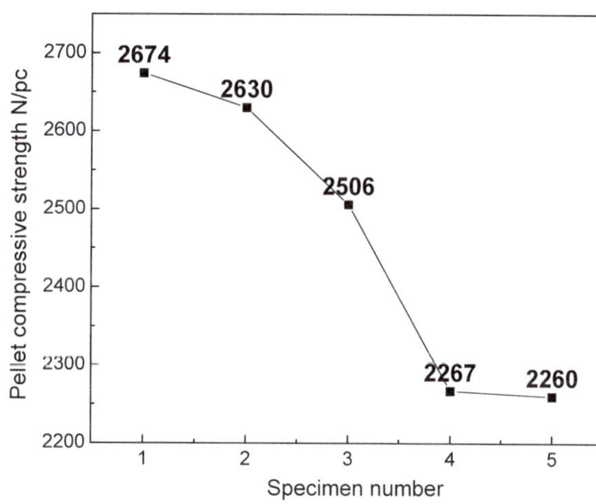

Fig. 6.11 Compressive strength curve of sintered pellets/N

important factor in pellet strength. Furthermore, the increasing FeO content in the finished pellets with higher ironsand proportions indicates insufficient oxidation and solid-phase bonding, contributing to the gradual decline in compressive strength. Nonetheless, all tested compositions still meet the standard for first-grade pellets, with compressive strength exceeding 2000 N.

6.4 High-Temperature Metallurgical Properties of Ironsands Pellets

6.4.1 Low Temperature Reduction Pulverization Properties of Ironsands Pellets

Experimental method for low-temperature reduction pulverization of ironsands pellet

The Reduced Pulverization Index (RDI) reflects the degree of disintegration of iron ore pellets after low-temperature reduction. Following the rotary drum test, the sample is subjected to sieving. The mass percentages of material retained on sieves with apertures larger than 6.30 mm, larger than 3.15 mm, and smaller than 500 μm are calculated relative to the initial sample mass before the test. These values characterize the pellet's resistance to pulverization under low-temperature reduction conditions and are denoted as $RDI_{+6.3}$, $RDI_{+3.15}$, and $RDI_{-0.5}$, respectively. The corresponding equations are given as Eqs. 6.1–6.3:

$$\text{Low-temperature reduced strength index: } RDI_{+6.3} = \frac{m_{D1}}{m_{D0}} \times 100\% \tag{6.1}$$

$$\text{Low-temperature reduction pulverization index: } RDI_{+3.15} = \frac{m_{D1} + m_{D2}}{m_{D0}} \times 100\% \tag{6.2}$$

$$\text{Low-temperature reduction anti-wear index: } RDI_{-0.5} = \frac{m_{D0} - (m_{D1} + m_{D2} + m_{D3})}{m_{D0}} \times 100\% \tag{6.3}$$

where:

m_{D0} Mass of the specimen before the rotary drum test, g;
m_{D1} Mass of the material retained on the 6.30 mm sieve after the test, g;
m_{D2} Mass of the material retained on the 3.15mm sieve after the test, g;
m_{D3} Mass of the material retained on the 500 μm sieve after the test, g.

In evaluating experimental results, $RDI_{+3.15}$ is typically adopted as the primary performance indicator for low-temperature reduction pulverization. The $RDI_{+6.3}$ and $RDI_{-0.5}$ values are considered supplementary reference indices.

Experimental results of low-temperature reduction pulverization of ironsands pellet

The experimental results for the low-temperature reduction strength index ($RDI_{+6.3}$), low-temperature reduction pulverization index ($RDI_{+3.15}$), and low-temperature reduction anti-wear index ($RDI_{-0.5}$) are presented in Table 6.9 and Fig. 6.12.

As the proportion of ironsand increases, the low-temperature reduction pulverization resistance of the pellets improves. Specifically, both $RDI_{+3.15}$ and $RDI_{+6.3}$

Table 6.9 Low-temperature reduction strength and low-temperature reduction abrasion resistance of pellets

Serial number	RDI + 6.3/%	RDI − 0.5/%
1	94.82	1.74
2	96.86	2.07
3	98.55	1.33
4	98.74	0.74
5	99.68	0.32

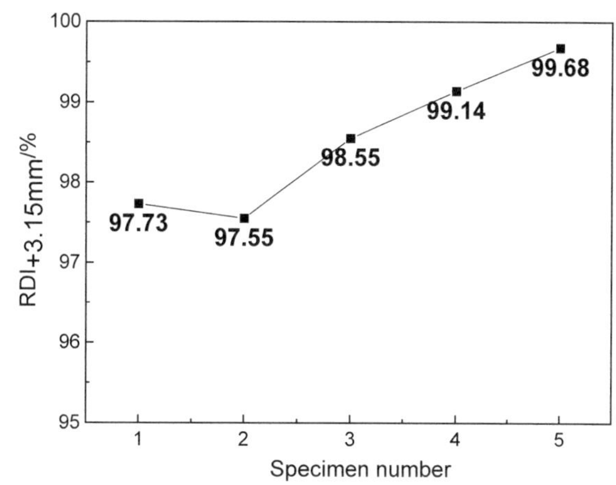

Fig. 6.12 Low-temperature reduction pulverization test results of pellets

show a clear upward trend. When the ironsand content is 10 wt%, $RDI_{+3.15}$ reaches 97.73% and $RDI_{+6.3}$ is 94.82%. In the case of pellets prepared entirely from ironsand, $RDI_{+3.15}$ increases to 99.68%. After reduction at 500 °C followed by the drum test, the particle size of the reaction products was predominantly larger than 6.3 mm, indicating almost no pulverization.

These results demonstrate that with increasing ironsand content, the pellets exhibit significantly enhanced resistance to pulverization under low-temperature reduction conditions, suggesting that ironsand is beneficial for improving the structural integrity of reduced pellets.

Mechanism analysis of the influence of ironsands on the low-temperature reduction pulverization rate of pellet

During blast furnace operation or direct reduction processes, pellets are exposed to low-temperature environments in the range of 400–600 °C, particularly in the upper part of the furnace or shaft. In this temperature range, pellets exhibit a tendency to disintegrate into fines. The primary mechanism responsible for this low-temperature reduction pulverization is the phase transformation associated with the reduction of hematite to magnetite, which induces crystal structure changes and internal stress, leading to pellet degradation.

Ironsand pellets primarily contain complex solid solutions of hematite–ilmenite. During low-temperature reduction, titanium-bearing phases such as titanium hematite are reduced to titanomagnetite, and pseudobrookite transforms into titanocrystalline phases. However, these transformations generally occur at elevated temperatures.

Specifically, below 600 °C, the phase composition of vanadium–titanium-bearing pellets remains largely unchanged. As the temperature increases from 600 to 750 °C, titanium hematite begins to reduce to titanomagnetite, triggering crystalline transformation and the formation of microcracks. In the present study, the experimental temperature was maintained at 500 °C. At this temperature, the complex solid solutions in ironsand pellets have not yet entered the reduction phase, and no significant crystalline transformation occurs.

As a result, the addition of ironsand inhibits the phase transformation responsible for pulverization at this stage, thereby enhancing the low-temperature reduction pulverization resistance of the pellets and improving the relevant RDI indices.

6.4.2 *Reduction Properties of Ironsands Pellets*

Experimental methods for reduction properties of ironsands pellet

The reduction behavior of finished ironsand pellets was investigated using a fixed-bed reactor. A reducing gas mixture consisting of 30 vol% ± 0.5 vol% CO and 70 vol% ± 0.5 vol% N_2 was used. The isothermal reduction experiments were conducted at a constant temperature of 900 °C. The reduction process was maintained for 3 h, after which the degree of reduction was calculated to assess the reducibility of the pellets. The experimental setup is shown in Fig. 6.13, and the formula used to calculate the reduction degree is provided in Eq. 6.4.

Reduction degree formula:

$$R_t = \left[\frac{0.11w_1}{0.43w_2} + \frac{m_1 - m_t}{m_0 \times 0.43w_2} \times 100\right] \times 100 \tag{6.4}$$

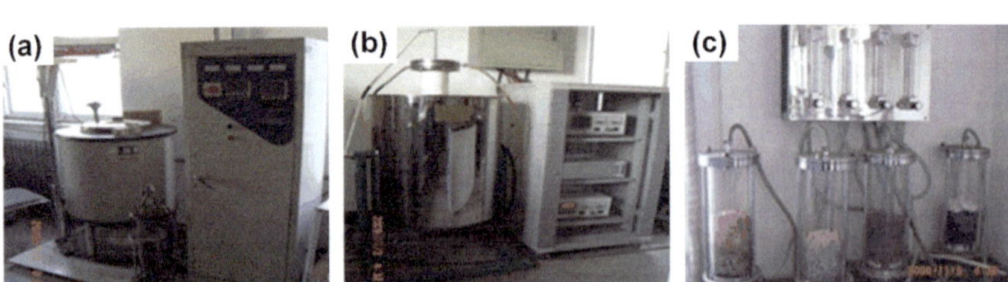

Fig. 6.13 Reduction testing equipment. **a** Medium-temperature tube furnace; **b** gas generator; **c** gas purification system

where:

R_t degree of reduction of the pellet at time t, %;
m_0 mass of the specimen at the beginning of reduction, g;
m_1 mass of the specimen when heated to 900°C before reduction, g;
m_t mass of the specimen after reduction at time t, g;
w_1 content of FeO in the specimen before reduction, wt%;
w_2 content of total iron in the specimen before reduction, wt%.

Experimental results and mechanism analysis of reduction performance of ironsands pellet

Based on the experimentally measured specimen mass after reduction (m_t), the initial specimen mass (m_0), and the specimen mass before reduction (m_1), the reduction performance index (reducibility index, RI) of the ironsand pellets was calculated. The results are presented in Fig. 6.14.

The reduced performance is further reflected in the fact that the finished pellets containing ironsand failed to meet the industry standard of > 65% reduction ratio. This decline in reducibility is primarily attributed to changes in the phase composition of the pellets after preheating and roasting [2–4]. The dominant phases in the ironsand-containing pellets include hematite, titanomagnetite, and pseudobrookite. Hematite undergoes reduction according to the sequence: $Fe_2O_3 \rightarrow Fe_3O_4 \rightarrow FeO \rightarrow Fe$, while pseudobrookite follows a more complex reduction path: $Fe_2TiO_5 \rightarrow FeTiO_4 \rightarrow FeTiO_3 + Fe \rightarrow FeTiO_5 + Fe \rightarrow Fe + TiO_2$ [5–8].

As the proportion of ironsand increases, the content of Fe_2TiO_5 also increases. Since Fe_2TiO_5 has a lower reducibility than Fe_2O_3, primarily due to the hindering effect of titanium on the migration of iron and oxygen atoms during the reduction process, the overall reducibility of the pellet is restricted. Consequently, a higher ironsand content results in reduced reduction performance.

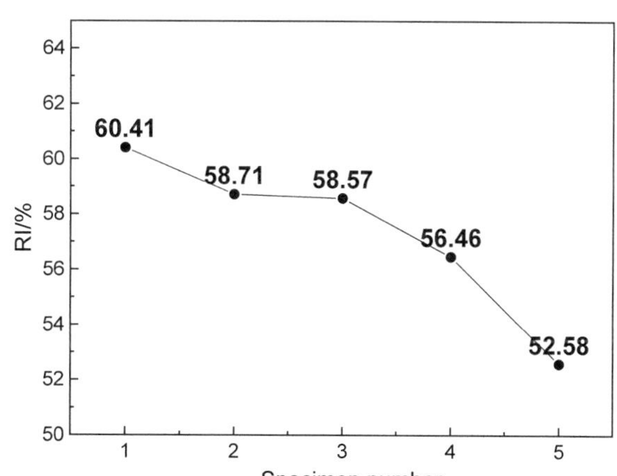

Fig. 6.14 Reduction index of finished ironsands pellets

6.4.3 Swelling Properties of Ironsands Pellets

Experimental methods for evaluating the reduction and swelling properties of ironsand pellets

Finished pellets with a particle size range of 10.0–12.5 mm were selected for testing. Isothermal reduction experiments were conducted at 900 °C using a reducing gas mixture composed of 30 vol% ± 0.5 vol% CO and 70 vol% ± 0.5 vol% N_2. During the reduction process, the pellets were allowed to expand freely. The Reduction Swelling Index (RSI) was determined by measuring the change in pellet volume before and after reduction, serving as a key parameter for evaluating the pellet's reduction-induced expansion behavior.

The RSI is calculated according to Eq. 6.5:

$$RSI = \frac{V_1 - V_0}{V_0} \times 100\% \tag{6.5}$$

In the formula:

V_1 Volume of the pellet after reduction, ml;
V_0 Volume of the pellet before reduction, ml.

Experimental results and mechanism analysis of the reduction and swelling behavior of ironsand pellets

The volume of the pellet before reduction (V_0) and after 1 h of reduction (V_1) was calculated based on buoyancy measurements obtained by immersing the pellets in water. The results of the reduction swelling behavior are presented in Fig. 6.15.

As shown in Fig. 6.15, the RSI decreases with the increasing proportion of ironsand in the pellet composition. The highest RSI value of 16.14% was observed at

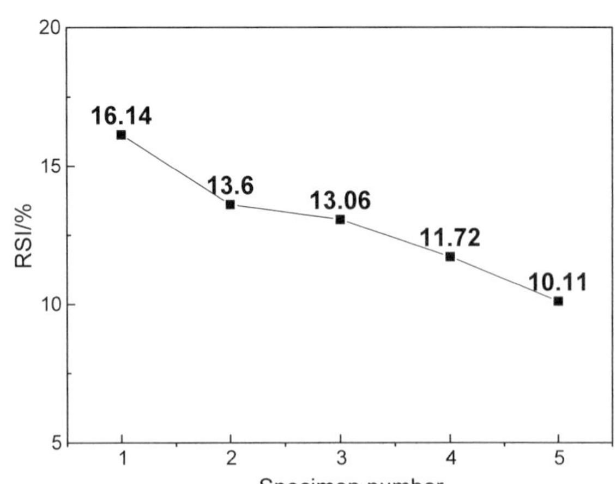

Fig. 6.15 Swelling behavior of ironsands pellets after reduction

10 wt% ironsand, while a moderate RSI of 13.06% was recorded at 50 wt% ironsand. When the pellets were prepared entirely from ironsand (100 wt%), the RSI decreased further to 10.11%, which falls within the first-grade pellet standard (RSI < 15%). This indicates that increasing the proportion of ironsand improves the swelling resistance of the pellet during reduction.

During the reduction process, pellets undergo crystalline phase transformation, particularly the conversion of Fe_2O_3 to Fe_3O_4, as well as the formation of metallic iron whiskers through the reduction of iron oxides. Both phenomena contribute to the volumetric expansion of the pellets. However, when ironsand is introduced into the pellet composition, the presence of titanium–iron solid solutions partially replaces hematite, thereby mitigating the expansion caused by the crystalline transformation [9].

In blast furnace reduction, the volume expansion of pellets is a critical indicator of their behavior and influence on furnace operation. When the expansion rate is less than 20%, it is classified as normal expansion, under which conditions blast furnace production can proceed smoothly. An expansion rate between 20 and 40% is considered abnormal, often leading to deterioration in furnace permeability and impaired gas flow. When the expansion exceeds 40%, it is classified as severe (or vicious) expansion, which can significantly disrupt furnace operations. In this case, the permeability of the burden bed is drastically reduced, dust generation increases sharply, and severe operational issues such as burden hanging, material collapse, and unstable burden descent may occur. These phenomena can cause process disturbances, leading to reduced productivity, increased coke consumption, and in extreme cases, furnace shutdown. Therefore, controlling pellet volume expansion is essential for maintaining stable and efficient blast furnace operation.

Therefore, increasing the ironsand content in the pellet formulation improves its reduction swelling resistance, which is beneficial for maintaining stable and efficient blast furnace operation.

6.5 Industrial Testing of Ironsands Pellet Production

6.5.1 Raw Material Conditions Using Indonesian Ironsands

The Indonesian ironsands used in this experiment have an iron grade of 42.79 wt%, with SiO_2 content of 9.26 wt%, TiO_2 content of 20.96 wt%, and FeO content of 15.01 wt% (Table 6.10).

The initial particle size of the Indonesian ironsands is relatively coarse, with only 2.80 wt% passing through a − 200 mesh, while particles larger than 100 mesh account for 40.33 wt%. Therefore, the material was first subjected to grinding in a ball mill. After fine grinding, the proportion of − 200 mesh particles increased to

Table 6.10 Chemical composition of Indonesian ironsands/wt%

TFe	FeO	SiO_2	Al_2O_3	CaO	MgO	MnO	P	TiO_2
42.79	15.01	9.26	2.34	1.65	4.12	0.55	0.043	20.96
K_2O	Na_2O	PbO	As_2O_3	ZnO	CuO	Cl	S	
0.04	0.14	0.001	0.001	0.041	0.01	0.006	0.014	

Table 6.11 Particle size distribution of Indonesian ironsands

Scope/object	> 100	100–120	120–140	140–160	160–180	180–200	– 200
Original particle size/wt%	40.33	15.67	18.30	17.64	1.88	3.38	2.80
Particle size after fine grinding/wt%	0.34	1.35	2.57	3.25	5.78	4.41	82.30

Table 6.12 Chemical composition of mineral powders/wt%

Name	TFe	FeO	SiO_2	CaO	Al_2O_3	MgO	– 200 mesh
Peruvian powder	69.27	29.28	2.11	0.51	0.35	0.53	88.51
Processed powder	67.57	29.71	3.19	0.77	0.61	1.02	80.30
Hainan powder	62.12	4.27	6.81	0.80	1.12	0.48	92.61

82.30 wt%, particles coarser than 100 mesh decreased to 0.34 wt%, and the 160–200 mesh fraction accounted for 10.19 wt%, making the material suitable for pellet formation (Table 6.11).

The Peruvian ore powder used in the experiment exhibits a relatively high iron grade of 69.27 wt% and a SiO_2 content of 2.11 wt%. In contrast, the processed ore powder has a SiO_2 content of 3.19 wt%, and the Hainan powder contains 6.81 wt% SiO_2. The detailed chemical compositions of these materials are presented in Table 6.12.

6.5.2 *Test Results Using Indonesian Ironsands*

Raw ball quality

Balling experiments were conducted using finely ground Indonesian ironsands. The raw material proportioning schemes for these experiments are detailed in Table 6.13. Schemes 1 and 4 served as baseline experiments, utilizing a blend of 85 wt% Peruvian ore powder, 10 wt% self-processed powder, 5 wt% Hainan powder, and 1.3 wt% magnesium oxide (MgO) powder in Scheme 1, while Scheme 4 excluded MgO. Both schemes employed 1.7 wt% bentonite as a binder.

Table 6.13 Pelletizing test proportioning scheme and pelletizing index results

Program number	Mineral powder name and proportion/wt%				Bentonite ratio/wt%	Magnesium oxide powder ratio/wt%	Number of raw ball drops/times	Raw ball compression/N	Water content of the pellet/wt%
	Peruvian powder	Hainan powder	Self-processing powder	Indonesian Ironsands					
1-based	85	5	10		1.7	1.3	5.2	8.04	9.0
2	85		10	5	1.7	1.3	5.3	9.31	8.9
3	75		10	15	1.7	1.3	5.0	8.88	9.0
1-based	85	5	10		1.7	–	5.1	8.53	9.2
5	85		10	5	1.7	–	4.9	7.43	8.9
6	80		10	10	1.7	–	5.1	7.28	9.0
7				100	1.7	–	5.1	11.34	9.2

Schemes 2 and 3 excluded Hainan powder and incorporated 5 wt% and 15 wt% Indonesian ironsands, respectively. Schemes 5 and 6 similarly included 5 wt% and 10 wt% Indonesian ironsands but without any MgO addition. Scheme 7 used 100 wt% Indonesian ironsands for ball-making.

According to the results, the average drop number (falling from 0.5 m) for Schemes 1 and 4 was 5.1 and 5.2 times, respectively, indicating consistent baseline pellet strength. When 5 wt% and 15 wt% finely ground Indonesian ironsands were added (Schemes 2 and 3), the drop numbers were 5.3 and 5.0 times, respectively—showing negligible variation in drop strength compared to the baseline.

For Scheme 7, using 100 wt% Indonesian ironsands, the pellets achieved a drop number of 5.1 times and demonstrated relatively high compressive strength, reaching 11.34 N, indicating good raw ball mechanical integrity even without additional ore blending.

Physical and chemical properties of finished balls

Roasting experiments were conducted using seven different raw pellet formulations, as detailed in Table 6.13. For Schemes 1–3, the maximum roasting temperature was set at 1280 °C, primarily due to the inclusion of magnesium oxide (MgO) powder. For Schemes 4–7, which excluded MgO, the highest roasting temperature was 1260 °C. The roasting duration for all schemes was 52 min.

The compressive strength results of the roasted pellets are presented in Table 6.14, and the corresponding chemical composition analysis is shown in Table 6.15.

The compressive strength data indicate that all pellet formulations exhibited relatively high mechanical strength. In the benchmark experiment (Scheme 1), the compressive strength of the roasted pellets reached 2936 N. For pellets containing 5 wt% and 15 wt% Indonesian ironsands (Schemes 2 and 3), the compressive strengths were 2948 N and 2810 N, respectively.

In Scheme 4, which served as a baseline without MgO, the compressive strength was 2918 N. For Schemes 5 and 6, which included 5 wt% and 10 wt% Indonesian ironsands without MgO, the compressive strengths were 2955 N and 2939 N, respectively. Notably, in Scheme 7, where 100 wt% Indonesian ironsands were used, the

Table 6.14 Compressive strength of roasted balls/N

Serial number	Average compressive strength N	Maximum compressive strength N	Minimum value of compressive strength N
1	2936	4105	1485
2	2948	4241	1363
3	2810	3916	1379
4	2918	4105	1485
5	2955	4254	1218
6	2939	4172	1332
7	2951	4321	1340

Table 6.15 Chemical composition of roasted balls/wt%

Serial number	TFe	FeO	SiO_2	Al_2O_3	CaO	MgO	TiO_2	K_2O	Na_2O	V_2O_5
1	65.14	0.18	3.94	0.69	0.55	1.66	0.14	0.080	0.12	0.003
2	63.73	0.22	4.13	0.77	0.64	1.86	1.37	0.077	0.15	0.030
3	61.08	0.29	4.92	1.03	0.80	2.35	3.80	0.072	0.12	0.072
4	65.79	0.34	4.02	0.61	0.60	0.61	0.10	0.089	0.13	0.004
5	64.43	0.29	4.17	0.79	0.42	0.86	1.31	0.073	0.13	0.021
6	63.52	1.19	4.39	0.89	0.51	1.01	2.05	0.075	0.12	0.034
7	43.35	1.15	9.76	2.46	1.70	4.79	18.14	0.072	0.13	0.300

pellets still achieved a high compressive strength of 2951 N, demonstrating excellent roasting performance and mechanical integrity across all formulations.

The experimental pellets exhibited relatively high SiO_2 content, primarily due to the elevated SiO_2 level in the Peruvian ore powder used in the formulation, which contained 2.11 wt% SiO_2 and had an iron grade of 69.31 wt%.

For the baseline pellet containing 1.3 wt% MgO powder, the SiO_2 content was 3.94 wt%, the iron grade was 65.14 wt%, and the MgO content reached 1.66 wt%. When 5 wt% Indonesian ironsands were incorporated into the mix (with 1.3 wt% MgO), the iron grade decreased to 63.73 wt%, and the TiO_2 content increased to 1.37 wt%. With a further increase to 15 wt% ironsands, the iron grade dropped to 61.08 wt%, while the TiO_2 content rose to 3.80 wt%.

In the absence of MgO powder, the pellet with 5 wt% Indonesian ironsands had an iron grade of 64.43 wt% and TiO_2 content of 1.31 wt%. When the ironsands proportion increased to 10 wt%, the grade further declined to 63.52 wt%, and the TiO_2 content increased to 2.05 wt%.

Reduced expansion behavior of roasted pellets

The reduced expansion rates of pellets containing 5, 10, and 100 wt% Indonesian ironsands were evaluated. All three types of roasted pellets exhibited expansion rates below 20%, indicating acceptable reduction behavior. Notably, the pellet containing 10 wt% Indonesian ironsands showed a reduced expansion rate approximately 7 percentage points higher than that of the 5 wt% sample. It is worth mentioning that all experimental pellets had SiO_2 contents exceeding 4 wt%, which contributed to suppressing excessive expansion during reduction (Table 6.16).

Summary of findings

① The Indonesian ironsands used in this study had an iron grade of 42.79 wt% and a high TiO_2 content of 20.96 wt%, with the two components collectively accounting for 63.75 wt%.

② The original particle size of the Indonesian ironsands was relatively coarse, with only 2.80 wt% passing through a 200-mesh sieve, while 40.33 wt% was retained

Table 6.16 Metallurgical properties of roasted balls

Serial number	Mineral powder name and proportion/wt%				Bentonite ratio/ wt%	Reduced expansion rate/%	Pelletized ore composition/ wt%	
	Peruvian powder	Hainan powder	Self-processing powder	Indonesian Ironsands			SiO_2	TiO_2
1	85		10	5	1.7	12.70	4.17	1.31
2	80		10	10	1.7	19.74	4.39	2.05
3				100	1.7	17.2	9.76	18.14

above 100 mesh. After laboratory ball milling, 82.30 wt% of the particles passed through the 200 mesh, improving suitability for balling.

③ The use of finely ground Indonesian ironsands had minimal impact on balling behavior and raw ball quality. Even with 100 wt% Indonesian ironsands, the drop number reached 5.1 times/0.5 m, and the compressive strength of the green balls reached 11.34 N, demonstrating good balling performance.

④ Adding 5–15 wt% Indonesian ironsands had negligible effects on the roasting temperature and compressive strength of the fired pellets. Under a roasting temperature of 1260 °C, the compressive strength of the pellet containing 100 wt% Indonesian ironsands reached 2951 N, comparable to that of the baseline pellet.

⑤ The Peruvian ore powder used in the experiments had a relatively high SiO_2 content of 2.11 wt% and a high grade of 69.31 wt%, resulting in elevated SiO_2 levels in the final pellets. With 5 wt% Indonesian ironsands, the pellet grade decreased by 1.4 percentage points and TiO_2 content increased to 1.3 wt%. At 10 wt%, the grade dropped by 2.27 percentage points, and TiO_2 rose to 2.05 wt%. At 15 wt%, the grade decreased by 4.06 percentage points, with TiO_2 reaching 3.8 wt%.

⑥ The pellet roasted using only Indonesian ironsands had a grade of 43.35 wt% and a TiO_2 content of 18.14 wt%.

⑦ When the proportion of Indonesian ironsands was $\leq$ 10 wt%, the reduction expansion rate remained below 20%, which is within the acceptable limit for blast furnace operation. The high SiO_2 content (> 4 wt%) of the Peruvian powder contributed to controlling the expansion. A decrease in SiO_2 content may lead to increased expansion and deteriorated pellet performance.

⑧ The reduction expansion rate of pellets made with 100 wt% Indonesian ironsands was 17.2%, meeting the requirements for blast furnace charging.

6.5.3 *Raw Material Conditions Using New Zealand Ironsands*

Chemical composition of New Zealand ironsands

The chemical composition of New Zealand ironsands was first analyzed, and the results are presented in Table 6.17. The New Zealand ironsands exhibit a relatively high iron grade of 52.49 wt%, with FeO content at 1.54 wt% and a significant TiO_2 content of 15.36 wt%.

Particle size distribution of New Zealand ironsands

The original particle size of New Zealand ironsands is relatively coarse, with 62.55 wt% of particles retained above 100 mesh, and only 21.69 wt% passing through a 200-mesh sieve. Therefore, ball milling was employed to refine the particle size. Two different levels of grinding were used. A finer grind, in which 95 wt% of the particles were below 200 mesh. A coarser grind, where 84 wt% of the particles passed through the 200-mesh sieve. Detailed particle size distributions are provided in Table 6.18.

Physical and chemical properties of additional mineral powders used in the experiment

The Peruvian ore powder used in the experiment has a relatively high SiO_2 content of 2.11 wt%, a grade of 69.17 wt%, and a − 200 mesh content of 88.51 wt%. The processed powder used in the experiments contains 3.19 wt% SiO_2, while the Hainan ore powder has a SiO_2 content of 6.81 wt%. The detailed chemical compositions of these materials are summarized in Table 6.19.

Table 6.17 Chemical composition of New Zealand ironsands/wt%

TFe	FeO	SiO_2	Al_2O_3	CaO	MgO	MnO
52.49	1.54	3.04	2.88	0.39	0.62	0.26
TiO_2	K_2O	Na_2O	ZnO	S	V_2O_5	P
15.36	0.019	1.47	0.042	0.22	0.28	0.002

Table 6.18 Particle size analysis results of New Zealand ironsands mineral powder

Particle size fan/mesh	> 100	100–120	120–140	140–160	160–180	180–200	< 200
Primary ore/wt%	62.55	3.21	3.01	6.13	0.76	2.65	21.69
Fine grinding/wt%	1.17	0.47	0.33	0.67	0.18	1.44	95.74
Rough grinding/wt%	1.51	1.34	0.89	2.07	4.39	5.23	84.57

Table 6.19 Chemical composition of mineral powder/wt%

Name	TFe	FeO	SiO_2	CaO	Al_2O_3	MgO	< 200 mesh
Peruvian powder	69.17	29.28	2.11	0.51	0.35	0.53	88.51
Processed powder	67.57	29.71	3.19	0.77	0.61	1.02	80.30
Hainan powder	62.12	4.27	6.81	0.80	1.12	0.48	92.61

6.5.4 Test Results Using New Zealand Ironsands

Ball-making experiments were conducted using both finely ground and coarsely ground New Zealand ironsands. The raw material proportions and corresponding results are presented in Table 6.20. Based on the experimental results, when finely ground New Zealand ironsands were used, the green balls exhibited severe agglomeration, hindering proper pellet formation. The balling process also encountered difficulties in water addition, and the drop strength of the green balls was notably low.

In contrast, when coarsely ground New Zealand ironsands were used, the pellet formation process improved significantly. The green balls rolled more easily without sticking to each other, water could be added more readily, and the drop strength was higher than that of the finely ground samples. Specifically, the drop strength reached 4.6 drops with 5 wt% coarsely ground New Zealand ironsands and 4.3 drops with 10 wt%, although both were still slightly lower than that of the baseline experiment, which achieved 5.1 drops.

Roasting experiment and physical and chemical properties of roasted pellets

Based on the results of the ball-making experiments, green pellets containing 5 wt%, 10 wt% coarsely ground New Zealand ironsands, and 100 wt% New Zealand ironsands were subjected to roasting in crucible cups. The raw material ratios used in the roasting experiments are provided in Table 6.21.

The compressive strength results of the roasted pellets are presented in Table 6.22. The compressive strength of the pellet with 5 wt% New Zealand ironsands was 2895 N, while that with 10 wt% was 2934 N, both showing negligible difference compared to the benchmark pellet, indicating that within the 5–10 wt% addition range, New Zealand ironsands have minimal impact on pellet compressive strength.

The chemical composition of the roasted pellets is shown in Table 6.23. With 5 wt% New Zealand ironsands, the Fe grade of the pellets decreased by 0.74 percentage points, and TiO_2 content reached 1.03 wt%. When the New Zealand ironsands content was increased to 10 wt%, the Fe grade declined by 1.36 percentage points, and the TiO_2 content rose to 1.56 wt%.

Table 6.20 Balling experiment proportion and raw ball indexes

Programmatic	Mineral powder name and proportion/wt%			New Zealand ironsands/wt%		Bentonite/wt%	Raw ball drop/times	Raw ball compression/N	Raw pellet moisture/wt%
	Peruvian powder	Self-processing powder	Hainan powder	Fine grind	Rough grinding				
1	85	10	5	–	–	1.7	5.1	8.53	9.2
2	85	10		5		1.7	3.8	6.74	8.3
3	75	10		15		1.7	3.9	9.26	8.2
4				100		1.7	6.1	10.99	8.5
5	85	10			5	1.7	4.6	7.67	9.1
6	80	10			10	1.7	4.3	7.65	9.2

Table 6.21 Ratio of raw balls in roasting experiment/wt%

Serial number	Peruvian powder	Processed powder	Hainan powder	Coarse grinding New Zealand ironsands	Bentonite
1	85	10	5		1.7
2	85	10		5	1.7
3	80	10		10	1.7
4				100	1.7

Table 6.22 Compressive strength of roasted ball/N

Program number	Average compressive strength	Maximum compressive strength	Minimum compressive strength
1	2918	4105	1485
2	2895	4352	1276
3	2934	4212	1412
4	2915	4109	1233

Table 6.23 Chemical composition of roasted ball/wt%

Serial number	TFe	FeO	SiO_2	Al_2O_3	CaO	MgO	TiO_2	K_2O	Na_2O	V_2O_5
1	65.79	0.34	4.02	0.61	0.60	0.61	0.10	0.09	0.13	0.004
2	65.05	0.29	3.66	0.77	0.38	0.70	1.03	0.09	0.11	0.019
3	64.43	0.43	3.77	0.89	0.40	0.64	1.56	0.10	0.13	0.029

6.6 Summary

(1) The original particle size of ironsands is relatively coarse, with the majority of particles concentrated in the 0.125–0.212 mm range. The particles are predominantly near-spherical in shape, indicating smooth surfaces and low roughness, which makes them less conducive to ball formation. Therefore, milling is required. After 35 min of grinding, the proportion of ironsands concentrate with a particle size below 0.074 mm reaches 85.63 wt%, and the specific surface area is 2741 cm^2/g, meeting the particle size and surface area requirements for the pelletizing process.

(2) Regarding the drop strength of green pellets, the highest drop resistance was observed at 30 wt% ironsands content, reaching 4.3 drops. The compressive strength of green pellets exceeded 8 N across all proportions of ironsands tested, indicating that the addition of ironsands does not significantly affect green pellet formation and can meet production requirements. Under the conditions

of preheating at 950 °C, roasting at 1250 °C, with preheating and roasting durations of 15 min each, the compressive strength of the roasted pellets exceeded 2200 N, fully satisfying industrial performance standards.

(3) Due to the slow development of complex solid solutions in ironsand pellets during reduction in the low-temperature zone, and the limited degree of crystalline transformation, the RDI of the finished pellets is improved with the addition of ironsands. Both $RDI_{+3.15}$ and $RDI_{+6.3}$ increased, and for pellets composed entirely of ironsands, the $RDI_{+3.15}$ and $RDI_{+6.3}$ values reached as high as 99.68%. However, the presence of titanium, particularly in the form of Fe_2TiO_5, hinders the migration of Fe and O atoms during reduction. As a result, Fe_2TiO_5 shows poorer reducibility compared to Fe_2O_3, which limits the overall reducibility of ironsand pellets. The RI of the pellets decreases significantly with increasing ironsands content. The maximum RI achieved was 60.41% at 10 wt% ironsands addition, which still falls short of the 65% industry standard. On the other hand, the substitution of hematite by the titanium–iron solid solution phase substantially reduces the volume expansion associated with phase transformation. Consequently, the reduction expansion behavior of the pellets improves with increasing ironsands content. The maximum reduction expansion index was 16.14% at 10 wt% ironsands, while the value decreased to 10.11% for pellets made entirely from ironsands. These results indicate that increasing the proportion of ironsands can effectively suppress excessive volume expansion during reduction.

References

1. Dang J, Hu X, Zhang G, et al. Kinetics of reduction of titano-magnetite powder by H_2. High Temp Mater Process (London). 2013;32(3):229–36.
2. Wang YM, Yuan ZF. Reductive kinetics of the reaction between a natural ilmenite and carbon. Int J Miner Process. 2006;81(3):133–40.
3. Kucukkaragoz CS, Eric RH. Solid state reduction of a natural ilmenite. Miner Eng. 2006;19(3):334–7.
4. Paunova R. Thermodynamic study of the reduction of titanium magnetite concentrate with solid carbon. Metall Mater Trans B-Proc Metall Mater Proc Sci. 2002;33(4):633–8.
5. Park E, Ostrovski O. Effects of preoxidation of titania-ferrous ore on the ore structure and reduction behavior. ISIJ Int. 2004;44(1):74–81.
6. Park E, Ostrovski O. Reduction of titania-ferrous ore by carbon monoxide. ISIJ Int. 2003;43(9):1316–25.
7. Park E, Ostrovski O. Reduction of titania-ferrous ore by hydrogen. ISIJ Int. 2004;44(6):999–1005.
8. Zboril R, Mashlan M, Petridis D. Iron(III) oxides from thermal processes-synthesis, structural and magnetic properties, Mossbauer spectroscopy characterization, and applications. Chem Mater. 2002;14(3):969–82.
9. Bold T, Kuziak R. Phase transformations during oxidizing roasting of titanomagnetite pellets in vanadium recovery process. 1990;167–78.

Chapter 7
Pre-oxidation Treatment of Ironsands

The gas-based reduction of ironsands using various reductants—including hydrogen, carbon monoxide, solid carbonaceous agents, and methane—has been extensively studied [1–6]. Across reductant types, a consistent finding is the relatively poor reduction kinetics of ironsands compared to hematite and magnetite [7–9]. Pre-oxidation treatment has thus been proposed and demonstrated to enhance the reduction performance of ironsands [10]. Similar techniques have also been applied in the processing of ilmenite and high-titanium blast furnace slag [11]. Although some studies suggest that improved reducibility arises from phase transformation, the microstructural evolution and elemental migration mechanisms remain unclear [12]. This chapter aims to clarify these aspects and elucidate the role of pre-oxidation in enhancing the reduction behavior of ironsands.

7.1 Pre-oxidation Methodology

A fixed-bed vertical tubular resistance furnace was employed for the pre-oxidation study of ironsands, with the schematic diagram of the apparatus presented in Fig. 7.1. A 3 g sample of ironsands was evenly spread over a square corundum disk (5 cm × 5 cm) to ensure full contact between the mineral particles and the incoming gas. The corundum disk was positioned within the constant-temperature zone of the tubular furnace. Compressed air, pre-dehydrated with calcium sulfate and further dried using a 3-Å molecular sieve, was introduced at a flow rate of 1 L/min, regulated by a volumetric flow meter located in the control cabinet. The air was delivered directly to the surface of the ironsand particles through a corundum gas line, maintaining a consistent oxygen concentration at the reaction interface.

The furnace temperature was increased to the predetermined experimental setpoints at a heating rate of 5 °C/min, with the temperature ranging from 300 to 1200 °C. Samples were collected at 100 °C intervals. During sampling, the corundum

Z. Wang et al., *Smelting and Utilization Technology of Ironsands*,
https://doi.org/10.1007/978-981-95-2868-4_7

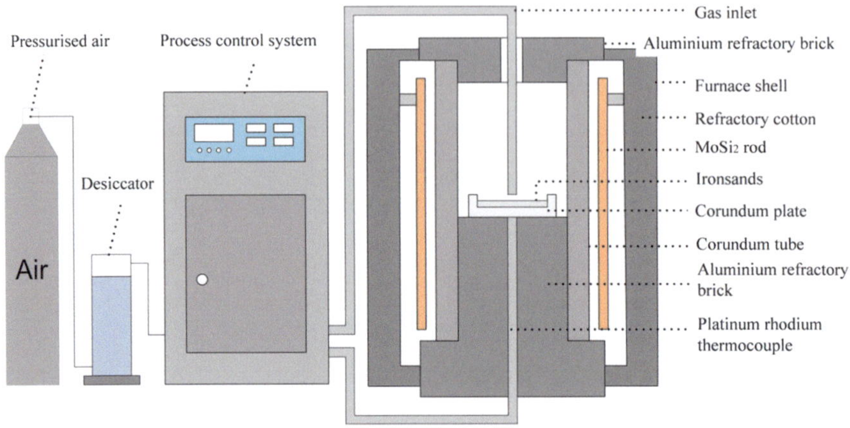

Fig. 7.1 Schematic diagram of the experimental setup for non-isothermal pre-oxidation of ironsands

disk containing the ironsands was removed from the furnace and cooled to room temperature under ambient atmospheric conditions. The physical phase transformations, lattice parameter changes, and patterns of elemental migration and enrichment occurring during the pre-oxidation process were investigated through subsequent sampling and analysis.

7.2 Physical Phase Transition During Pre-oxidation

X-ray diffraction (XRD) analysis was performed to investigate the physical phase transitions of ironsands after pre-oxidation at various temperatures. The analysis was conducted using a Rigaku Ultima IV diffractometer operating at 40 kV and 250 mA with a CuKα radiation source. The scanning rate was 0.03° 2θ/s. Pre-experimental calibration was carried out using pure silica powder, achieving an angular deviation of less than 0.01°, and the experimental reproducibility exceeded 0.0001. A total of ten ironsands samples subjected to different pre-oxidation temperatures were analyzed, and the resulting diffraction patterns were compared with standard reference cards from the International Centre for Diffraction Data (ICDD) [13]. The results, presented in Fig. 7.2, illustrate the phase transformations of ironsands during the non-isothermal pre-oxidation process.

As shown in Fig. 7.2, titanomagnetite (TTM) remains the dominant phase in ironsands up to 800 °C. Above this temperature, the relative intensities of TTM diffraction peaks gradually decline. Some crystallographic orientations exhibit high thermal stability; for instance, the (220) peak of TTM persists up to 1200 °C. In contrast, the (222) and (400) TTM peaks vanish at 1000 °C and 1200 °C, respectively. Due to peak overlap among several crystallographic planes—specifically (311), (511),

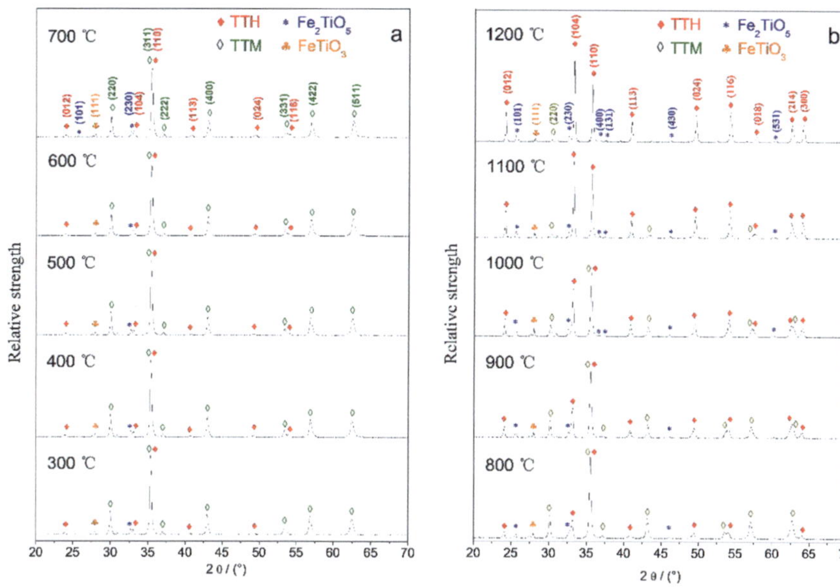

Fig. 7.2 X-ray diffraction (XRD) patterns of ironsands during heating from room temperature at a rate of 5 °C/min to various target temperatures: **a** 300–700 °C; **b** 800–1200 °C

(433), and (331)—accurate identification of specific phase transition temperatures in these directions is challenging.

During pre-oxidation heating, TTM with a cubic spinel structure progressively transforms into titanhematite (TTH), which possesses a hexagonal crystal structure. Between 300 and 700 °C, no notable changes were observed. However, beginning at 800 °C, the emergence and intensification of TTH characteristic peaks—corresponding to the (012), (104), (113), (024), (116), and (300) planes—suggest progressive crystal growth with increasing temperature and time. Due to overlapping diffraction angles of certain TTM and TTH peaks [e.g., (018), (214), and (110)], the precise temperature at which TTH first forms cannot be definitively determined. Titanomagnetohematite, an unstable intermediate phase reported in previous studies [10], was not detected, likely due to its rapid transformation into TTH.

Pseudobrookite (PSB), a titanium–iron oxide with a higher oxidation state than both TTM and TTH, was first detected at 700 °C as a result of further oxidation of TTH. PSB initially forms and grows along the (101) crystallographic direction and continues to develop up to 1200 °C. Diffraction peak intensities corresponding to the (400) and (131) planes of PSB also increase between 1000 and 1200 °C, albeit remaining weak and showing no further development. This suggests that the transformation of TTH to PSB is limited, requiring higher temperatures and longer oxidation durations.

Ilmenite ($FeTiO_3$), originally present in the raw ironsands, exhibits dynamic behavior during pre-oxidation. Crystal growth along the (111) direction does not commence until PSB formation begins at 1000 °C. After a brief increase, the ilmenite content decreases and is completely absent at 1200 °C. Thus, ilmenite follows a trend of initial growth, subsequent decline, and eventual disappearance. The chemical oxidation of ilmenite spinel within TTM to ilmenite in TTH is represented by Eq. 7.1, while the oxidation of TTH to PSB is shown in Eq. 7.2. In these equations, δ denotes the degree of oxidation, ranging from TTH (δ = x) to PSB and ultimately to hematite (δ = 0).

$$Fe_{3-x}Ti_xO_4 + \frac{9}{4}O_2 = [(1.5 - x)Fe_2O_3 \cdot x\,FeTiO_3] \tag{7.1}$$

$$Fe_{2-x}Ti_xO_3 + \frac{1}{2}\left(\frac{x-\delta}{2-3\delta}\right)O_2 = \left(\frac{2-3x}{2-3\delta}\right)Fe_{2-\delta}Ti_\delta O_3 + 2\left(\frac{x-\delta}{2-3\delta}\right)Fe_2TiO_5 \tag{7.2}$$

where δ represents the degree of oxidation, ranging from TTH (δ = x) to complete oxidation into PSB and hematite (δ = 0).

Equation 7.1 indicates that the oxidation of titanomagnetite initially leads to an increase in ilmenite content, which, as shown in Eq. 7.2, subsequently undergoes further oxidation and decomposes into hematite and pseudobrookite. Moreover, the oxidation process and the associated changes in elemental valence states are closely linked to the migration and enrichment of titanium and iron within the ironsands. During this process, these elements become progressively separated and concentrated in different mineral phases—iron is ultimately enriched in hematite, while titanium is predominantly enriched in pseudobrookite [10].

7.3 Lattice Parameters Variation

JADE 6.5 (Materials Data Incorporated) is a widely used XRD analysis software that employs the least-squares refinement method to accurately determine crystallographic parameters [13–16]. Based on the XRD test results discussed above, the crystallographic parameters (including cell dimensions, unit cell volume, and associated calculation errors) of ten ironsands samples containing different physical phases—titanomagnetite (TTM), titanhematite (TTH), pseudobrookite (PSB), and ilmenite—under various pre-oxidation temperature conditions are summarized in Tables 7.1, 7.2, 7.3 and 7.4.

As shown in Table 7.1, the unit cell parameter *a* of TTM (cubic crystal system) decreases from 8.4126 Å at 300 °C to 8.3679 Å at 1200 °C, with a corresponding reduction in unit cell volume from 596.13 to 585.94 $Å^3$. This trend indicates that the crystallographic parameters of TTM decrease progressively with increasing pre-oxidation temperature and duration, deviating from the ideal TTM crystal values (a = 8.4237 Å, V = 597.74 $Å^3$; ICDD card No. 75-1373). This reduction is primarily attributed to the substitution of Fe^{2+} (larger ionic radius) by Fe^{3+} (smaller ionic

Table 7.1 TTM lattice parameters and unit cell volume during pre-oxidation of ironsands at 300–1200 °C

Temperature	TTM (cubic crystal system)		
	a/Å	±	Volumetric/Å^3
Raw ore	8.4165	0.0006	596.20
300 °C	8.4162	0.0007	596.13
400 °C	8.4159	0.0005	596.08
500 °C	8.4157	0.0004	596.04
600 °C	8.4096	0.0009	594.74
700 °C	8.4026	0.0008	593.25
800 °C	8.3829	0.0004	589.09
900 °C	8.3812	0.0006	588.74
1000 °C	8.3772	0.0002	587.89
1100 °C	8.3729	0.0008	586.98
1200 °C	8.3679	0.0006	585.94
Ideal crystal	8.4237	–	597.74

Table 7.2 TTH lattice parameters and unit cell volume during pre-oxidation of ironsands at 300–1200 °C

Temperature	TTH (hexagonal crystal system)				
	a/Å	±	*c*/Å	±	Volumetric/Å^3
Raw ore	5.0947	0.0008	14.0451	0.0003	315.71
300 °C	5.0939	0.0003	14.0449	0.0002	315.61
400 °C	5.0835	0.0010	13.9781	0.0011	312.83
500 °C	5.0790	0.0010	13.9449	0.0004	311.53
600 °C	5.0788	0.0005	13.9177	0.0006	310.90
700 °C	5.0763	0.0008	13.8547	0.0012	309.19
800 °C	5.0581	0.0011	13.8341	0.0009	306.52
900 °C	5.0548	0.0007	13.7605	0.0002	304.49
1000 °C	5.0448	0.0006	13.7591	0.0009	303.26
1100 °C	5.0387	0.0005	13.7544	0.0005	302.42
1200 °C	5.0348	0.0004	13.7471	0.0003	301.79
Ideal crystal	5.0342	–	13.7460	–	301.70

radius) during the oxidation process. The incorporation of additional oxygen atoms into the crystal structure results in lattice shrinkage, particularly along the *a*-axis, as the oxidation state of TTM increases.

Table 7.2 shows that, in the raw ironsands, the lattice parameters of TTH significantly deviate from those of the ideal crystal (a = 5.0342 Å, c = 13.7460 Å, V = 301.70 Å^3). However, as the degree of pre-oxidation increases, the unit cell parameters of TTH gradually approach their ideal values. Specifically, the *a* and *c* parameters

Table 7.3 PSB lattice parameters and unit cell volume during pre-oxidation of ironsands at 300–1200 °C

Temperature	PSB (orthogonal crystal system)						
	a/Å	±	b/Å	±	c/Å	±	Volumetric/$Å^3$
700 °C	9.8380	0.0006	9.9387	0.0009	3.7194	0.0009	363.67
800 °C	9.8323	0.0003	9.9489	0.0004	3.7206	0.0002	363.95
900 °C	9.8216	0.0004	9.9592	0.0013	3.7227	0.0004	364.14
1000 °C	9.8111	0.0008	9.9663	0.0007	3.7244	0.0003	364.17
1100 °C	9.8025	0.0011	9.9709	0.0005	3.7267	0.0001	364.25
1200 °C	9.7958	0.0007	9.9768	0.0003	3.7299	0.0007	364.53
Ideal crystal	9.7965	–	9.9805	–	3.7301	–	364.71

Table 7.4 Changes in ilmenite lattice parameters and unit cell volume at 100 °C intervals during pre-oxidation of ironsands from 300 to 1200 °C

Temperature	Ilmenite (rhodochrosite system)				
	a/Å	±	c/Å	±	Volumetric/$Å^3$
Raw ore	5.0793	0.0002	13.9461	0.0007	311.60
300 °C	5.0787	0.0003	13.9437	0.0002	311.47
400 °C	5.0758	0.0002	13.9417	0.0003	311.07
500 °C	5.0738	0.0008	13.9405	0.0007	310.80
600 °C	5.0700	0.0007	13.9400	0.0006	310.32
700 °C	5.0689	0.0005	13.9285	0.0007	309.93
800 °C	5.0587	0.0003	13.9283	0.0003	308.68
900 °C	5.0569	0.0008	13.9190	0.0006	308.25
1000 °C	5.0558	0.0005	13.8896	0.0007	307.47
1100 °C	5.0474	0.0002	13.7848	0.0007	304.14
1200 °C	5.0347	0.0008	13.7634	0.0004	302.14
Ideal crystal	5.0561	–	13.9115	–	307.99

decrease from 5.0939 Å and 14.0449 Å to 5.0348 Å and 13.7471 Å, respectively. These changes suggest that the early-stage TTH crystallization in raw ironsands is incomplete—likely due to rapid cooling during the initial liquid-to-solid transition, followed by sluggish oxidation and limited ionic migration. With further oxidation, TTH crystallizes more completely, and its cell parameters progressively align with those of the ideal structure.

The extent of TTH crystallinity is influenced by its formation temperature and mineralization time, both of which affect the solid solution oxygen content within the phase. Specifically, higher formation temperatures and shorter mineralization times generally correspond to lower oxygen incorporation, resulting in smaller unit cell sizes. Therefore, the crystallographic parameters of TTH may serve as useful indicators for inferring the formation conditions and mineralization history of ironsands.

Pseudobrookite was not detected in the temperature range from room temperature to 600 °C. Its crystallographic parameters, measured between 700 and 1200 °C, are presented in Table 7.3. With increasing pre-oxidation temperature and duration, pseudobrookite—belonging to the orthorhombic crystal system—exhibited contraction along the *a*-axis and expansion along the *b*- and *c*-axes. This anisotropic deformation led to an overall increase in unit cell volume and a gradual convergence toward the ideal crystallographic dimensions.

The variation in ilmenite cell parameters as a function of pre-oxidation temperature is summarized in Table 7.4. Compared to the ideal ilmenite structure (ICDD Card No. 75-1212), the lattice parameters of ilmenite in raw ironsands deviate significantly, a discrepancy attributed to the non-equilibrium crystallization conditions during the early stages of mineralization. As the pre-oxidation temperature increases, the ilmenite lattice exhibits a gradual contraction along both the *a*- and *c*-axes. At 1000 °C, the ilmenite unit cell dimensions (a = 5.0558 Å, c = 13.8896 Å) closely approximate those of the ideal structure (a = 5.0561 Å, c = 13.9115 Å), and continued contraction is observed with further temperature increase. This lattice contraction is primarily driven by the progressive substitution of Ti^{4+} ions (ionic radius = 0.53 Å) with Fe^{3+} ions (ionic radius = 0.49 Å), as described by the reaction in Eq. 7.2. Accordingly, the unit cell dimensions of ilmenite decrease more rapidly beyond 700 °C as oxidation and cation exchange intensify.

7.4 Elemental Migration and Enrichment

Under an air flow rate of 1 L/min, ironsands samples were heated to 300 °C, 900 °C, and 1200 °C at a rate of 5 °C/min. Scanning electron microscopy (SEM) and energy-dispersive X-ray spectroscopy (EDS) were employed to investigate the evolution of microstructural morphology and elemental distribution as a function of pre-oxidation temperature. The results are presented in Fig. 7.3. Elemental surface scans for calcium and silicon are not included here, as calcium showed no significant variation, and silicon predominantly exists in the form of discrete vein-like particles.

As shown in Fig. 7.3, lamellar structures precipitated on a relatively homogeneous matrix exhibit widths of approximately 0.1–2 μm. Within these lamellae, titanium and magnesium are significantly enriched, whereas iron, aluminum, and oxygen are concentrated primarily within the substrate phase.

The formation of titanomagnetite (TTM) is primarily driven by the progressive substitution of iron atoms in the B-site of the spinel lattice by titanium. The degree of this substitution depends on the titanium content and the specific mineralization conditions. TTM solid solutions are generally represented as $Fe_{3-x}Ti_xO_4$, where x reflects the extent of titanium incorporation. As pre-oxidation progresses, changes in crystal structure, lattice size, and physical phase occur. Specifically, the cubic spinel structure of TTM ($Fe_{3-x}Ti_xO_4$, $0 < x < 1$) transforms into the hexagonal

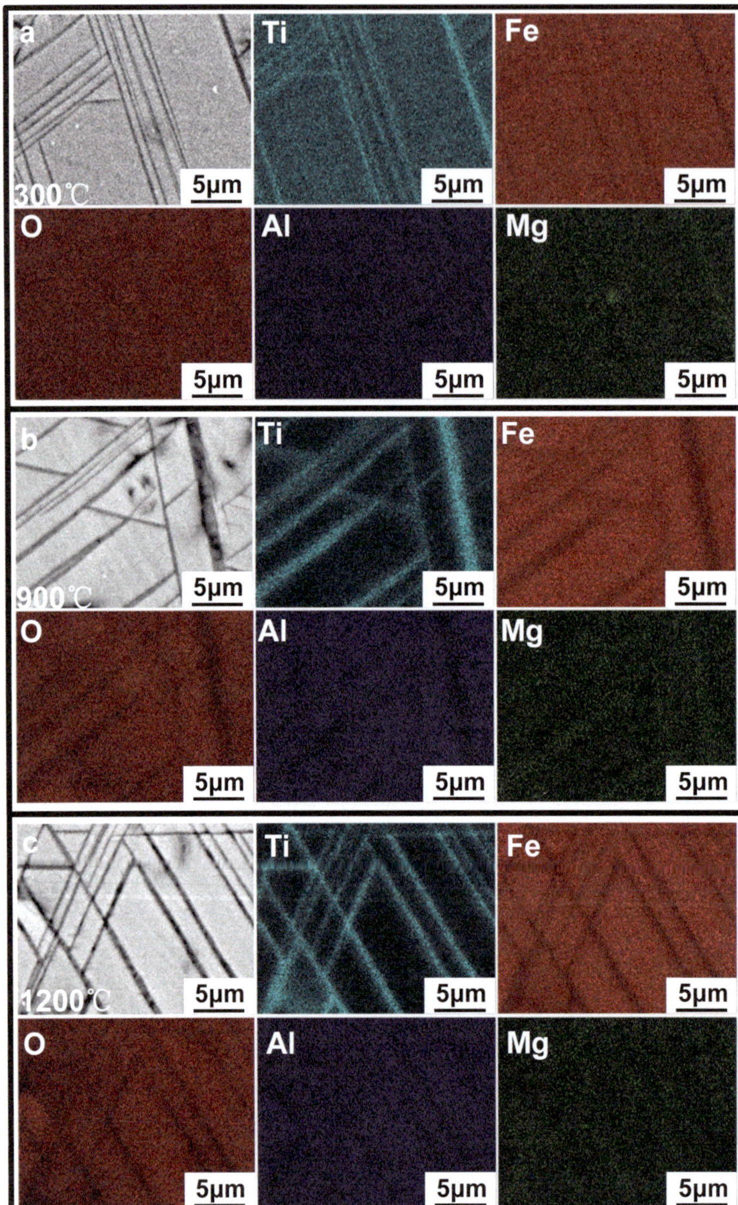

Fig. 7.3 Scanning electron microscope (SEM) micrographs and elemental surface mapping (five elements) of ironsands subjected to different pre-oxidation temperatures

structure of titanhematite (TTH, $Fe_{2-y}Ti_yO_3$, $0 < y < 1$), and ultimately evolves into the pseudobrookite solid solution system.

During this transformation, titanium becomes increasingly enriched within the lamellar structures, while iron and oxygen concentrations in these regions gradually decline. The final enrichment of titanium leads to the formation of the pseudobrookite phase. Consequently, with increasing pre-oxidation temperature and duration, the spatial distribution boundaries between iron, titanium, and oxygen become increasingly distinct. Titanium is predominantly concentrated in the lamellar regions, whereas iron and oxygen are mainly retained in the homogeneous matrix. Aluminum is similarly concentrated in the matrix, incorporated into the TTM phase as aluminum-bearing spinel. In the early stages of oxidation, magnesium is relatively dispersed; however, its segregation and enrichment become more pronounced at higher pre-oxidation temperatures.

This elemental partitioning reduces the inhibitory effect of titanium on oxygen migration during the subsequent reduction of ironsands. Therefore, pre-oxidation facilitates the formation of discrete elemental phases and enhances the reducibility of ironsands.

In summary, the non-isothermal pre-oxidation of ironsands in the temperature range of 300–1200 °C is characterized by sequential phase transformations from titanomagnetite (TTM) to titanhematite (TTH), and eventually to pseudobrookite (PSB). Titanium migrates into the PSB phase, while iron is preferentially enriched in the hematite phase. Ilmenite undergoes generation and subsequent decomposition during this process. Titanium becomes more concentrated in lamellar structures, whereas iron, aluminum, and oxygen accumulate in the homogeneous matrix. Pre-oxidation promotes the separation of titanium and oxygen, thereby reducing the adverse effect of titanium on oxygen migration during reduction and ultimately improving the reducibility of ironsands.

7.5 Summary

As discussed in Sect. 3.2, the oxidation of magnetite represents the primary reaction during the early stages of ironsands pre-oxidation, particularly at 304 °C and 532 °C. At approximately 686 °C, significant formation of the TTH phase begins, while the generation of ilmenite becomes the dominant reaction at 910 °C. When the temperature exceeds 1052 °C, the formation of pseudobrookite (PSB) initiates. The five sub-reactions identified from the derivative differential thermogravimetric (DDTG) curves are consistent with the phase evolution observed in XRD analysis and theoretical thermodynamic calculations.

However, even at 1200 °C, only about 80.8 wt% of the ironsands sample was fully oxidized, underscoring the structural stability and oxidation resistance of certain mineral phases within the ironsands. With increasing pre-oxidation degree, the TTM lattice gradually contracts due to the substitution of Fe^{2+} (with a larger ionic radius) by Fe^{3+} (with a smaller ionic radius). Although the TTH phase in the raw ironsands is initially poorly crystallized, its unit cell parameters progressively approach those of the ideal structure as pre-oxidation proceeds. The orthorhombic unit cell of pseudobrookite contracts along the *a*- and *c*-axes while elongating along the *b*-axis. For ilmenite, the cell dimensions are closest to their ideal values at 1000 °C, after which further shrinkage is observed with increasing temperature.

Before 800 °C, TTM remains the dominant phase in ironsands. As the temperature rises, the cubic spinel solid solution system of TTM is progressively transformed into the hexagonal TTH structure. Subsequently, a portion of TTH is further converted into pseudobrookite. During this transformation sequence, titanium becomes increasingly concentrated in the pseudobrookite phase, while iron is enriched in the hematite phase. This elemental segregation is closely associated with the phase evolution from titanomagnetite to titanhematite, and eventually to pseudobrookite.

The observed changes in physical phase composition and lattice parameters are key contributors to the improved reducibility of ironsands following pre-oxidation. The redistribution and selective enrichment of iron and titanium into distinct phases—hematite and pseudobrookite, respectively—not only modify the microstructure but also facilitate enhanced oxygen migration and reduction kinetics during subsequent processing.

References

1. Park E, Ostrovski O. Reduction of titania-ferrous ore by carbon monoxide. ISIJ Int. 2003;43(9):1316–25.
2. Wang YM, Yuan ZF. Reductive kinetics of the reaction between a natural ilmenite and carbon. Int J Miner Process. 2006;81(3):133–40.
3. Paunova R. Thermodynamic study of the reduction of titanium magnetite concentrate with solid carbon. Metall Mater Trans B-Proc Metall Mater Proc Sci. 2002;33(4):633–8.
4. Schwebel GL, Filippou D, Hudon G, et al. Experimental comparison of two different ilmenites in fluidized bed and fixed bed chemical-looping combustion. Appl Energy. 2014;113:1902–8.
5. Longbottom RJ, Ostrovski O, Zhang JQ, et al. Stability of cementite formed from hematite and titanomagnetite ore. Metall Mater Trans B-Proc Metall Mater Proc Sci. 2007;38(2):175–84.
6. Park E, Ostrovski O. Reduction of titania-ferrous ore by hydrogen. ISIJ Int. 2004;44(6):999–1005.
7. Mcdma G, GD M. Instability of titanium-rich ironsands in reducing gases. Ironmakg Steelmakg 1974;1(3):138–50.
8. McAdam G, Dall R, Marshall T. Direct reduction of New Zealand ironsand concentrates. 1. Gaseous Reduction. New Zealand J Sci. 1969;12(4):649.
9. McAdam G, Dall R, Marshall T. Direct reduction of New Zealand ironsand concentrates. 2. Solid reductants. N Z J Sci 1969;12(4):669.
10. Park E, Ostrovski O. Effects of preoxidation of titania-ferrous ore on the ore structure and reduction behavior. ISIJ Int. 2004;44(1):74–81.
11. Wei X, Lu XG, Zou XL, et al. Phase transitions, micro-morphology and its oxidation mechanism in oxidation of ilmenite ($FeTiO_3$) powder. Trans Nonferrous Metals Soc China. 2013;23(8):2439–45.
12. Zhang L, Zhang L, Wang M, et al. Dynamic oxidation of the Ti-bearing blast furnace slag. ISIJ Int. 2006;46(3):458–65.

13. Chen L, Li JW, Rye RO, et al. Mineralogical, chemical, and crystallographic properties of supergene jarosite-group minerals from the Xitieshan Pb-Zn sulfide deposit, northern Tibetan Plateau, China. Miner Petrol. 2013;107(4):487–99.
14. Singh Y, Viswanathan R, Parihar PS, et al. X-ray crystallography of uraninites associated with the albitite belt of western India: evidence for the high-temperature origin of uranium and associated mineralisation. J Geol Soc India. 2013;81(1):79–90.
15. Gu XP, Xie XD, Wu XB, et al. Ferrisepiolite: a new mineral from Saishitang copper skarn deposit in Xinghai County, Qinghai Province, China. Eur J Mineral. 2013;25(2):177–86.
16. Wall AJ, Mathur R, Post JE, et al. Cu isotope fractionation during bornite dissolution: An in situ X-ray diffraction analysis. Ore Geol Rev. 2011;42(1):62–70.

Chapter 8
Direct Reduction of Ironsands

The phase transition process during the hydrogen reduction of ironsands remains unclear. One of the primary objectives of this chapter is to elucidate the phase transitions occurring during the reduction process, with a focus on reaction time and solid solubility. Furthermore, the experimental parameters of temperature and hydrogen concentration are systematically varied to investigate their effects on the reduction mechanism of ironsands. This approach aims to identify reduction parameters that optimize both reaction rate and economic feasibility. Finally, the microscopic morphology of ironsands during the hydrogen reduction process is presented, and three distinct reduction pathways characterized by different morphological features are summarized.

8.1 Influence of Gas-Based Reduction Parameters

Due to the progressive depletion of high-grade hematite resources, ironsands have attracted increasing attention as a supplementary alternative to conventional iron ores. Their advantages—such as abundant reserves and relatively low mining costs—have made ironsands a focal point of research in various industrial applications, including magnetic separation [1, 2], sintering [3–6], and pelletizing [7–9]. In particular, hydrogen-based direct reduction of ironsands has emerged as a key research direction, owing to its favorable reaction kinetics and low environmental impact [1, 10–19].

Eungyeul [17] investigated the influence of temperature and hydrogen concentration on the gas-based reduction of ironsands, demonstrating that under a hydrogen–argon atmosphere and temperatures above 900 °C, the reduction pathway follows: $Fe_{3-x}Ti_xO_4 \rightarrow FeO + Fe_{3-x-\delta}Ti_{x+\delta}O_4 \rightarrow Fe + Fe_{3-x-\delta}Ti_{x+\delta}O_4 \rightarrow Fe + xTiO_2$. In contrast, Jie [20], in a kinetic study of hydrogen-reduced sepiolite, suggested a dual-pathway mechanism: the simultaneous conversion of titanomagnetite to

Z. Wang et al., *Smelting and Utilization Technology of Ironsands*,
https://doi.org/10.1007/978-981-95-2868-4_8

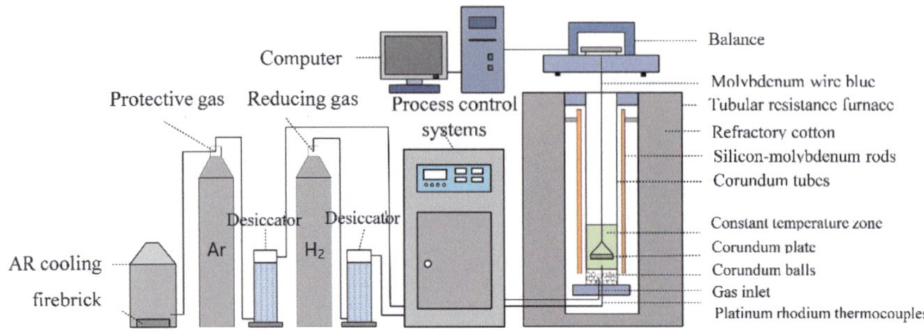

Fig. 8.1 Schematic diagram of the hydrogen–argon gas mixture reduction apparatus for ironsands

pumice and ilmenite, and the subsequent transformation of pumice and ilmenite into high-titanium-bearing phases.

In this chapter, the reduction behavior of ironsands is studied within a temperature range of 800–1100 °C and hydrogen concentrations ranging from 10 to 100 vol%. The investigation includes optimization of reduction parameters, analysis of physical phase transformations, solid solubility changes associated with titanium and iron separation and enrichment, as well as examination of microstructural evolution.

The hydrogen–argon mixed-gas reduction experiments were conducted using a fixed-bed vertical tubular resistance furnace, as shown in Fig. 8.1. Hydrogen and argon gases were first dehydrated using calcium sulfate and a 3-Å molecular sieve, then regulated by gas flow meters in the control cabinet. A total gas flow rate of 5 L/min was maintained, with hydrogen volume fractions ($H_2/(H_2 + Ar)$) set at 10, 20, 30, 40, 50, and 100 vol%, respectively. For each experiment, 3 g of ironsands was evenly distributed on a 40 mm diameter corundum disk to ensure uniform gas–solid contact.

The temperature of the furnace's constant-temperature zone was precisely controlled between 800 °C and 1100 °C, in 50 °C intervals. Each sample was preheated under an argon atmosphere for 30 min to reach the target temperature. Subsequently, the hydrogen–argon reduction mixture was introduced via a bottom gas inlet, preheated and evenly dispersed by corundum balls, initiating the reduction reaction immediately.

The ironsands samples and corundum disks were suspended on molybdenum wires attached to a high-precision balance, allowing continuous mass measurement at 6-s intervals. The data were recorded and stored via a computer connected to the balance. To analyze the phase transformations and microstructural evolution during reduction, samples were extracted at predefined time intervals, rapidly cooled under an argon atmosphere, and then subjected to XRD analysis for phase identification. SEM (FEI Quanta 250) and EDS (EDAX) were employed to investigate morphological changes and elemental distribution within the reduced samples.

8.1.1 Temperature Effects

The reduction curves of ironsands under varying temperatures (800–1100 °C) and hydrogen concentrations (10–100 vol%) are presented in Fig. 8.2.

As shown in Fig. 8.2, at hydrogen concentrations of 10 vol% and 20 vol%, ironsands undergo only partial reduction within a 1-h reaction time, making it difficult

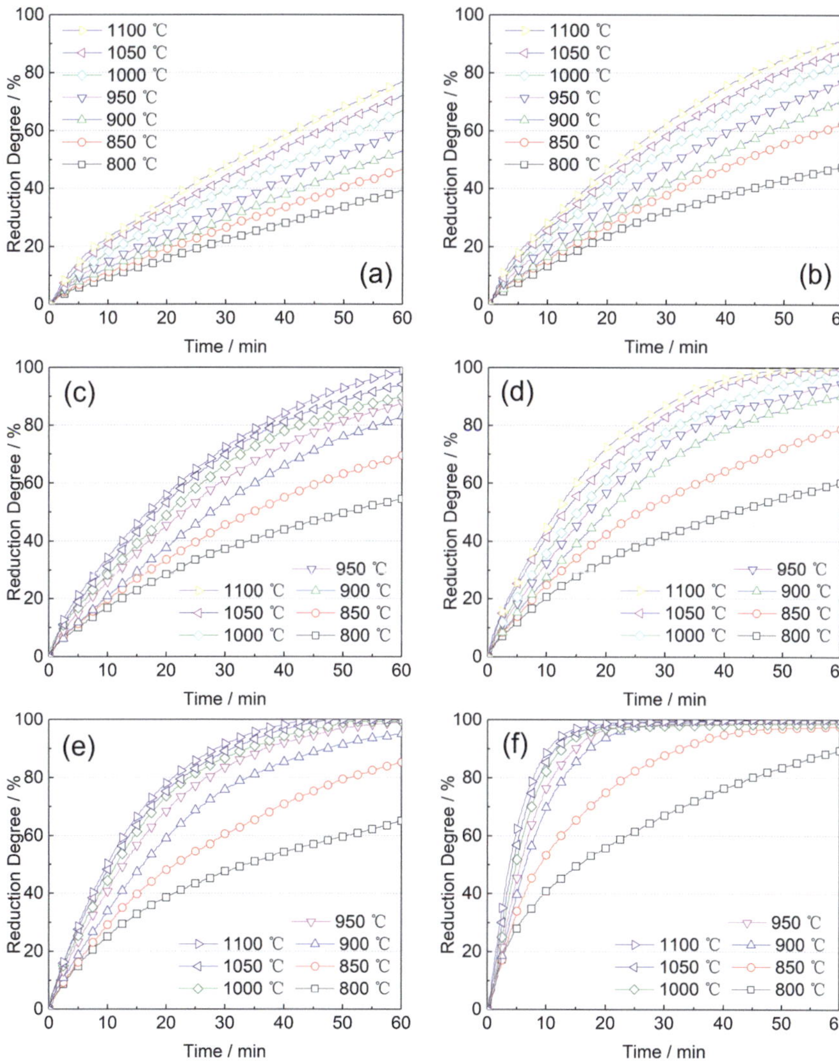

Fig. 8.2 Effect of temperature on the reduction behavior of ironsands under varying $H_2/(H_2 + Ar)$ volume ratios. **a** 10 vol%; **b** 20 vol%; **c** 30 vol%; **d** 40 vol%; **e** 50 vol%; **f** 100 vol%

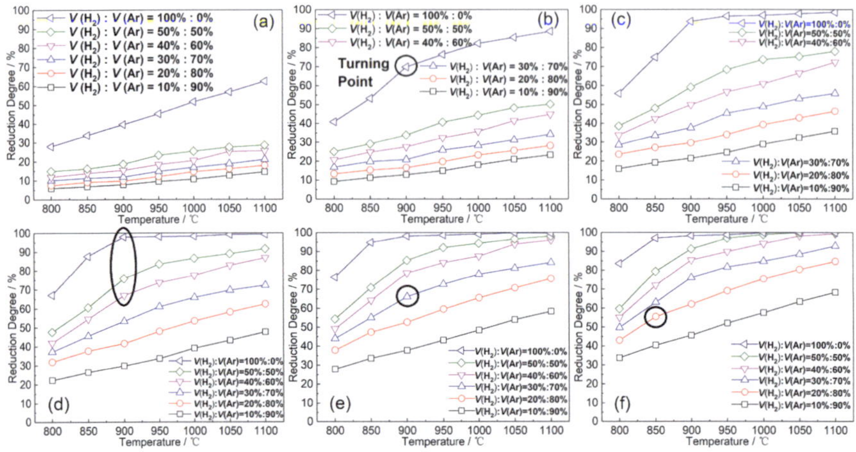

Fig. 8.3 Promotion of reduction over time at a fixed temperature. **a** 5 min; **b** 10 min; **c** 20 min; **d** 30 min; **e** 40 min; **f** 50 min

to achieve complete reduction under these conditions. However, when the hydrogen partial pressure is increased to 0.3 bar or higher, complete reduction of ironsands can be achieved within 1 h, depending on the temperature.

For a fixed hydrogen concentration and constant reaction time, both the rate and extent of reduction increase with rising temperature. However, the magnitude of this increase is not uniform across temperature intervals. For instance, at 100 vol% hydrogen and 10 min of reaction time, increasing the temperature from 850 to 900 °C results in an increase in the degree of reduction from 53.27 to 69.81%, representing a 16.54% improvement. In contrast, raising the temperature from 900 to 950 °C under the same conditions yields a smaller increase of only 6.54%.

To further evaluate and quantify the effect of temperature on the reduction process, the degree of reduction was plotted as a function of temperature for a fixed reaction time, as shown in Fig. 8.3.

To characterize the incremental effect of temperature on the reduction behavior of ironsands, an index representing the increase in the degree of reduction per 50 °C temperature rise was defined, as shown in Eq. 8.1:

$$\Delta R_{T,V_{\mathrm{H_2}}} = R_{T,V_{\mathrm{H_2}}} - R_{T-50\,^{\circ}\mathrm{C},V_{\mathrm{H_2}}} \tag{8.1}$$

where:

R the degree of reduction, %;
T the reaction temperature, °C;
V_{H2} the hydrogen volume fraction, vol%;
ΔR the increase in reduction percentage per 50 °C increment, %.

As shown in Fig. 8.3a, during the initial stage of the reduction reaction (at 5 min), the incremental reduction rate (ΔR) remains nearly constant across the temperature range of 800–1100 °C for each specific hydrogen concentration. However, with increasing hydrogen content in the reducing gas, ΔR also increases. This trend becomes more pronounced at 10 and 20 min, as illustrated in Fig. 8.3b, c. At these time points, a distinct turning point—defined as the temperature at which a noticeable decrease in the slope of the reduction curve occurs—emerges at 900 °C under a pure hydrogen atmosphere. This indicates that the temperature's contribution to promoting the reduction reaction diminishes beyond 900 °C when operating under pure hydrogen.

At 30 min, turning points appear at 900 °C for hydrogen concentrations of 40, 50, and 100 vol%. As the reaction progresses to 40–50 min, the turning point for 30 vol% hydrogen shifts to 900 °C, while for 20 vol% hydrogen, it appears at 850 °C. This progression suggests that as the hydrogen concentration decreases from 100 to 20 vol%, the emergence of the turning point is significantly delayed—from occurring at 10 min to as late as 50 min.

In contrast, under 10 vol% hydrogen conditions, ΔR remains essentially unaffected by variations in temperature and reaction time. Consequently, the reduction increment per 50 °C temperature rise is approximately constant, and no turning point is observed within the experimental temperature and time range.

In summary, taking into account both the reduction kinetics and the cost-efficiency of temperature increase, a temperature of 900 °C is identified as an optimal condition for the pre-reduction of ironsands.

8.1.2 Reducing Gas Concentration Effects

To investigate the influence of hydrogen concentration in the reducing gas on the reduction behavior of ironsands, the reduction degree under various conditions is compared in Fig. 8.4. At temperatures of 800 and 850 °C, hydrogen concentrations ranging from 10 to 50 vol% are insufficient to achieve a reduction degree greater than 90% within 60 min. However, when the temperature increases to 900 and 950 °C, a hydrogen concentration of 50 vol% enables the reduction degree of ironsands to exceed 90% within 60 min. Moreover, at 1000 °C and higher, achieving a 90% reduction becomes significantly easier across the tested hydrogen concentrations.

To further quantify the effect of hydrogen concentration on the reduction process, Fig. 8.5 illustrates the trend of increasing reduction degree with rising hydrogen content under various temperature conditions. At 5 min, the increase in reduction degree as hydrogen concentration rises from 10 to 50 vol% is relatively uniform and steady. However, at 10 min, a marked acceleration in the reduction degree is observed when hydrogen concentration increases from 30 to 40 vol%, particularly at 1050 and 1100 °C. This inflection point is also evident in the 20- and 30-min reduction profiles across the temperature range of 900–1100 °C.

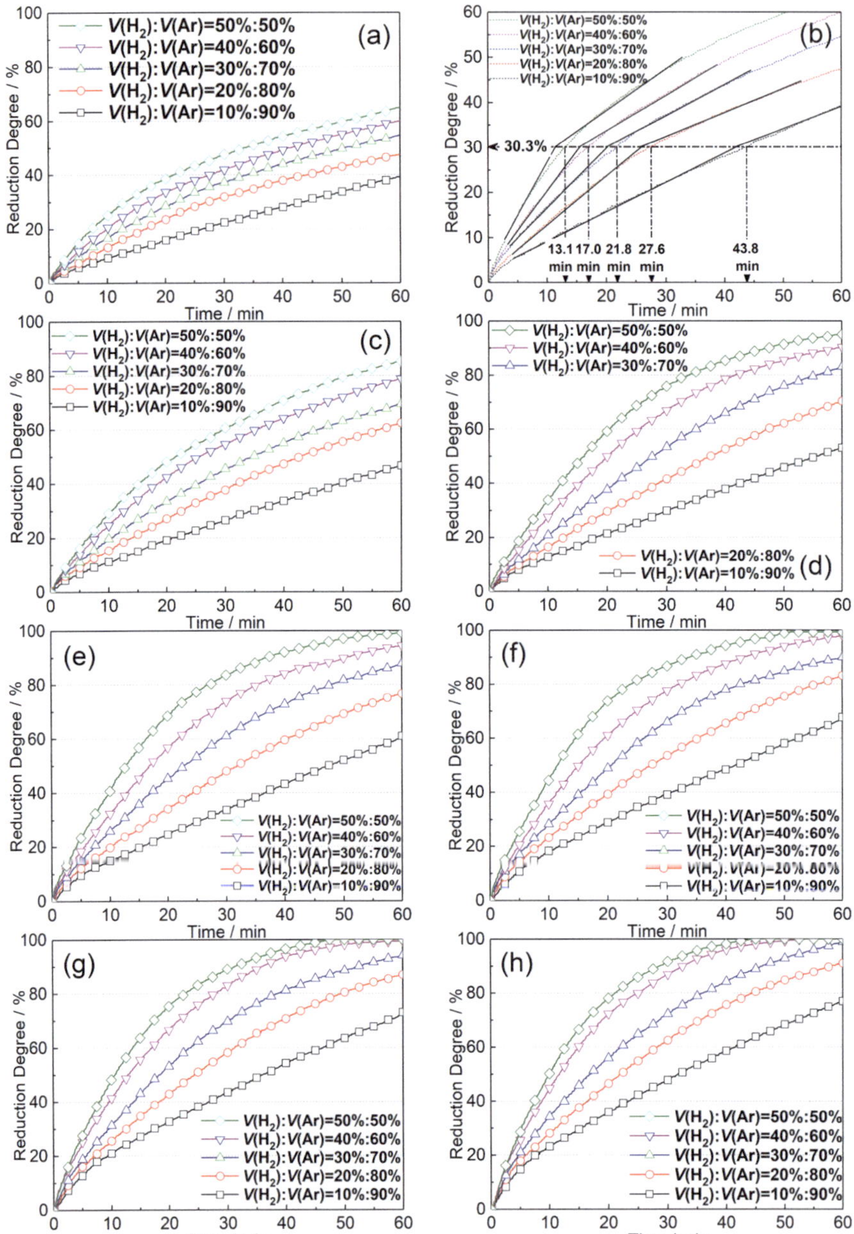

Fig. 8.4 Effect of hydrogen concentration on the reduction degree of ironsands at different temperatures. **a**, **b** 800 °C; **c** 850 °C; **d** 900 °C; **e** 950 °C; **f** 1000 °C; **g** 1050 °C; **h** 1100 °C

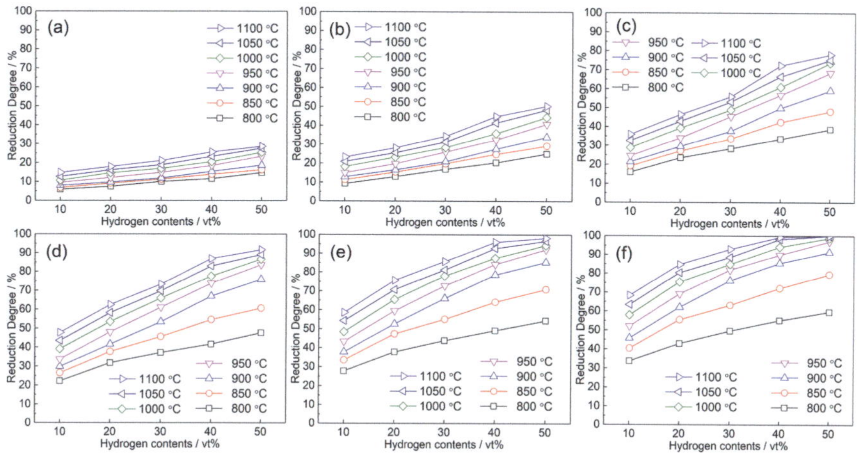

Fig. 8.5 Influence of hydrogen concentration changes on the reduction degree at fixed durations. **a** 5 min; **b** 10 min; **c** 20 min; **d** 30 min; **e** 40 min; **f** 50 min

At later stages—40 and 50 min—the transition from 40 to 50 vol% hydrogen continues to promote the reaction, but the incremental gain in reduction degree diminishes compared to the earlier increase from 30 to 40 vol%, or from lower concentrations. For instance, at 900 °C and 40 min, increasing the hydrogen concentration from 30 to 40 vol% results in a 12.5% increase in reduction degree, whereas a further increase from 40 to 50 vol% yields only a 6.8% improvement.

Therefore, considering both the reduction efficiency and the economic implications of increasing hydrogen concentration, a hydrogen content of approximately 40 vol% is deemed optimal for the reduction of ironsands.

8.1.3 Mechanistic Analysis of Temperature and Gas Concentration

The reduction rate during the hydrogen-based reduction of ironsands varies due to changes in the rate-controlling mechanisms, which manifest as inflection points in the reduction curve. These inflection points, referred to as turning points (TPs), are characterized by noticeable changes in the slope of the reducibility curve and can be identified through linear fitting and breakpoint analysis, as illustrated in Fig. 8.4b.

Typically, each turning point is defined by two key parameters: the degree of reduction at its onset, denoted as R_{TP}, and the time of occurrence, denoted as t_{tp}. According to the unreacted core model for gas–solid reduction, both R_{TP} and t_{tp} are strongly influenced by the reaction temperature and the concentration of the reducing gas.

A schematic diagram illustrating the mechanisms by which temperature and hydrogen concentration influence the turning point is presented in Fig. 8.6. In this

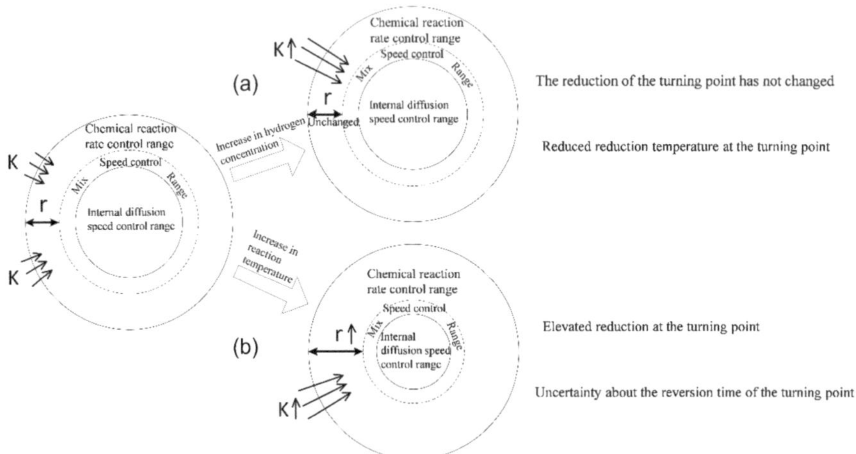

Fig. 8.6 Conceptual illustration showing the effect of temperature and gas concentration on the reduction of ironsands and the appearance of critical transition (tipping) points

model, r denotes the radius of the region controlled by the chemical reaction within the unreacted core, which directly determines R_{TP}, while k represents the reaction rate or the rate of product layer formation, which governs t_{tp}.

As shown in Fig. 8.6a, when the hydrogen concentration is increased while maintaining a constant reaction temperature, the molecular thermal mobility remains unchanged; thus, r stays constant. However, the increased partial pressure of hydrogen enhances the reaction rate constant K, thereby accelerating the reaction and reducing t_{tp}, while R_{TP} remains unchanged.

Conversely, Fig. 8.6b illustrates that under constant hydrogen concentration, an increase in temperature promotes both faster molecular motion and an expansion in the reaction-controlled zone, leading to simultaneous increases in both r and K. As a result, R_{tp} increases. However, the effect on t_{tp} is not straightforward, as it depends on the relative magnitudes of the changes in r and K.

The relationships between these two turning point parameters (R_{tp} and t_{tp}) and the reaction conditions are further detailed in Fig. 8.6 and Table 8.1. The characterization equations derived from the unreacted core model, where chemical reaction is the rate-controlling step, are provided in Eq. 8.2.

$$t_{\text{Chemical complaint}} = \frac{\rho_{Ironsands} \cdot r}{\mathrm{M}_{Ironsands} k_{\text{Chemical reaction}} c_{H2}} \tag{8.2}$$

where:

- t time required to complete the reaction within the chemical reaction rate-controlled regime, s;
- ρ density of ironsands particles, kg/m^3;

Table 8.1 Reductions R_{TP} and time t_{TP} at the appearance of the turning point at different temperatures (800–1100 °C) and hydrogen concentrations (10–100 vol%)

Hydrogen concentration $V(H_2)/[V(H_2) + V(Ar)]$/vol%	10	20	30	40	50	100	Turning point reduction R_{TP}/%
Temperature/°C	Turning point occurrence time t_{TP}/min						
800	43.8	27.6	21.8	17.0	13.1	5.8	30.3
850	59.2	38.2	30.7	22.3	18.8	7.8	46.8
900	–	56.3	41.7	30.5	26.4	9.6	67.6
950	–	56.1	41.1	30.1	22.8	9.5	73.9
1000	–	49.5	37.0	28.3	21.0	8.3	75.1
1050	–	45.3	35.3	25.6	20.4	7.9	76.4
1100	–	41.5	33.5	23.0	19.4	7.3	77.0

r radius of the reaction-controlled zone in the unreacted core model, m;
M molar mass of ironsands, g/mol;
k chemical reaction rate constant, mol/s;
c hydrogen concentration, mol/L.

As shown in Eq. 8.2, when the reaction temperature is unchanged, r does not change because the molecular thermal mobility is unchanged, which means that R_{TP} is unchanged, and at this time, increasing the hydrogen concentration, k and c will increase, which will make the reaction time t in the controlled rate range of the chemical reaction decrease which means that t_{TP} will be shortened. While when the reaction temperature is increased and the hydrogen concentration remains constant, the R_{TP} is increased with increasing temperature because the increase in temperature increases both the chemical reaction rate-control range, r, and the reaction rate, k. However, the reaction time t (t_{TP}) in the rate-controlled range of the chemical reaction shows uncertainty, and t_{TP} exhibits an increasing trend when the increase in the rate-controlled range of the chemical reaction, r (R_{TP}), is predominant and the increase in the reaction rate, k, is of secondary importance; When the increase in reaction rate k predominates and the increase in chemical reaction rate control range r (R_{TP}) is secondary, t_{TP} shows a decreasing trend.

As shown in Eq. 8.2, when the reaction temperature remains constant, the radius of the rate-controlled zone r does not change due to unchanged molecular thermal mobility. Consequently, the reduction degree at the turning point (R_{TP}) remains constant. In this case, increasing the hydrogen concentration results in an increase in both the chemical reaction rate constant k and the hydrogen concentration c, which together lead to a decrease in the reaction time t within the rate-controlled regime—i.e., the time to reach the turning point (t_{TP}) becomes shorter.

In contrast, when the reaction temperature increases while the hydrogen concentration remains constant, both r and k increase due to enhanced molecular motion. As a result, R_{TP} increases with rising temperature. However, the behavior of t_{TP} becomes uncertain and depends on the relative magnitude of changes in r and k. If the increase in r (and thus R_{TP}) dominates, while the increase in k is secondary, t_{TP} tends to increase. Conversely, if the increase in k outweighs that of r, t_{TP} decreases.

As mentioned earlier, t_{TP} depends on both the reaction rate and the chemical reaction rate control range such that a higher reaction temperature and a lower R_{TP} correspond to a shorter t_{TP} and vice versa. As shown in Table 8.1, at the same temperature, accompanied by an increase in hydrogen concentration which is the increase in chemical reaction rate K, t_{TP} strictly exhibits a decreasing trend, mainly due to the unchanged R_{TP} and higher chemical reaction rate K. However, at the same hydrogen concentration, when the reaction temperature is increased, the reaction rate K increases, and at the same time, the R_{TP} is increased. When the reaction temperature is increased from 800 to 900 °C, the effect of temperature on R_{TP} will be greater than that on the reaction rate K. Therefore, t_{TP} increases gradually from 800 °C to 900 °C and reaches a great value at 900 °C. When the temperature continued to increase from 900 to 1100 °C, although the chemical reaction rate-control range gradually increased (confirmed by the gradually increasing R_{TP}), the growth of the reaction rate played a major role in this temperature band, so t_{TP} gradually decreased from 900–1100 °C.

As discussed earlier, t_{TP} is influenced by both the reaction rate and the extent of the rate-controlled regime, such that a higher temperature and lower R_{TP} generally correspond to a shorter t_{TP}, and vice versa. As shown in Table 8.1, at a fixed temperature, increasing the hydrogen concentration leads to a consistent decrease in t_{TP}, primarily because R_{TP} remains constant while the reaction rate K increases. However, under constant hydrogen concentration, when the temperature is raised, both K and R_{TP} increase. In the temperature range from 800 °C to 900 °C, the increase in R_{TP} dominates over that of K, resulting in a gradual increase in t_{TP}, reaching its maximum at 900 °C. Beyond 900 °C, from 900 °C to 1100 °C, although R_{TP} continues to increase, the growth in reaction rate becomes the dominant factor, leading to a progressive decrease in t_{TP}.

8.2 Phase Transitions During Gas-Based Reduction

The XRD patterns of raw ironsands and samples at various reduction stages (2.5–60 min) are presented in Fig. 8.7. To eliminate the interference of other variables in examining phase transformations, a hydrogen concentration of 100 vol% was selected as the reducing atmosphere. The XRD analysis revealed that a total of eight distinct phases emerged during the reduction process of ironsands. Based on these XRD results and corresponding reduction data, the phase compositions at each reduction stage are summarized in Table 8.2.

Titanomagnetite (TTM) is identified as the dominant phase in the raw ironsands. Additionally, a small amount of titanhematite (TTH) is detected, primarily formed from the supersaturation of titanium during ironsands formation and the long-term oxidation of TTM. A weak ilmenite ($FeTiO_3$) peak is observed near $2\theta = 32.72°$, and

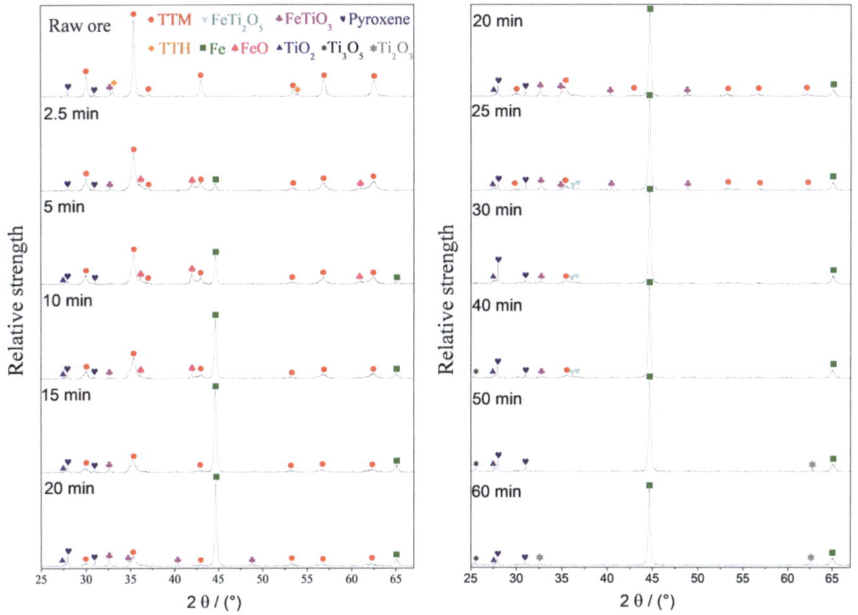

Fig. 8.7 XRD patterns of ironsands reduced at 900 °C under 100 vol% H_2 for 60 min from raw ore

Table 8.2 Physical phases and reduction degrees of ironsands at 900 °C and 100 vol% hydrogen from initial reduction to 60 min

Reduction time/min	Reduced phase	Reduction degree/%
0	TTM; TTH; $FeTiO_3$; pyroxene	0
2.5	TTM; Fe; FeO; $FeTiO_3$; pyroxene	24.98
5	TTM; Fe; FeO; TiO_2; pyroxene	51.71
10	Fe_2TiO_4; Fe; FeO; $FeTiO_3$; TiO_2; pyroxene	82.18
15	Fe_2TiO_4; Fe; $FeTiO_3$; TiO_2; pyroxene	94.12
20	Fe_2TiO_4; Fe; $FeTiO_3$; TiO_2; pyroxene	96.97
25	Fe_2TiO_4; Fe; $FeTiO_3$; $FeTi_2O_5$; TiO_2; pyroxene	97.58
30	Fe_2TiO_4; Fe; $FeTiO_3$; $FeTi_2O_5$; TiO_2; pyroxene	98.20
40	Fe_2TiO_4; Fe; $FeTiO_3$; $FeTi_2O_5$; TiO_2; Ti_3O_5; pyroxene	99.19
50	Fe; TiO_2; Ti_3O_5; Ti_2O_3; pyroxene	100
60	Fe; TiO_2; Ti_3O_5; Ti_2O_3; pyroxene	100

oxides of silicon, magnesium, aluminum, and manganese occur as chain pyroxenes, persisting throughout the reduction process.

2.5 min: After 2.5 min of exposure to the reducing atmosphere, the reduction degree reaches 24.98%. At this stage, the TTH phase is fully reduced, primarily due to its low concentration and the rapid reduction of Fe_2O_3 (a component of TTH) to Fe_3O_4. The Fe_3O_4 present—including that from TTM and reduced TTH—is partially reduced to wüstite (FeO) and metallic iron (Fe).

5 min: By 5 min, $FeTiO_3$ originally present in the raw ironsands is fully reduced to Fe and TiO_2, with the reduction degree exceeding 50%. However, XRD patterns between 5 and 10 min reveal the reappearance of $FeTiO_3$. This phenomenon is attributed to the initial reduction of another spinel solid solution phase, Fe_2TiO_4 (a component of TTM), as well as residual $FeTiO_3$ from TTH. In this process, additional metallic iron is generated through the reduction of Fe_2TiO_4 in TTM.

15 min: At 15 min, the FeO peak completely disappears, indicating the full conversion of Fe_3O_4 (including the intermediate FeO) in TTM to metallic iron. The corresponding reduction degree of 94.12% confirms that the majority of iron oxides have been reduced by this point.

20 min: Between 15 and 20 minutes, no new phases are detected, though the reduction degree increases further to 96.97%.

25 min: Subpseudobrookite ($FeTi_2O_5$), an intermediate phase in the reduction pathway from $FeTiO_3$ to Fe and TiO_2, does not appear until 25 min into the reduction. This suggests that the $FeTiO_3$ formed between 5 and 20 min begins to gradually reduce to $FeTi_2O_5$ only after accumulating to a detectable level.

30 min: Between 25 and 30 min, the phase composition remains unchanged. However, the TTM content continues to decrease, as evidenced by the disappearance of its characteristic diffraction peaks at $2\theta = 53.4°$, $56.6°$, and $62.6°$.

40 min: In the 40–50 min interval, low-valence titanium oxides begin to emerge. Ti_3O_5 is first detected at 40 min, followed by Ti_2O_3 at 50 min. Although TiO_2 had already appeared as early as 5 min into the reduction, low-valence titanium oxides are delayed in their formation. This delay may be attributed to the presence of iron oxides early in the reaction, which reduces the overall reduction potential for titanium. When the reduction degree reaches 99.19% at 40 min, the diminishing concentration of iron oxides allows for the reduction of titanium oxides, resulting in the appearance of lower-valent titanium species. Nevertheless, complete reduction has not yet been achieved at this stage, as minor amounts of Fe_2TiO_4, $FeTiO_3$, and $FeTi_2O_5$ persist.

50 min: By 50 min, $FeTiO_3$ and $FeTi_2O_5$ are fully reduced to TiO_2 and metallic Fe, and their diffraction peaks disappear. Additionally, the primary diffraction peak of TTM at $2\theta = 35.3°$ vanishes entirely, indicating the complete reduction of Fe_2TiO_4 in the TTM phase.

After 50 min of reduction under a 100 vol% hydrogen atmosphere, the final physical phases of the ironsands consist of metallic iron, titanium oxides in three different

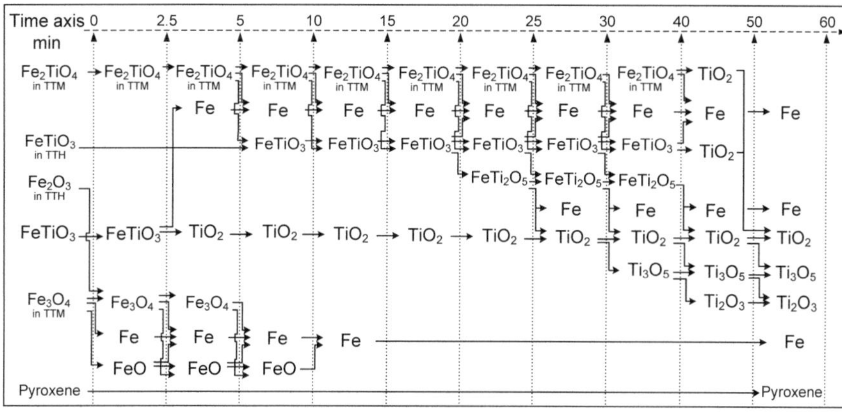

Fig. 8.8 Physical phase transformation of ironsands after 60 min of reduction at 900 °C under 100 vol% hydrogen

valence states, and chain pyroxenes. Based on the preceding analysis, the evolution of the physical phases throughout the gas-based reduction process is illustrated in Fig. 8.8.

The phase transitions and solid solubility variations of titanomagnetite (TTM) and titanohematite (TTH) during the reduction process are depicted in Fig. 8.9. The corresponding chemical reactions for the solid solution components in TTM and TTH are expressed by Eqs. 8.3 and 8.4, respectively.

$$\mathrm{TTM}(\mathrm{Fe}_{3-x}\mathrm{Ti}_x\mathrm{O}_4)\begin{cases}(1-x)\mathrm{Fe}_3\mathrm{O}_4:(1-x)\mathrm{Fe}_3\mathrm{O}_4+(4-4x)\mathrm{H}_2\\ \quad=(3-3x)\mathrm{Fe}+(4-4x)\mathrm{H}_2\mathrm{O}\\ x\mathrm{Fe}_2\mathrm{TiO}_4:\mathrm{Fe}_{2x}\mathrm{Ti}_x\mathrm{O}_{4x}+\delta\mathrm{H}_2\\ \quad=\mathrm{Fe}_{2x-\delta}\mathrm{Ti}_x\mathrm{O}_{4x-\delta}+\delta\mathrm{Fe}+\delta\mathrm{H}_2\mathrm{O}(0<\delta<2x)\end{cases} \tag{8.3}$$

$$\mathrm{TTH}\left(\mathrm{Fe}_{2-y}\mathrm{Ti}_y\mathrm{O}_3\right)\begin{cases}(1-y)\mathrm{Fe}_2\mathrm{O}_3:(1-y)\mathrm{Fe}_2\mathrm{O}_3+(3-3y)\mathrm{H}_2\\ \quad=(2-2y)\mathrm{Fe}+(3-3y)\mathrm{H}_2\mathrm{O}\\ y\mathrm{FeTiO}_3:\mathrm{Fe}_y\mathrm{Ti}_y\mathrm{O}_{3y}+\varepsilon\mathrm{H}_2\\ \quad=\mathrm{Fe}_{y-\varepsilon}\mathrm{Ti}_y\mathrm{O}_{3y-\varepsilon}+\varepsilon\mathrm{Fe}+\varepsilon\mathrm{H}_2\mathrm{O}(0<\varepsilon<y)\end{cases} \tag{8.4}$$

where:

x solid solubility of the Fe_3O_4-Fe_2TiO_4 system;
y solid solubility of the Fe_2O_3-$FeTiO_3$ system;
δ degree of reduction of Fe_2TiO_4;
ε degree of reduction of $FeTiO_3$.

According to the atomic proportions of TTM and TTH in the raw ironsands (as shown in Table 8.2), the initial solid solubility values for TTM are $x = 0.26$ and 0.12 (corresponding to homogeneous and heterogeneous particles, respectively), and for

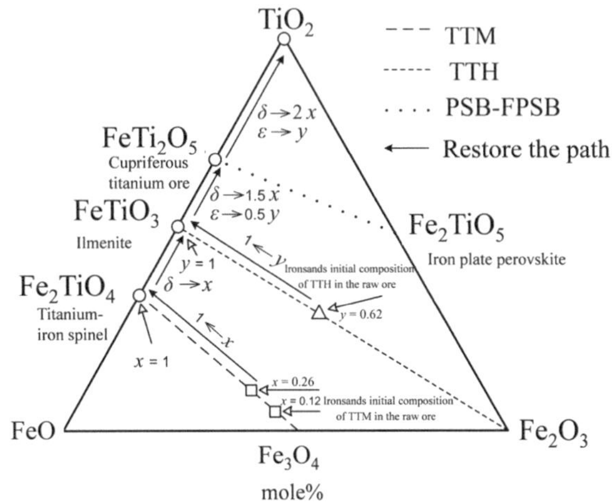

Fig. 8.9 FeO-Fe_2O_3-TiO_2 ternary system: phase evolution and solubility changes of titanium and iron during ironsands reduction (TTM and TTH phases)

TTH, $y = 0.62$. These initial values are marked with square and triangular symbols on the Fe_3O_4–Fe_2TiO_4 and Fe_2O_3–$FeTiO_3$ solid solubility curves in Fig. 8.9.

As the reduction proceeds, the Fe_3O_4 and Fe_2O_3 contents within TTM and TTH decrease, shifting the compositional points progressively toward the FeO–TiO_2 solid solution line. Concurrently, the values of x and y in Eqs. 8.3 and 8.4 increase gradually and ultimately approach 1, indicating the complete transformation of Fe_3O_4 and Fe_2O_3.

During this process, the reduction of Fe_2TiO_4 to $FeTiO_3$ in TTM causes δ to approach x. Further reduction of $FeTiO_3$ (originating from both TTH and Fe_2TiO_4) to $FeTi_2O_5$ leads δ and ε to gradually approach $1.5x$ and $0.5y$, respectively. When δ and ε reach $2x$ and y, the iron oxides are fully reduced, thereby completing the phase separation between iron and titanium oxides.

8.3 Micromorphological Evolution During Gas-Based Reduction

In this section, SEM and EDS were employed to analyze the morphological and compositional evolution of ironsands laminae and homogeneous regions during the reduction process. Under identical external reduction conditions, three distinct types of ironsands particles exhibiting different reduction mechanisms and pathways were identified, as shown in Figs. 8.10, 8.11, and 8.12.

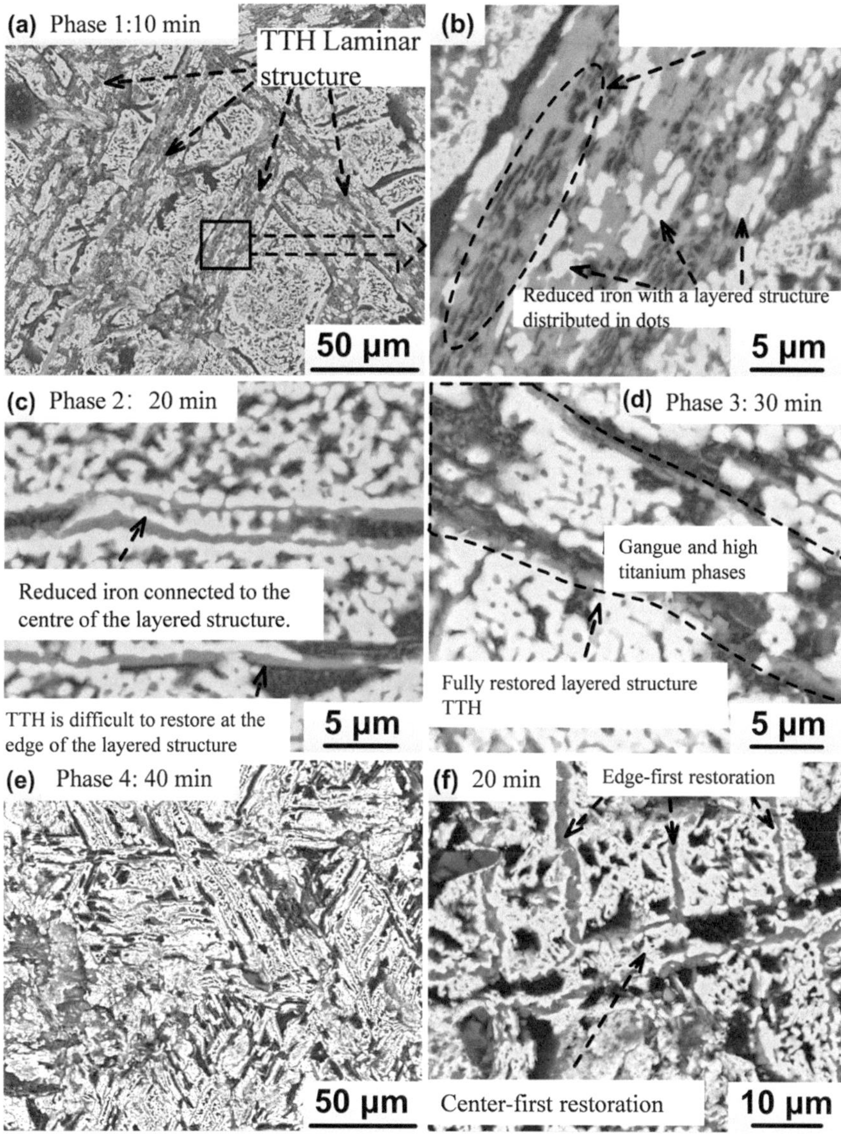

Fig. 8.10 Microstructural evolution of non-homogeneous ironsands particles (with central lamellar TTH precipitation) at various stages of reduction at 900 °C in 100 vol% H_2

8.3.1 Nonhomogeneous Particles of Ironsands

Figures 8.10a–f illustrate the changes in the microscopic morphology of the laminar structures of ironsands particles at various stages of reduction at 900 °C. As shown

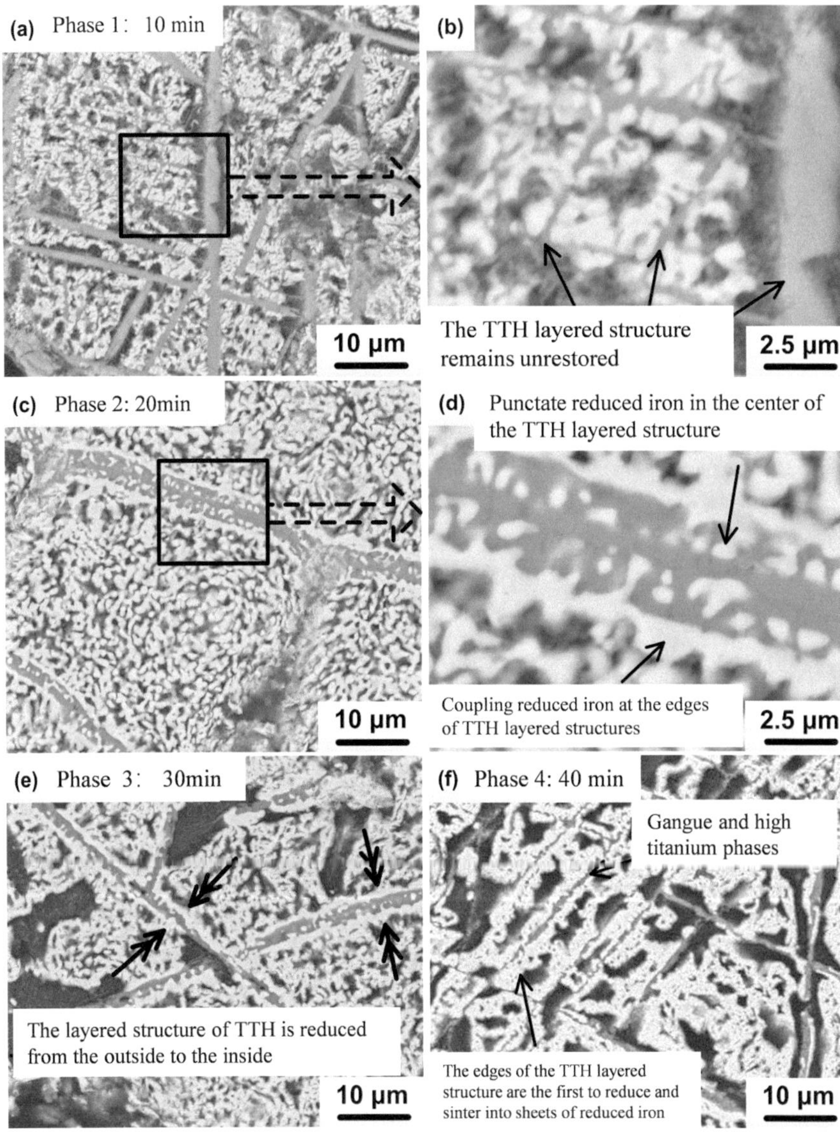

Fig. 8.11 Microstructural evolution of non-homogeneous ironsands particles (with edge-initiated TTH precipitation) during reduction at 900 °C under 100 vol% H_2

in Fig. 8.10a, while iron whiskers formed in the basal regions (TTM) of the particles—and in some cases, sintered metallic iron aggregates were observed—only sparsely distributed metallic iron dots appeared within the laminar (TTH) regions. This suggests that the laminar TTH phase exhibits comparatively lower reducibility. Furthermore, subcircular and needle-like pores, primarily concentrated in the central

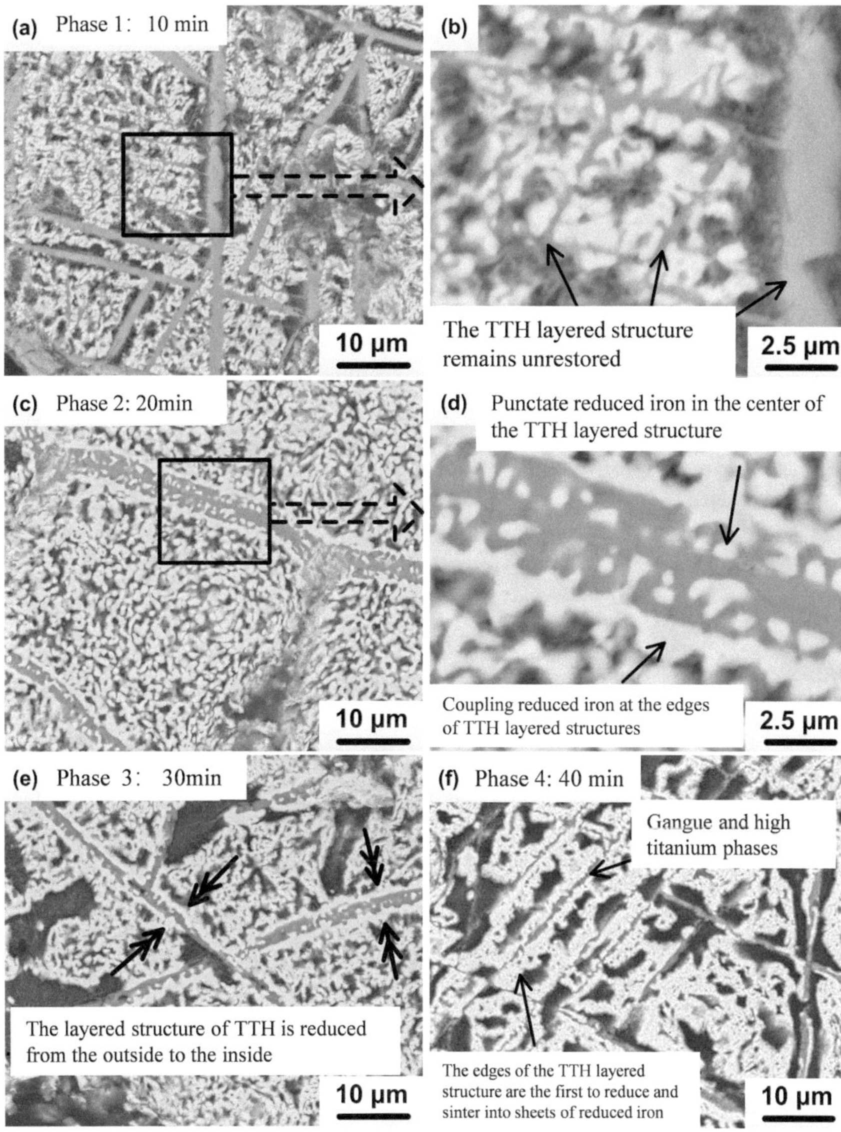

Fig. 8.12 Microstructure of homogeneous ironsands particles at 5 min of reduction under 900 °C and 100 vol% H_2, showing reduction progression from surface inward along dark phase bands

regions of the laminae, were observed in conjunction with the dotted metallic iron, as shown in Fig. 8.10b.

During the second and third reduction stages, the point-like metallic iron and associated pores gradually coalesced into lamellar metallic iron structures and slit-like voids (Fig. 8.10c, d), indicating a center-outward reduction pathway within the laminae. Notably, even after 40 min of reduction, when the heterogeneous particles were nearly fully reduced, the original lamellar architecture and the phase boundary at the basal site remained discernible (Fig. 8.10e).

At the 20-min mark, a second reduction pattern of the laminar structure emerged, as illustrated in Fig. 8.10f. While the first pathway initiates reduction from the laminar center (as previously described), this newly observed pathway begins at the laminar edge, which is explored in detail in Figs. 8.11a–f.

Figure 8.11a, b demonstrate that the laminar structures within heterogeneous particles exhibit greater reduction inertia compared to the basal regions. Despite the presence of substantial metallic iron aggregates in the basal area within 10 min, the laminar regions remain largely unreduced. It is only during the second reduction stage (10–20 min) that dot-like metallic iron begins to appear at the laminar center (Fig. 8.11c, d); meanwhile, at the laminar edges, the formation of associated metallic iron is already evident. This suggests a preferential reduction from the laminar edge toward the center—opposite to the center-outward pattern observed in Fig. 8.10.

In the third reduction stage (20–30 min), the TTH laminae clearly exhibit an outward-to-inward reduction sequence (Fig. 8.11e). By the fourth stage (after 40 min), the TTH laminar structures evolve into a morphology characterized by veinstone formations and high-titanium phases, with the edges enveloped by lamellar metallic iron aggregates, as shown in Fig. 8.11f.

8.3.2 Homogeneous Particles of Ironsands

Figure 8.12 characterizes the reduction behavior of a homogeneous TTM particle, which exhibits a smooth surface morphology. It presents the microstructural features and compositional distributions at different sites after reduction at 900 °C for 5 min under a 100 vol% hydrogen atmosphere. On the surface of the homogeneous TTM particle, four distinct regions—labeled A, B, C, and D in Fig. 8.12b—were analyzed using EDS.

As Fe, Ti, Mg, Al, Si, and O account for over 99 wt% of the total composition of the ironsands, the atomic percentages of these six major elements were normalized to a total of 100 at% to facilitate consistent comparative analysis among the four regions. The results of the EDS and normalization calculations are summarized in Table 8.3. In this analysis, oxygen required to stabilize chalcopyrite-forming elements was first subtracted, and the remaining oxygen was attributed to FeO. The metallic iron content was then determined by subtracting the iron in FeO from the total iron. The metallization rate was calculated as the mass fraction of metallic iron divided by the total iron mass fraction

The reduction of homogenized TTM pellets proceeds in an outward-to-inward manner, with a distinct boundary separating the reduced and unreduced regions. Within the reduced region, the brightly colored areas (labeled A) primarily consist of metallic iron and exhibit a metallization rate approaching 100 wt%. However, some partially reduced zones (labeled B) remain, with a slightly lower metallization rate of

Table 8.3 Atomic percent content in four regions of homogeneous TTM particles reduced for 5 min at 900 °C under 100 vol% hydrogen atmosphere

Shore	Relative atomic weight/at%						Metallization rate/wt%
	Fe	Ti	O	Si	Mg	Al	(metallic iron/all iron) wt%
A	77.76	2.97	13.87	1.51	1.90	1.98	99.96
B	58.39	6.34	26.18	1.55	3.20	4.33	98.80
C	52.65	2.39	39.86	0.99	2.49	1.61	46.45
D	46.68	4.58	40.63	1.15	2.83	4.12	56.85

approximately 98.80 wt%. Notably, region B contains significantly higher concentrations of magnesium, silicon, and especially aluminum and titanium compared to region A.

In the unreduced portion, two distinct microstructural regions exhibiting different reduction characteristics are observed. Bright gray, streak-like areas (labeled C) appear alternately with darker gray matrix areas (labeled D). The compositional differences between regions C and D closely mirror those between regions A and B, with variations primarily in the concentration of chalcopyrite-associated elements.

At the interface between the reduced and unreduced zones, the striated bright gray region C in the unreduced portion is directly connected to the metallic region A in the reduced portion. This spatial correlation suggests that reduction preferentially propagates along region C, facilitating the selective formation of metallic iron in region A. This behavior is attributed to the heterogeneous distribution of gangue-forming elements such as Mg, Si, Al, and Ti. During the early formation of ironsands, rapid cooling of volcanic lava resulted in a non-uniform distribution of these elements, impeding their homogenization.

Consequently, areas enriched in chalcopyrite-forming phases—such as magnesium–aluminum spinel, iron spinel, and titanium–iron spinel—are more resistant to reduction, while areas deficient in these phases are more readily reduced. This elemental heterogeneity leads to preferential, uneven reduction behavior and results in the alternating, streak-like distribution of regions A, B, C, and D throughout the reduction process.

8.4 Comparison of Raw and Pre-oxidized Ironsands During Gas-Based Reduction

Based on the process parameters of the well-established Middlesex method for gas-based reduction [21, 22], the experimental conditions were selected within a temperature range of 800–1000 °C and a hydrogen volume fraction (H_2 / (H_2 + Ar) × 100 vol%) between 10–100 vol%, as detailed in Table 8.4. According to Dalton's law of partial pressures, for both ideal gases and real gases under atmospheric pressure, the partial pressure of a gas component in a mixture is proportional to its

Table 8.4 Experimental conditions for gas-based reduction of raw and pre-oxidized ironsands

Hydrogen partial pressure conditions		Reducing temperature	
T/°C	P_{H2}/P_{sum} (H_2 vol%)	P_{H2}/P_{sum}	T/°C
900	0.1 (10 vol%) 0.2 (20 vol%) 0.3 (30 vol%) 0.4 (40 vol%) 0.5 (50 vol%) 1.0 (100 vol%)	0.4	800 850 900 950 1000

volume fraction. Accordingly, the hydrogen volume fractions were converted into partial pressures for subsequent calculations, as shown in Table 8.4.

The conversion rate α, which varies with time, is a key indicator of the extent of reaction progress [23]. In this study, α was determined based on the mass loss of the sample during reduction, as expressed in Eq. 8.5:

$$\alpha = \frac{m_{Ironsands}|_{t=0} - m_{Ironsands}|_{t=t}}{m_{\text{Theoretical Oxygen Loss}}|_{t=0}} \tag{8.5}$$

where:

$m_{ironsands}|_{t=0}$ the initial mass of the ironsands sample before the reaction, g;

$m_{ironsands}|_{t=t}$ the mass at time t during the reduction, g;

$m_{ironsands}|_{t=0}-m_{ironsands}|_{t=t}$ the actual oxygen loss, g;

$m_{Theoretical\ Oxygen\ Loss}|_{t=0}$ the theoretical maximum oxygen loss upon complete reduction, g.

It is important to note that $m_{theoretical\ oxygen\ loss}|_{t=0}$ differs between raw and pre-oxidized ironsands. During pre-oxidation, additional oxygen enters the ironsands, converting magnetite phases into hematite, thereby increasing the oxygen content. Consequently, the theoretical oxygen loss varies depending on the pre-oxidation temperature and duration. As shown in Fig. 8.13 and discussed in Chap. 7, the mass gain of ironsands during pre-oxidation at 900 °C is approximately 1.74 wt%.

The phase transformation pathway during pre-oxidation follows the sequence with a corresponding gradual increase in theoretical oxygen loss: TTM ($Fe_{3-x}Ti_xO_4$) → TTH ($Fe_{2-y}Ti_yO_3$) → PSB ($Fe_{2-z}Ti_zO_5$). Considering an initial theoretical oxygen loss of 20.55 wt% for raw ironsands, the adjusted value for the pre-oxidized sample at 900 °C is calculated as: (20.55 wt% + 1.74 wt%)/(100 wt% + 1.74 wt%) × 100% = 21.9 wt%.

By applying Eq. 8.5 and measuring the mass loss of both raw and pre-oxidized samples, the respective conversion curves under different reduction temperatures and hydrogen partial pressures are plotted in Fig. 8.14.

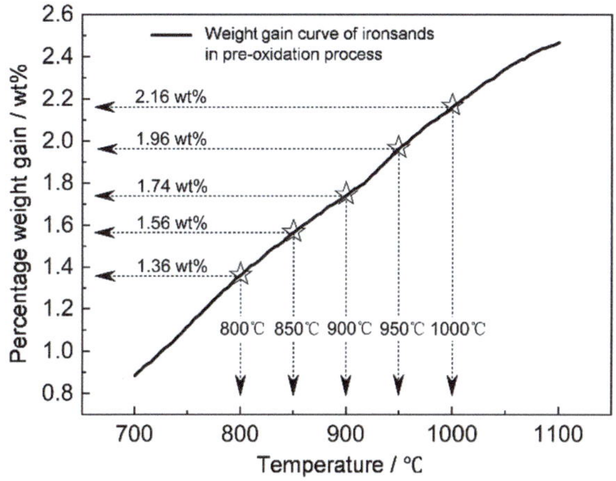

Fig. 8.13 Weight gain percentage of ironsands during the pre-oxidation process

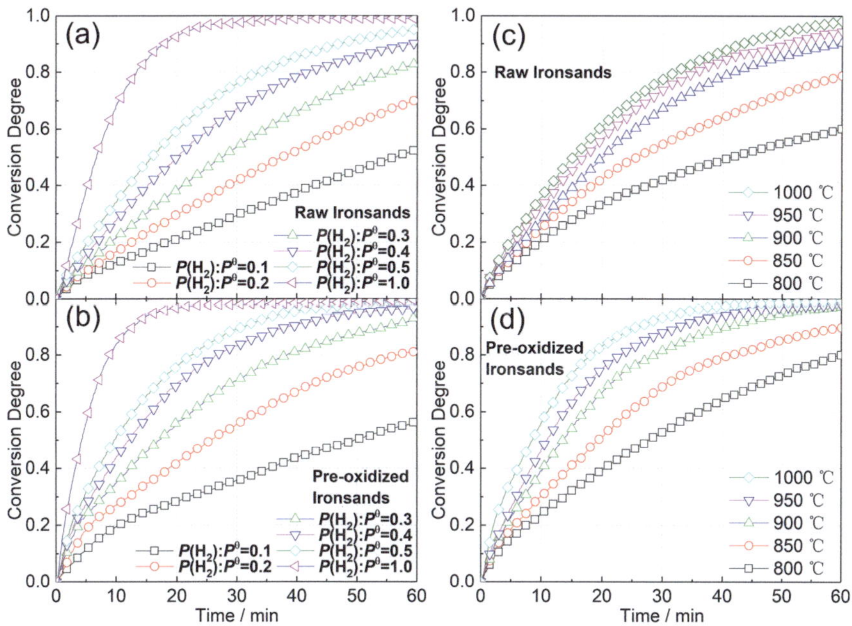

Fig. 8.14 Temperature-dependent conversion rates of raw and pre-oxidized ironsands under different atmospheres: **a**, **b** Conversion versus hydrogen partial pressure; **c**, **d** Conversion versus reduction temperature

In both raw and pre-oxidized ironsands, the increase in conversion per 0.1 bar increment of hydrogen partial pressure diminishes progressively within the 0.1–0.5 bar range. Similarly, in the 800–1000 °C temperature interval, the incremental increase in conversion per 50 °C also decreases with temperature. In particular, when the hydrogen partial pressure exceeds 0.4 bar or the temperature surpasses 900 °C, the enhancing effect becomes significantly less pronounced.

By comparing Fig. 8.14a, b, as well as Fig. 8.14c, d, it is evident that the conversion rate of pre-oxidized ironsands is, on average, approximately 32% higher than that of raw ironsands under identical reduction conditions and durations. This enhancement is primarily attributed to the phase transformation induced by pre-oxidation, which promotes the separation of titanium from the Fe–Ti–O solid solution. This phase separation reduces the inhibitory effect of titanium on the reduction and diffusion of iron and oxygen, thereby facilitating a more efficient reduction process.

8.5 Kinetic Analysis of Pre-oxidation-Promoted Ironsands in Gas-Based Reduction

8.5.1 Hydrogen Partial Pressure and Reaction Level Fitting

In the field of reaction kinetics, the rate of change of the conversion degree α with respect to time (dα/dt) serves as a fundamental descriptor of the reaction process. This rate typically depends on three key parameters: the reaction temperature T, the conversion degree α, and the partial pressure P of the gaseous reactants, as expressed in Eq. 8.6:

$$\frac{d\alpha}{dt} = k(T) \times f(\alpha) \times h(P) \tag{8.6}$$

In Eq. 8.6, the partial pressure factor h(P) of the reactants is negligible under specific conditions, which include that the reactants and products do not include a gas phase, or that there is an excess of gaseous reactants, or that the gaseous products can be removed in a short time [24, 25]. In this study, the partial pressure factor h(P) is not negligible under the conditions of this experiment because the partial pressure of the hydrogen reductant varies over a wide range (0.1–1 bar) and the partial pressure of hydrogen will inevitably affect the reaction rate. The partial pressure factor [26, 27] can be expressed by Eq. 8.7.

Under certain conditions—such as when no gaseous reactants or products are involved, when gaseous reactants are present in large excess, or when gaseous products are rapidly removed—the influence of partial pressure, represented by the pressure function *h(P)*, can often be neglected [24, 25]. However, in the present study, the partial pressure function *h(P)* cannot be ignored, as the hydrogen partial pressure varies substantially (from 0.1 to 1 bar), which significantly impacts the reaction rate. The pressure dependency of the reaction rate can be described by the following empirical relation [26, 27]:

$$h(P) = P^n \tag{8.7}$$

where:

- P is the partial pressure of the gaseous reactant,
- n is the reaction order with respect to the gaseous reactant, reflecting the sensitivity of the reaction rate to changes in P.

Taking the natural logarithm of both Eqs. 8.6 and 8.7 leads to the following linearized form:

$$\ln(\frac{d\alpha}{dt}) = n \times \ln(P) + \text{constant} \tag{8.8}$$

To determine the reaction order n, the reaction rate (dα/dt) is calculated at specific values of the conversion degree α. In this study, four representative conversion levels were selected (α = 0.1, 0.2, 0.4, and 0.6) to account for different stages of the reduction process. Under identical reaction conditions (900 °C), the relationship between ln(dα/dt) and ln(P) was established for both raw and pre-oxidized ironsands. The results are plotted in Fig. 8.15, enabling the evaluation of the influence of hydrogen partial pressure on the reduction kinetics.

In Fig. 8.15, the slopes of the linear fits of ln(dα/dt) versus ln(P) represent the experimentally determined reaction order n. For the raw ironsands, the average value of n was 1.0002 (ranging from 0.98 to 1.02), with a correlation coefficient (R^2) of 0.9941. For the pre-oxidized ironsands, the average reaction order was 1.0086 (ranging from 1.00 to 1.01), with an R^2 of 0.9892.

Although the reaction order of the pre-oxidized ironsands is marginally higher than that of the raw sample by 0.0084, this difference falls within the margin of

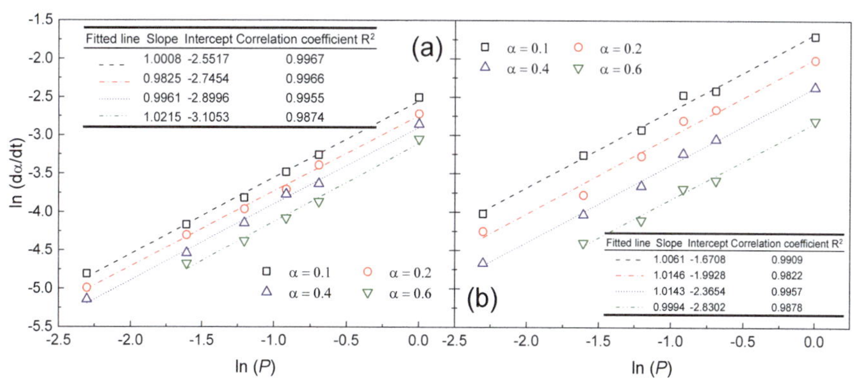

Fig. 8.15 Linear fitting of gas-phase reduction kinetics for: **a** raw ironsands; **b** pre-oxidized ironsands

experimental error. Therefore, it cannot be conclusively stated that the partial pressure of hydrogen exerts a greater influence on the reaction rate of pre-oxidized ironsands compared to raw ironsands.

Nevertheless, since the reaction order remains close to $n = 1$ for both materials over the primary reaction stage ($\alpha = 0.1$–0.6), this value can be reasonably and reliably applied in subsequent kinetic calculations.

8.5.2 Conversion Rate and Apparent Reaction Rate Constants

By integrating Eq. 8.6 under isothermal conditions, Eq. 8.9 is obtained:

$$\int_0^{\alpha} \frac{d\alpha}{f(\alpha)} = \int_0^{t} k(T) \cdot h(P) \cdot dt \tag{8.9}$$

To facilitate the calculation, the integral form $G(\alpha)$ of the conversion-dependent function $f(\alpha)$ is introduced, with its defining equation given in Eq. 8.10:

$$G(\alpha) = \int_0^{\alpha} \frac{d\alpha}{f(\alpha)} \tag{8.10}$$

Based on different assumptions regarding reaction mechanisms, several researchers [26–28] have proposed various kinetic models. However, preliminary analysis of the experimental data reveals that certain models deviate significantly from the measured values due to incompatibility with the actual reaction mechanism. Among all the models examined, seven show excellent agreement with the experimental data. The functional forms of their $f(\alpha)$ and $G(\alpha)$ expressions are summarized in Table 8.5.

In Table 8.5, solid-state reaction kinetic models are classified into four categories based on their underlying reaction mechanisms and process assumptions:

(1) Diffusion-controlled models—Models 1, 2, and 3;
(2) Nucleation and growth—Model 4;
(3) Chemical reaction control—Model 5;
(4) Phase boundary (interface) movement—Models 6 and 7.

In gas–solid reactions, gas diffusion occurs between the product nuclei and the unreacted substrate. When diffusion is hindered or its rate becomes lower than that of other rate-controlling processes, diffusion becomes the dominant rate-controlling step. Diffusion models 1, 2, and 3 are based on idealized assumptions regarding gas diffusion through planar, cylindrical, and spherical solid geometries, respectively.

At the onset of the reaction, product phases form as nuclei at active sites, and this nucleation process is influenced by external factors such as surface roughness,

Table 8.5 Reaction kinetics models

Serial number	Reaction model	Establishment of mechanism	$f(\alpha)$	$G(\alpha)$
1	One-dimensional diffusion	Proliferate	$1/2\alpha^{-1}$	α^2
2	Two-dimensional diffusion		$[-\ln(1-\alpha)]^{-1}$	$(1-\alpha)\ln(1-\alpha)+\alpha$
3	Three-dimensional diffusion		$3/2(1-\alpha)^{2/3}[1-(1-\alpha)^{1/3}]^{-1}$	$[1-(1-\alpha)^{1/3}]^2$
4	Power law	Nucleation and growth	$2/3\alpha^{-1/2}$	$\alpha^{3/2}$
5	Primary reaction	Chemical reaction	$1-\alpha$	$-\ln(1-\alpha)$
6	Spherical contraction	Phase transition	$3(1-\alpha)^{2/3}$	$1-(1-\alpha)^{1/3}$
7	Cylindrical contraction		$2(1-\alpha)^{1/2}$	$1-(1-\alpha)^{1/2}$

purity, and lattice defects of the reactants. Once the number of nuclei reaches a critical threshold, they begin to grow while consuming the reactant phase. The nucleation and growth process is affected by initial nucleus size, ion diffusion, and the supersaturation of the product phase. When these conditions deteriorate, nucleation and growth may become the rate-limiting steps. In this study, the nucleation and growth process is modeled using the power-law model (Model 4), which assumes a power-law dependence for nucleation and a constant rate for nucleus growth.

In some scenarios, the reaction rate is governed by the concentration or quantity of reactants, following a specific power-law dependence reflecting the number of reaction steps. Among the models considered, the chemical reaction control model with a single reaction order (Model 5) demonstrates strong agreement with experimental observations.

The geometrical contraction models (Models 6 and 7) assume that nucleation on the reactant surface occurs rapidly, such that the overall reaction is controlled by the movement of the interface between the reactant and product phases (i.e., phase boundary control). Depending on the assumed particle geometry, Models 6 and 7 describe three-dimensional spherical and two-dimensional cylindrical phase interface movement, respectively.

By combining Eqs. 8.7, 8.9, and Eqs. 8.10, 8.11 is derived:

$$G(\alpha) = k(T) \cdot P^n \cdot t \tag{8.11}$$

where t is the reaction time and $k(T)$ is the apparent rate constant, which is temperature-dependent. Based on the relationship between conversion α and time t, and considering the specific functional forms of $G(\alpha)$ listed in Table 8.5, fitting curves of $G(\alpha)$ versus t were generated at isothermal conditions (800, 850, 900, 950,

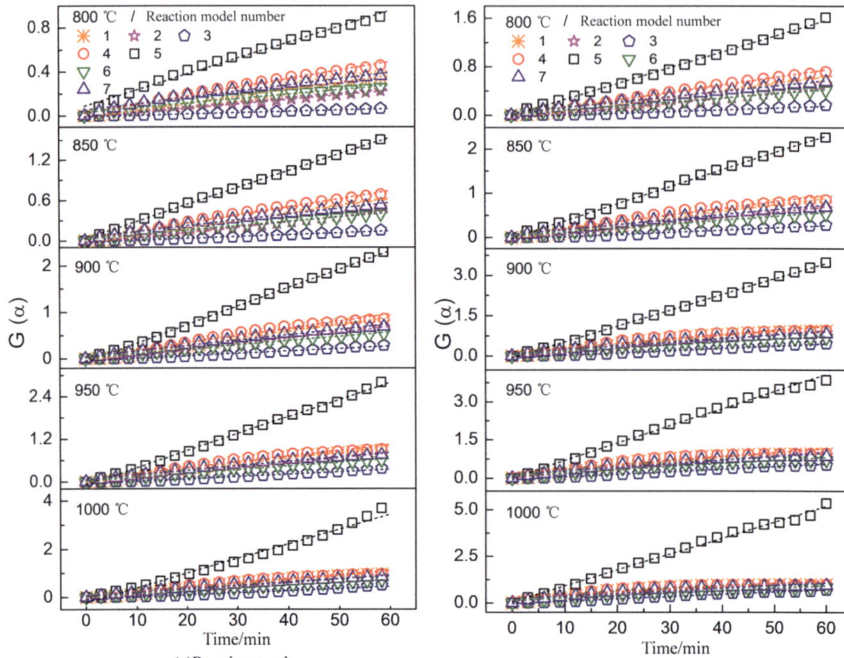

(a)Raw ironsands (b)Pre-oxidized ironsands

Fig. 8.16 Linear relationship between G(α) and time (t); slope used to determine the apparent reaction rate constant, *k(T)*

and 1000 °C), a hydrogen partial pressure of 0.4 bar, and a reaction order n = 1. The resulting curves are presented in Fig. 8.16, with the corresponding statistical data summarized in Table 8.6.

The goodness of fit between the linear regression model and the experimental data is typically evaluated using the coefficient of determination, R^2. As shown in Table 8.6, 84.3% of the R^2 values exceed 0.95, and 92.9% exceed 0.90. These high values indicate a strong correlation between the fitted and experimental data, thereby providing a reasonable degree of confidence and accuracy in the calculated apparent rate constants k. Furthermore, for each kinetic model, the apparent rate constant k generally increases with rising temperature, consistent with the expected thermally activated behavior of solid-state reactions. Notably, under identical temperature conditions, the rate constants for pre-oxidized ironsands are significantly higher than those for the raw ironsands, suggesting that the pre-oxidation treatment enhances the reducibility of the ironsands.

Table 8.6 Apparent reaction rate constants k(T) based on G(α)-t fitting results and corresponding statistical analysis

Model	Experimental materials	Raw ironsands				
	Temperature/°C	800	850	900	950	1000
1	Slope	0.0063	0.0112	0.0156	0.0168	0.0180
	R^2	0.9989	0.9982	0.9877	0.9785	0.9766
	k(*T*)/min^{-1}	0.0159	0.0279	0.0390	0.0421	0.0450
2	Slope	0.0041	0.0081	0.0128	0.0147	0.0168
	R^2	0.9975	0.9921	0.9909	0.9924	0.9950
	k(*T*)/min^{-1}	0.0103	0.0203	0.0319	0.0367	0.0419
3	Slope	0.0012	0.0028	0.0052	0.0067	0.0088
	R^2	0.9913	0.9716	0.9704	0.9736	0.9573
	k(*T*)/min^{-1}	0.0030	0.0070	0.0131	0.0168	0.0220
4	Slope	0.0078	0.0121	0.0156	0.0163	0.0170
	R^2	0.9894	0.9915	0.9749	0.9583	0.9530
	k(*T*)/min^{-1}	0.0195	0.0302	0.0391	0.0408	0.0426
5	Slope	0.0144	0.0251	0.0398	0.0479	0.0608
	R^2	0.9894	0.9993	0.9987	0.9974	0.9840
	k(*T*)/min^{-1}	0.0360	0.0627	0.0996	0.1197	0.1521
6	Slope	0.0041	0.0065	0.0092	0.0103	0.0118
	R^2	0.9799	0.9938	0.9940	0.9925	0.9981
	k(*T*)/min^{-1}	0.0102	0.0162	0.0231	0.0258	0.0296
7	Slope	0.0056	0.0086	0.0117	0.0126	0.0140
	R^2	0.9740	0.9876	0.9851	0.9801	0.9862
	k(*T*)/min^{-1}	0.0141	0.0215	0.0292	0.0316	0.0349
Pre-oxidized ironsands						
1	Slope	0.0116	0.0153	0.0174	0.0172	0.0163
	R^2	0.9940	0.9834	0.9456	0.8883	0.8223
	k(*T*)/min^{-1}	0.0288	0.0380	0.0434	0.0429	0.0405
2	Slope	0.0085	0.0125	0.0163	0.0171	0.0174
	R^2	0.9810	0.9911	0.9842	0.9491	0.9061
	k(*T*)/min^{-1}	0.0211	0.0309	0.0407	0.0425	0.0434
3	Slope	0.0029	0.0051	0.0086	0.0100	0.0124
	R^2	0.9538	0.9780	0.9873	0.9899	0.9866
	k(*T*)/min^{-1}	0.0073	0.0128	0.0214	0.0250	0.0310
4	Slope	0.0124	0.0152	0.0163	0.0158	0.0146
	R^2	0.9975	0.9682	0.9133	0.8494	0.7810
	k(*T*)/min^{-1}	0.0310	0.0378	0.0407	0.0394	0.0363
5	Slope	0.0261	0.0387	0.0586	0.0676	0.0855

(continued)

Table 8.6 (continued)

Model	Experimental materials	Raw ironsands				
	Temperature/°C	800	850	900	950	1000
	R^2	0.9974	0.9984	0.9992	0.9944	0.9939
	k(T)/min^{-1}	0.0649	0.0963	0.1459	0.1684	0.2131
6	Slope	0.0067	0.0089	0.0114	0.0121	0.0130
	R^2	0.9993	0.9900	0.9820	0.9546	0.9386
	k(T)/min^{-1}	0.0167	0.0223	0.0284	0.0302	0.0325
7	Slope	0.0089	0.0113	0.0134	0.0138	0.0139
	R^2	0.9968	0.9799	0.9592	0.9200	0.8894
	k(T)/min^{-1}	0.0221	0.0281	0.0334	0.0343	0.0347

8.5.3 Model Predictions of Apparent Activation Energy and Conversion Rate

The apparent reaction rate constant k(T) follows the Arrhenius equation, as expressed in Eq. 8.12:

$$k(T) = A\exp(-\frac{E}{RT}) \tag{8.12}$$

where:

A the pre-exponential factor, min^{-1};
E the apparent activation energy, kJ mol^{-1};
T the absolute temperature, K;
R the universal gas constant, 8.314 J mol^{-1} K^{-1}.

Applying the natural logarithm to the Arrhenius equation yields the linearized form shown in Eq. 8.13:

$$\ln k(T) = -\frac{E}{R}\cdot\frac{1}{T} + \ln A \tag{8.13}$$

Based on the data presented in Table 8.6, a linear regression was performed for lnk (T) versus 1/T. The apparent activation energy E was determined from the slope of the fitted line, while the pre-exponential factor A was derived from the intercept. The fitting process is illustrated in Fig. 8.17, and the corresponding results are summarized in Table 8.7.

As shown in Table 8.7, the various models are constructed based on different reaction mechanisms—such as diffusion, power law, or geometric contraction—leading to considerable variation in the resulting kinetic parameters. Notably, for each model, the kinetic parameters derived for the pre-oxidized ironsands are consistently lower than those for the raw ironsands. Since activation energy represents the minimum

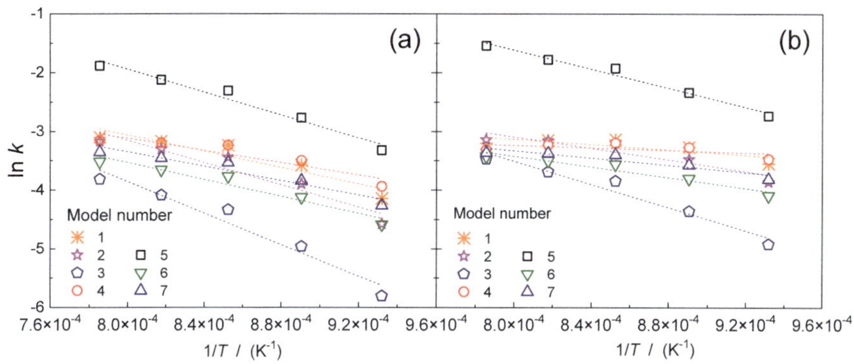

Fig. 8.17 Arrhenius plots used to determine activation energy and pre-exponential factor: **a** raw ironsands; **b** pre-oxidized ironsands

Table 8.7 Kinetic parameters of raw and pre-oxidized ironsands based on different reaction models

Model number	Raw ironsands		Pre-oxidized ironsands	
	E/kJ mol^{-1}	A/min^{-1}	E/kJ mol^{-1}	A/min^{-1}
Model 4	43.3	2.8	8.8	0.09
Model 7	50.7	4.6	25.6	0.42
Model 1	57.8	12.1	19.0	0.27
Model 6	59.8	9.2	37.7	1.2
Model 2	78.5	81.0	49.1	5.0
Model 5	80.9	346.3	67.2	127.6
Model 3	111.8	990.2	81.8	78.5

energy barrier that must be overcome for the reactants to be transformed into products, it can be inferred that pre-oxidation reduces this energy threshold, thereby accelerating the reduction and conversion processes of ironsands.

Table 8.7 also highlights the strong dependence of the kinetic parameter values on the choice of reaction model. In fact, within the framework of this kinetic approach (i.e., model fitting), factors such as the quality of the linear regression, the selection of kinetic data, and the model itself significantly influence the calculated values of the apparent activation energy E and the pre-exponential factor A. Therefore, due to inherent randomness and uncertainty in the fitting process, model-based kinetic analysis may not always yield accurate results. It remains essential to reintroduce the obtained kinetic parameters into the model to simulate conversion rates, compare them with experimental data, and thereby determine the most appropriate kinetic model and its corresponding parameters.

Using the number of reaction stages n and the apparent activation energy E obtained under various reaction model assumptions—along with the corresponding pre-exponential factor A—the predicted conversion rates of raw and pre-oxidized

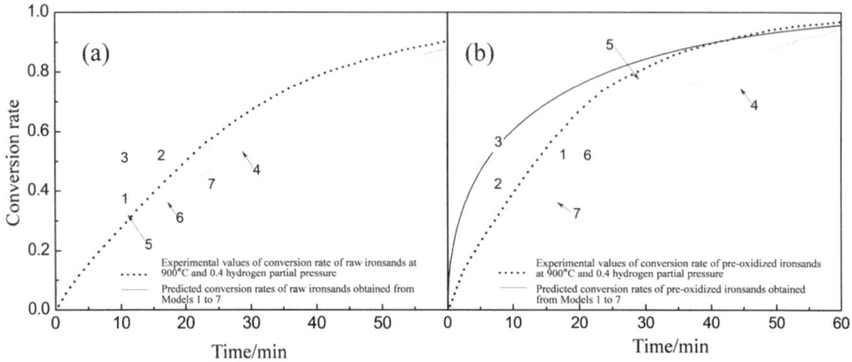

Fig. 8.18 Comparison between experimental and model-predicted conversion rates at 900 °C and 0.4 bar H_2 partial pressure: **a** raw ironsands; **b** pre-oxidized ironsands

ironsands were calculated using Eq. 8.14. These predicted values were then compared with the experimental data, as shown in Fig. 8.18.

$$G(\alpha) = A \cdot \exp(-\frac{E}{RT}) \cdot P^n \cdot t \tag{8.14}$$

As illustrated in Fig. 8.18, all seven models exhibit conversion trends that generally align with the experimental results. Among them, Model 5 and Model 3 show particularly strong agreement. Model 5 provides more accurate predictions during the early to mid-stages of the reaction (up to a conversion rate of approximately 0.5–0.6), while Model 3 more accurately captures the behavior in the later stages of the reaction. Despite these differences, the predicted conversion values across the models remain relatively close.

To quantitatively evaluate the fit of each model and provide a more robust basis for selecting the most appropriate reaction mechanism and kinetic parameters, two statistical metrics were introduced: the residual sum of squares and a comparison factor. The calculation methods for these indices are provided in Eqs. 8.15 and 8.16.

$$S_j^2 = \frac{1}{n}\sum_{i=1}^{n}(\alpha\Big|_j^{t_i} - \alpha\Big|_{\text{experiment}}^{t_i})^2 \tag{8.15}$$

$$F_j = \frac{S_j^2}{S_{\min}^2} \tag{8.16}$$

In the formula:

J Model Number;

I Data numbered from 1 to n;

$\alpha\Big|_j^{t_i}$ based on the predicted value of the conversion rate of model j at moment t_i;

Table 8.8 Residual sum of squares (S^2) and comparison factor (F) for conversion rate predictions using different reaction models

Model number	Raw ironsands		Pre-oxidized ironsands	
	S^2 (×1000)	F	S^2 (×1000)	F
1	3.33	5.26	8.14	19.03
2	5.86	9.25	4.59	10.72
3	14.63	23.11	12.51	29.25
4	7.71	12.17	29.46	68.90
5	0.63	1.00	0.43	1.00
6	6.49	10.24	13.76	32.18
7	13.03	20.58	29.46	68.89

$\alpha\big|_{\text{experiment}}^{t_i}$ The experimental value of the conversion rate at t_i moment;
S_j^2 Residual sum of squares of conversion rates based on model j;
$S_{\min}^2$ The smallest residual sum of squares in the model.

Therefore, the comparison of the correlation between the predicted and actual results of each model can be transformed into a comparison between the comparison factors, and the results are shown in Table 8.8.

The residual sum of squares S^2 and comparison factors F presented in Table 8.8 indicate that the first-order chemical reaction model (Model 5) provides the best fit for both raw and pre-oxidized ironsands across the entire range of reduction experiments. Using this model as the benchmark, the apparent activation energy E for raw ironsands is 80.9 kJ mol^{-1}, with a pre-exponential factor A of 346.3 min^{-1}. For pre-oxidized ironsands, E is reduced to 67.2 kJ·mol^{-1} and A to 127.6 min^{-1}. This represents an approximate 17% reduction in apparent activation energy following the pre-oxidation process.

As discussed previously, the pre-oxidation treatment lowers both the apparent activation energy and the pre-exponential factor during reduction, suggesting that the energy barriers associated with the transformation of titanium–iron solid solution oxides into metallic iron and titanium oxides are reduced [29]. Consequently, the reducibility of pre-oxidized ironsands is significantly enhanced compared to that of the raw ore.

8.5.4 Revalidation of Apparent Activation Energy by Equal Conversion Rate Method

The analyses in Sects. 8.5.2–8.5.3 are based on existing kinetic models, with relevant parameters derived by fitting experimental data to these models. However, real experimental processes may involve certain rate-controlling mechanisms not accounted for in currently established kinetic models. In this chapter, although Table 8.5 identifies the seven models with the highest correlation to the experimental data—selected from among all available reaction kinetics models—this selection does not guarantee that the most accurate or comprehensive model has been included. As Vyazovkin [26, 28] emphasized, even if the optimal kinetic model for a given reaction has not yet been developed or discovered, it is still possible to identify one or more models within the selected group that better describe and predict the reaction compared to others.

In this study, the first-order chemical reaction model (Model 5) was found to be the most suitable among the tested models, providing the best predictive performance. However, this conclusion is limited to a comparison among existing models. Consequently, while the model-fitting method can offer reasonable estimates of kinetic parameters and suggest likely reaction mechanisms, it cannot confirm whether a more accurate model exists beyond the current scope. Thus, the model-fitting approach identifies only the optimal model from the available set, not necessarily the true or most accurate one.

To reduce the influence of model selection on the calculated apparent activation energy E and pre-exponential factor A, the isoconversional method is applied in this section to re-evaluate the activation energy across the entire reaction process [30]. Taking the natural logarithm of Eq. 8.14 yields the core expression used in the isoconversional method, presented as Eq. 8.17:

$$\ln t_{\alpha} = \frac{E}{\mathrm{R}T} - n \ln P - \ln \frac{G(\alpha)}{A} \tag{8.17}$$

where tα denotes the time required to reach a specific conversion rate α. Assuming that the reaction model G(α) and the number of reaction stages n are independent of temperature, the apparent activation energy Eα can be obtained from the slope of the linear regression of $\ln(t_{\alpha})$ and 1/T at each given conversion rate α. Thus, by tracking changes in Eα as the reaction progresses, one can infer the kinetic mechanisms active at different reaction stages. Unlike the model-fitting method, which provides a single activation energy for the entire reaction, the iso-conversion rate method offers the advantage of yielding Eα values across the full reaction range. This advantage allows for a more detailed kinetic analysis. The results are presented in Fig. 8.19.

As shown in Fig. 8.19, for pre-oxidized ironsands, the apparent activation energy Eα remains relatively constant throughout the reduction process, fluctuating only slightly around the value of 67.20 $\mathrm{kJ{\cdot}mol^{-1}}$, as obtained from the first-order chemical reaction model (Model 5) in the range of $0.2 < \alpha < 0.6$. This indicates good agreement between the two methods. In contrast, for raw ironsands, Eα increases significantly with conversion, rising from approximately 49 $\mathrm{kJ{\cdot}mol^{-1}}$ at $\alpha = 0.2$ to 105 $\mathrm{kJ{\cdot}mol^{-1}}$ at $\alpha = 0.6$, whereas the model-fitting method gives an average activation energy of 80.94 $\mathrm{kJ\ mol^{-1}}$.

The apparent activation energy Eα of pre-oxidized ironsands exhibits minimal fluctuation across the experimental conversion range, indicating that the gas-based reduction process is primarily governed by a single kinetic model and mechanism—namely, a first-order chemical reaction.

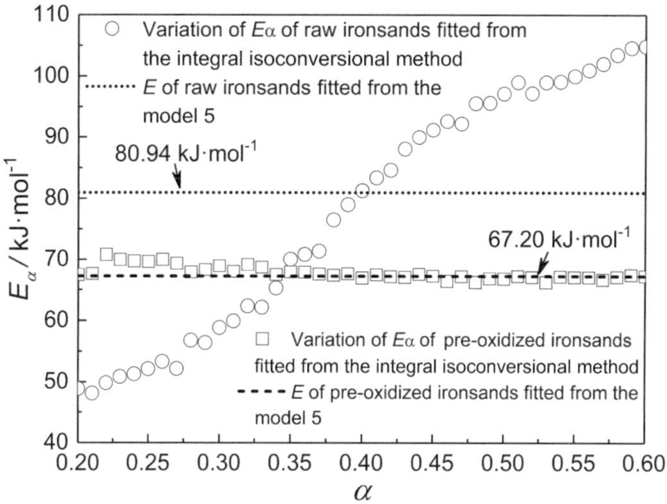

Fig. 8.19 Apparent activation energy ($E\alpha$) as a function of conversion rate (α) during reduction of raw and pre-oxidized ironsands; includes comparison with model-fitted results

In contrast, the Eα of raw ironsands increases progressively with the conversion rate. Notably, the activation energy of 80.94 kJ mol^{-1} obtained from the model-fitting approach closely matches the average of the activation energies over the reaction course. This suggests that multiple reaction models and mechanisms influence the hydrogen reduction process of raw ironsands. Consequently, a transition between different rate-controlling mechanisms occurs throughout the reaction. This implies that the activation energy observed at a given conversion rate α represents a superposition of the effects of several distinct mechanisms.

Combining these findings with results from the model-fitting method (Table 8.7; Fig. 8.18), it can be observed that Models 4 and 5 yield conversion predictions closely matching the experimental data during the early reaction stage ($\alpha \leq 0.2$). Their respective activation energies are 43.3 kJ·mol^{-1} and 80.9 kJ·mol^{-1}, respectively, which align well with the value obtained using the iso-conversion method (49 kJ·mol^{-1} at $\alpha \leq 0.2$). Therefore, it can be concluded that Model 4 (power-law nucleation) predominantly governs the reaction kinetics in this initial phase, with the process mainly controlled by the nucleation and growth of reduction products.

As the conversion rate increases to the 0.2–0.5 range, Eα rises from ~ 50 kJ·mol^{-1} ($\alpha = 0.2$) to ~ 80–90 kJ·mol^{-1} ($\alpha = 0.4$–0.5), gradually diverging from the behavior of Model 4 and approaching that of Model 5. During this stage, the reaction becomes increasingly governed by the first-order chemical reaction mechanism (Model 5). At higher conversion levels ($\alpha = 0.5$–0.6), Eα reaches approximately 105 kJ mol^{-1}, closely matching the activation energy of 111.8 kJ mol^{-1} predicted by Model 3, where three-dimensional diffusion becomes the dominant rate-controlling mechanism.

Thus, the rate-controlling mechanism of raw ironsands undergoes the following transitions during hydrogen reduction: Nucleation and growth of reaction products → First-order chemical reaction → Three-dimensional diffusion. In the early stage, the formation and crystal growth of metallic iron and titanium-iron oxides control the reaction rate. As the reaction progresses, the chemical reaction itself becomes the limiting step. Eventually, the gradual accumulation of the reaction products forms a dense product layer, which hinders the diffusion of gaseous reactants and products—making gas diffusion the dominant rate-controlling step in the later stage.

This sequential transition in rate-controlling mechanisms is not observed in the reduction of pre-oxidized ironsands. During pre-oxidation, the contraction of the main-phase crystal structure leads to the development of enlarged pores and micro-cracks (as detailed in Sect. 8.6). These structural changes provide ample nucleation sites and facilitate the formation of gas channels that enable efficient transport of gaseous reactants and products to and from the reaction interface. As a result, the limitations caused by nucleation and growth in the early stage and gas diffusion in the later stage—both of which are present in raw ironsands—are effectively eliminated in pre-oxidized ironsands.

In conclusion, hydrogen-based reduction of ironsands benefits significantly from pre-oxidation treatment, which lowers the apparent activation energy E—i.e., the energy barrier for converting titanium-iron solid solution into titanium oxides and metallic iron. For pre-oxidized ironsands, the entire reduction process is consistently governed by a chemical reaction mechanism. In contrast, raw ironsands undergo a complex, multi-stage kinetic evolution. Consequently, the reducibility of pre-oxidized ironsands is greatly enhanced compared to that of the untreated material.

8.6 Comparison of Raw and Pre-oxidized Ironsands in Microscopic Morphology During Gas-Based Reduction

Numerous studies have investigated the microscopic morphology and structural characteristics of ironsands. As early as 1964, Wright [31, 32] reported that the TTH (titanium–titanomagnetite–hematite) phase exists in a laminated structure within ironsands. More recently, Cruz-Sánchez [33] characterized the particle-level micromorphology and crystal structure of ironsands collected from the Rica coast. Additionally, the micromorphological changes during the reduction of ironsands using various reducing agents—such as carbon monoxide [10], hydrogen [17], methane [16, 34], and solid carbon [35]—have been examined to better understand the reduction mechanisms [18, 36].

However, owing to the structural complexity of titanium and iron oxides in ironsands, several microstructural features and their formation mechanisms during the reduction of both raw and pre-oxidized ironsands remain insufficiently reported. These include the relationship between microstructure and chemical composition, the migration and enrichment behavior of titanium, iron, and associated sulfides (e.g., chalcopyrite), and—most notably—the lack of a comprehensive summary describing the microstructural evolution of ironsands throughout the entire reduction process.

This section, therefore, aims to comparatively investigate the microstructural formation mechanisms of raw and pre-oxidized ironsands during reduction. Special emphasis is placed on the distinctive microstructures and reduction behaviors exhibited by pre-oxidized ironsands, which are largely attributed to the formation of microcracks during the pre-oxidation process. These features not only differentiate the reduction characteristics from those of raw ironsands but also serve as a focal point of this discussion.

8.6.1 Early Stage Reduction Morphology

Preferential appearance of dark grain-like structures prior to the formation of metallic iron phases

During the initial stage (1 min) of the ironsands reduction reaction, dark grain-like structures appear on a relatively homogeneous matrix (see area A in Fig. 8.20). These features emerge and develop rapidly, propagating from the particle edge toward the center before any metallic iron phase becomes detectable. As described in Sect. 8.3, the formation of these dark structures results from preferential reduction at specific sites (e.g., region B in Fig. 8.20). Importantly, their occurrence is not random but localized at regions with relatively low concentrations of chalcopyrite-associated elements. Thus, the heterogeneous distribution of chalcopyrite elements appears to significantly influence the formation and localization of these dark textures.

Furthermore, Sect. 8.3 confirms that metallic iron phases subsequently grow along the paths defined by these dark structures. This implies that metallic iron does not nucleate and grow arbitrarily within the mineral grains; rather, its nucleation sites are predetermined by the earlier-formed dark grain-like structures.

A three-layered nucleation structure derived from dark grain-like features

At the early stage of ironsands reduction (1 min), a distinct three-layered nucleated microstructure is observed, as shown in Fig. 8.21a. A subcircular central region, which retains its original composition and remains unreduced, is surrounded by a

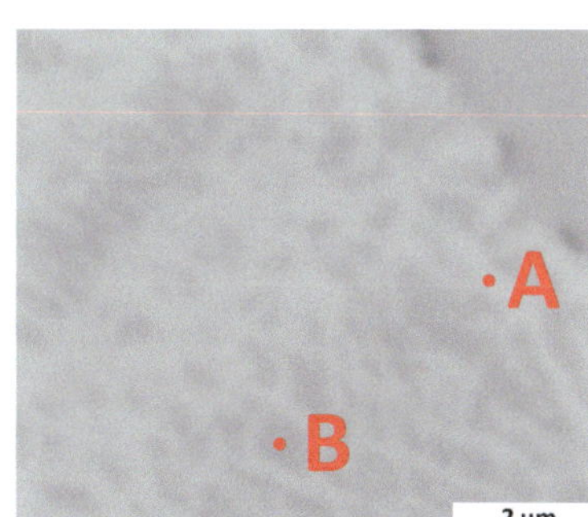

Fig. 8.20 Formation of dark granular structures at the base of ironsands particles during the early stage of reduction (1 min)

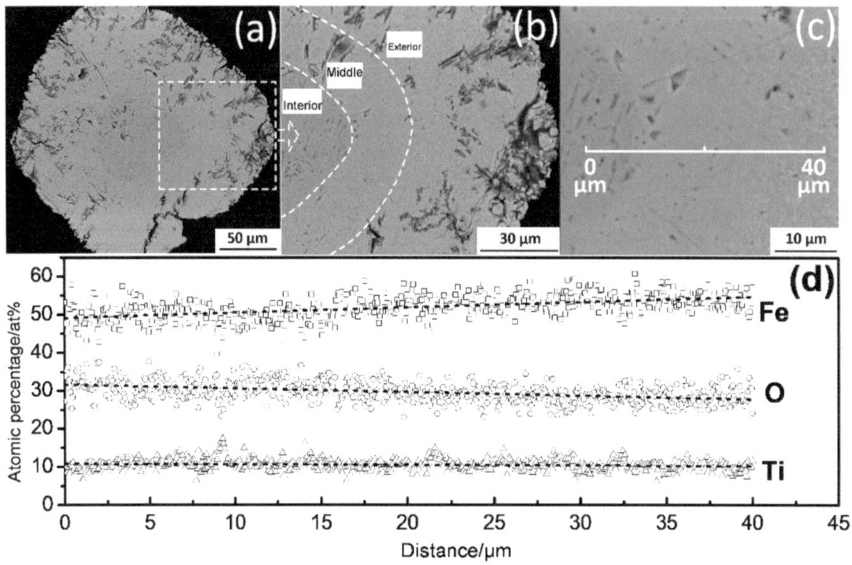

Fig. 8.21 Three-layered core structure observed at the onset of reduction: **a–c** progressive magnifications; **d** line scan showing atomic % of Fe, O, and Ti across image (**c**)

partially transformed matrix. To better visualize this structure, Fig. 8.21b provides a magnified view revealing three distinct layers: (1) an outer ring of preferentially reduced, dark stripe-like features; (2) an intermediate region representing a partially reduced matrix; and (3) a core that remains nearly unchanged from its original state.

To assess the compositional variation across these layers, EDS line scans were performed to quantify the atomic concentrations of Fe, O, and Ti, as shown in Fig. 8.21bc, d. From the center outward, the atomic fraction of iron increases slightly, while oxygen shows a corresponding decrease. Titanium content exhibits only minor fluctuations. These results indicate that reduction occurs progressively and selectively, not in a strictly layer-by-layer manner from the exterior inward. The early presence of dark grain-like structures at the edges of the particle, prior to metallic iron formation, demonstrates that the gas-phase reductant penetrates the interior and initiates reduction even at the central regions during the earliest stage.

8.6.2 *Intermediate-Stage Morphology*

Incompletely reduced reaction nuclei in raw ironsands

By the intermediate stage of reduction (5 min), raw and pre-oxidized ironsands exhibit markedly different microstructures. In raw ironsands, structures resembling incompletely reduced reaction nuclei are predominant, as illustrated in Fig. 8.22. These resemble the traditional unreacted core model; however, findings from Sect. 8.6.1 reveal that the core has, in fact, undergone partial reduction in the form of dark striated features. Therefore, the term incompletely reacted nucleus is more appropriate, reflecting that the reducing atmosphere has already altered the interior region.

Due to variation in initial composition among individual ironsand grains, different reduction microstructures develop externally. Figure 8.22a presents microstructures formed in Ti–rich particles, showing a peripheral reduced zone surrounding a central incompletely reacted nucleus. Figure 8.22b displays similar features for Fe-rich particles. Both samples represent homogeneous grains dominated by the TTM phase. In contrast, Fig. 8.22c shows a non-homogeneous grain (TTM–TTH dominant), where the TTH lamellar structure remains largely unreduced despite the adjacent basal phase transforming into titanium oxides and metallic iron. This highlights the lower reducibility of TTH lamellae relative to the basal TTM matrix.

To further investigate elemental distribution at the interface between the reduced outer zone and the incompletely reduced core, EDS elemental mapping was conducted (Fig. 8.22d). Fe is relatively evenly distributed in the core but becomes

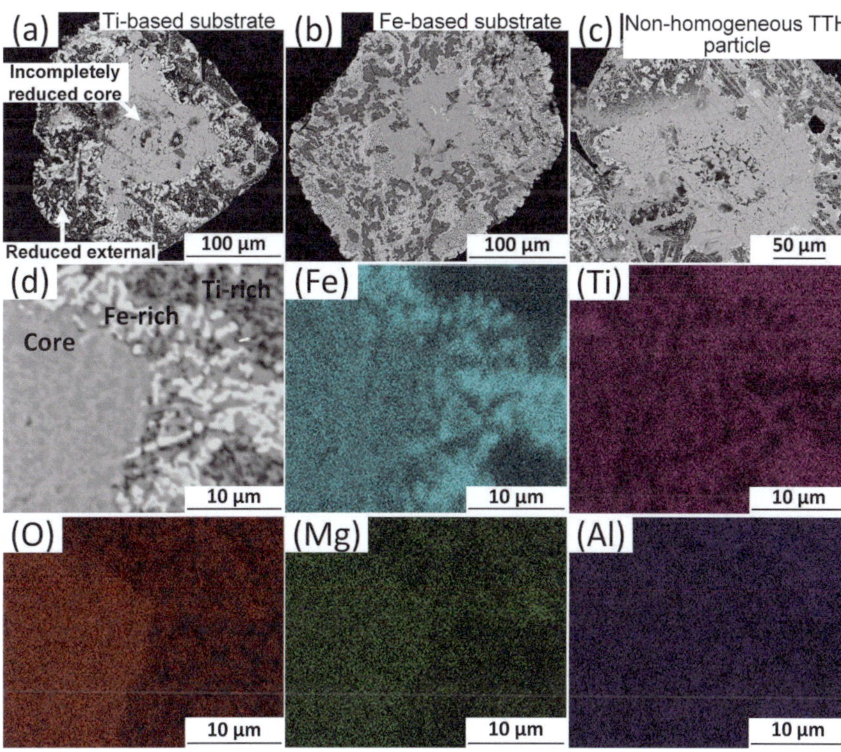

Fig. 8.22 Incomplete reduction nuclei observed on various substrate types: **a** high-Ti substrate; **b** high-Fe substrate; **c** TTM–TTH mixed substrate; **d** EDS mapping at nucleus–substrate interface

enriched in the reduced outer region. A similar pattern is observed for Ti, which transitions from a dotted, uniform distribution in the core to localized enrichment outside. This suggests both Ti and Fe undergo spatial migration and concentration during reduction. Concurrently, oxygen levels decrease significantly outside the core due to deoxygenation, where it predominantly binds with Ti and chalcopyrite-associated elements. Mg and Al are mainly enriched in the Ti–rich outer zone, likely due to their affinity for titanium oxides.

Disordered reduction structures in pre-oxidized ironsands

Unlike raw ironsands, which undergo outward reduction from a central incompletely reacted core, pre-oxidized ironsands exhibit a disordered reduction pattern characterized by multiple incompletely reduced nuclei distributed randomly at both particle edges and centers (Fig. 8.23a). As described in Sect. 7.3, the pre-oxidation process induces unit cell contraction in the main mineral phases—for instance, cubic titanomagnetite contracts along the *a*-axis, and orthorhombic titanomagnetite along both *a*- and *c*-axes. This contraction leads to an expansion of pore structures and the formation of microcracks (discussed in detail in Sect. 8.6.5), disrupting the original compact microstructure and facilitating deep penetration of the gas-phase reductant into the particles. Consequently, a disordered distribution of incompletely reduced reaction nuclei emerges, initiating reduction at multiple locations simultaneously.

In this disordered reduction pattern of pre-oxidized ironsands, reduced regions predominantly exhibit streak-like enrichment structures. Figure 8.23b, d represent high-Ti grains characterized by dark, Ti–rich streaks, whereas Fig. 8.23c, e show Fe-rich grains featuring bright, Fe-enriched streaks.

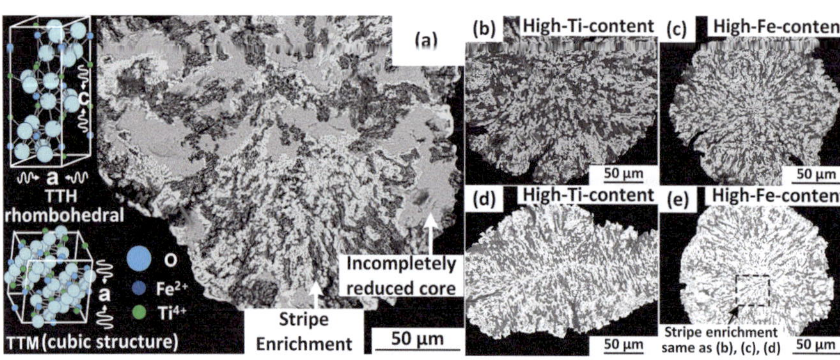

Fig. 8.23 **a** Microcracks induced by pre-oxidation result in disordered reduction pathways and random nucleation of reduced zones, which later cause streak-like Fe/Ti enrichment; **b**, **d** High-Ti substrates; **c**, **e** High-Fe substrates

8.6.3 Late-Stage Morphology

Titanium and iron enrichment in block structures

During the intermediate stage of reduction in raw ironsands, as shown in Fig. 8.22, the peripheral regions of the mineral particles exhibit block-like enrichment structures. As the reduction progresses to a later stage, illustrated in Fig. 8.24a, titanium and iron become enriched in distinct dark and light blocky regions, respectively. Thus, the structure in Fig. 8.24a can be interpreted as an evolved form of the microstructure observed in Fig. 8.22.

Local magnification images (Fig. 8.24b, c) reveal that the dark, titanium-rich blocks are not completely solidified by high-titanium phases. Instead, they consist of embedded phases interspersed with numerous cavities. These cavities are primarily attributed to the loss of oxygen during reduction and the formation of vacancies caused by elemental migration and enrichment.

To compare the composition of the iron-rich and titanium-rich block structures, EDS analysis was conducted on zones A and B in Fig. 8.24. The results are presented in Fig. 8.25. In zone A, iron is the dominant element, with other constituents each contributing less than 7 wt%. In contrast, the embedded phases in the black, titanium-rich region of zone B show titanium as the primary component, exceeding 25 wt%. Additionally, elements such as magnesium, aluminum, and vanadium are more likely to concentrate in the titanium-rich embedded phases than in the iron-rich areas.

Compared to the elemental distribution in unreduced ironsands, the major elements—particularly iron and titanium—have undergone significant migration and localized enrichment, eventually forming distinct block-like enriched regions by the end of the reduction process.

Dotted iron-rich regions distributed in a reticulated titanium-rich structure

In addition to the block-like enrichment structure, another distinctive morphology appears in the later stages of raw ironsands reduction—characterized by bright, dotted phases distributed within a gray reticulated matrix, as illustrated in Fig. 8.26a. Closer examination reveals that the reticulated structure itself comprises two distinct

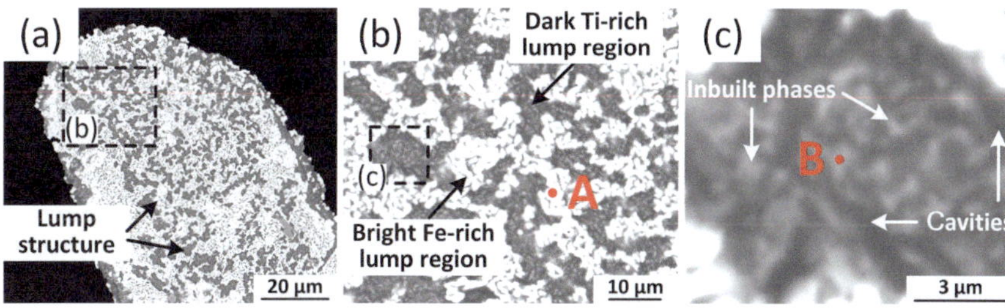

Fig. 8.24 Formation of interphase block structures during the later stage of reduction: **a–c** stepwise zoomed-in views; compositions of zones A and B are presented in Fig. 8.25

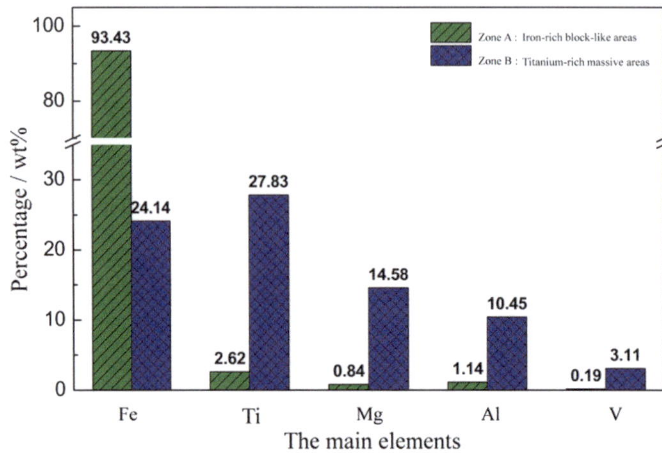

Fig. 8.25 Major elemental composition of zones A and B from Fig. 8.24

patterns: one exhibits a uniform distribution of point-like iron-rich phases across the reticulation, while the other shows a concentrated clustering of these bright dots, leaving parts of the reticulated structure visibly exposed.

Elemental mapping (Fig. 8.26c, d) shows that the bright, dotted phases are primarily composed of metallic iron, whereas titanium, oxygen, magnesium, and aluminum are mainly concentrated within the reticulated matrix—regardless of whether the distribution is homogeneous or heterogeneous. To better understand the formation mechanisms of these two reticulated morphologies, quantitative EDS analysis was conducted on two areas with uniformly distributed point-like phases (Fig. 8.26c) and one area with non-uniform distribution (Fig. 8.26d). The corresponding results are shown in Fig. 8.27, where region A represents the point-like iron-rich phase, and regions B and C represent the uniformly and non-uniformly distributed reticulated structures, respectively.

As shown in Fig. 8.27, region A consists predominantly of metallic iron, with a content approaching 95 wt%. In contrast, the reticulated regions (B and C) are rich in titanium (around 50 wt%) and show a higher concentration of magnesium compared to aluminum and vanadium among the chalcopyrite elements. This titanium content is notably higher than that observed in the blocky titanium-rich structures discussed earlier in this section, suggesting that such high Ti concentration may favor the formation of a reticulated titanium-rich matrix instead of massive structures at the final stage of reduction. Furthermore, region C exhibits higher titanium and lower iron contents than region B, which may explain the tendency for point-like metallic iron to cluster in isolated domains within the exposed reticulated matrix.

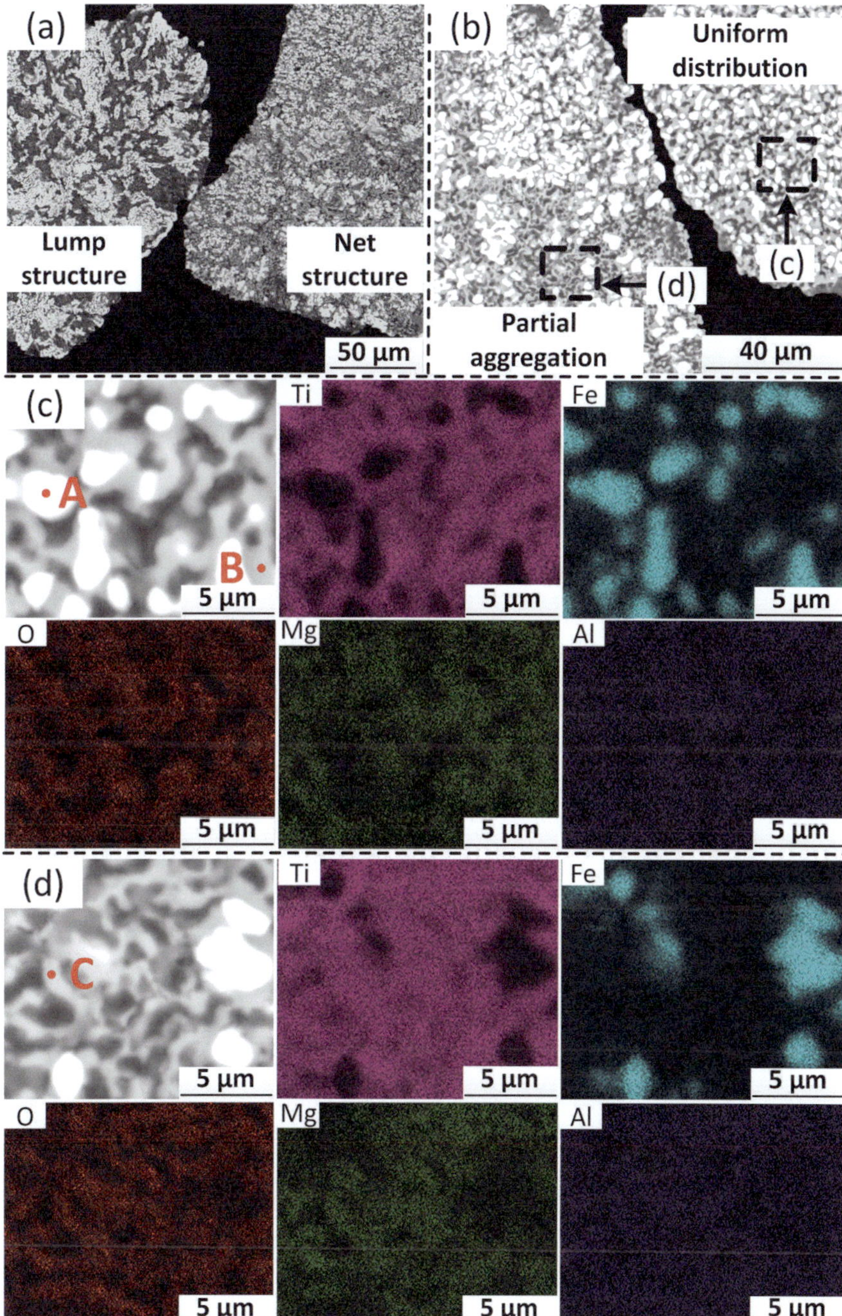

Fig. 8.26 Two reticulated structures and corresponding elemental maps: **a** Comparison of block and mesh morphologies; **b** Uniform versus enriched distributions of metallic iron; **c**, **d** Localized EDS mapping of regions in **b**

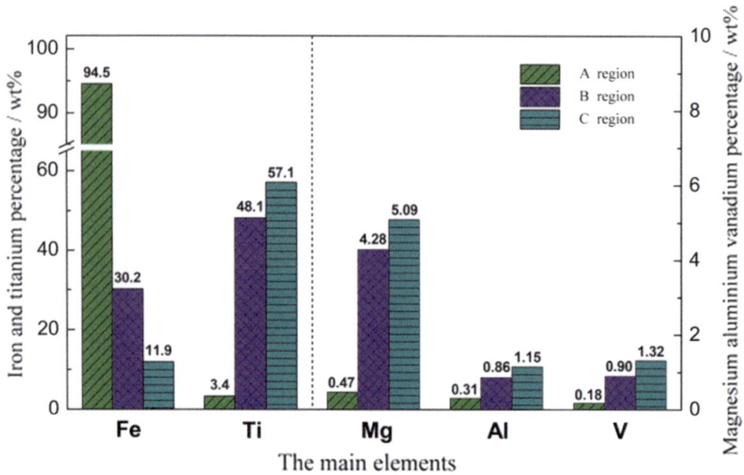

Fig. 8.27 Major element contents in regions A, B, and C of Fig. 8.26c, d

Interwoven structure of quartz-based veinstone phases with titanium-iron oxides

Ironsands particles are primarily composed of solid solutions of titanium-iron oxides. However, quartz phases, unlike magnesium–aluminum elements, exhibit weak bonding with titanium oxides. The quartz-based mineral phase appears in two principal forms: as isolated quartz grains and as an interwoven structure of quartz-based veinstone with useful mineral phases. This is demonstrated in Fig. 8.28, with Fig. 8.28a–c providing progressively magnified views. Although these two phases are embedded within the same mineral grain, a distinct physical boundary between them remains evident. As shown in Fig. 8.28c, two distinguishable subphases, labeled B and C, exist within the quartz-based material phase. These, along with region A (representing the average composition of the quartz-based veinstone), were subjected to EDS analysis, with results presented in Fig. 8.29.

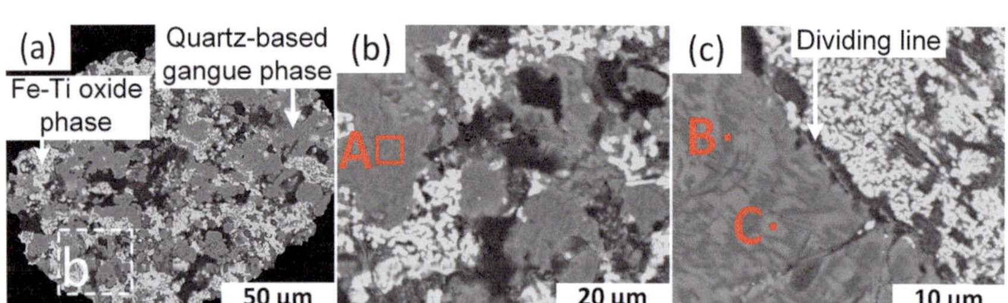

Fig. 8.28 Ironsands particles featuring quartz-rich gangue phases and interwoven Ti–Fe oxide structures

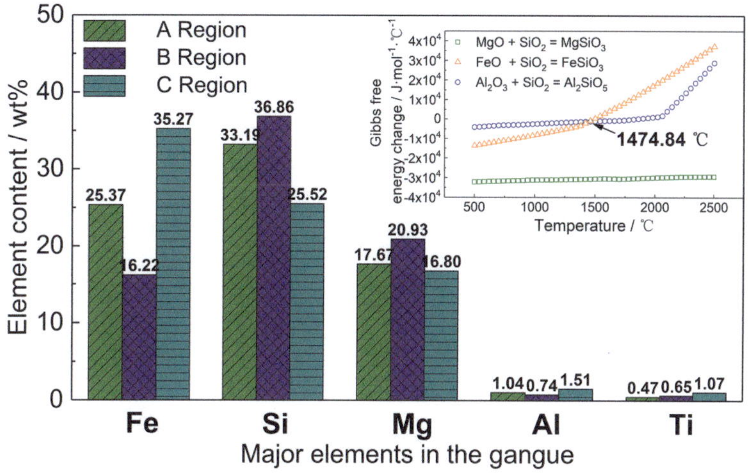

Fig. 8.29 Chemical composition of zones A, B, and C in Fig. 8.28, and calculated Gibbs free energy changes for reactions involving MgO, FeO, and Al_2O_3 with SiO_2

According to Fig. 8.29, region A primarily contains silica and magnesium oxide. Notably, even within the quartz-rich material, iron is still present—constituting 16.22 wt% and 35.27 wt% in regions B and C, respectively. However, aluminum and titanium are virtually absent in the quartz-based phase, likely due to the thermodynamic stability constraints of their compound formation.

Thermodynamic analysis of the Gibbs free energy changes associated with the binding reactions of MgO, Al_2O_3, and FeO with SiO_2 (shown in Fig. 8.29) reveals that MgO most readily reacts with SiO_2 to form silicates, compared to FeO and Al_2O_3. This tendency plays a key role during the transformation of ironsands from volcanic lava to solid mineral phase. If we define the elemental enrichment coefficient as the ratio of the current elemental content to the average content in the raw ironsands, magnesium shows an enrichment coefficient of 7.87 in the veinstone phase, which far exceeds that of iron (0.46) and aluminum (0.58). This suggests that magnesium is more prone to enrichment in quartz-based gangue than either iron or aluminum.

Similarly, TiO_2 and SiO_2 exhibit limited reactivity and do not form crystalline products. They may only coexist in amorphous states. This explains why SiO_2 is rarely incorporated into titanium-rich phases (unlike Al_2O_3 and MgO) and instead exists predominantly as standalone quartz gangue or as an interwoven structure with titanium-iron oxides.

8.6.4 *TTH Layered Phases Throughout the Reduction Process*

This section presents the reduction behavior of non-homogeneous ironsands particles containing laminated structural material phases, specifically the TTH phase, under complete reduction conditions.

Figure 8.30a and the magnified view in Fig. 8.30b correspond to the microstructure after 1 min of reduction. At this stage, while the matrix phase of the non-homogeneous particles begins to exhibit reduction features—evident in the dark striated structural regions—the TTH laminated structure remains unchanged, indicating no observable reduction. Only when the reduction time reaches 2.5 min, as shown in Fig. 8.30c, does the TTH layered structure begin to display initial signs of reduction, visible as a dark granular morphology. By this point, the bright metallic iron phase and the dark titanium-rich massive phase have already partially developed within the surrounding matrix. These observations indicate that the reduction of TTH lamellae is delayed compared to the matrix phase, both in non-homogeneous particles and in the homogeneous particulate phase.

Figure 8.30d–g depict the microstructural evolution of the non-homogeneous ironsands particles containing TTH lamellae after 60 min of reduction, approaching the endpoint of the reduction process. At this stage, the TTH laminar structure transforms into two distinct morphologies: a rod-like structure (labeled A and B), which largely retains the original position and orientation of the TTH lamellae and is a product of reduction-induced contraction; and a point-like phase (labeled C and D), scattered within the matrix.

The chemical compositions of these rod-like and point-like residual phases are presented in Fig. 8.31. The point-like phases (C and D) exhibit lower iron content and

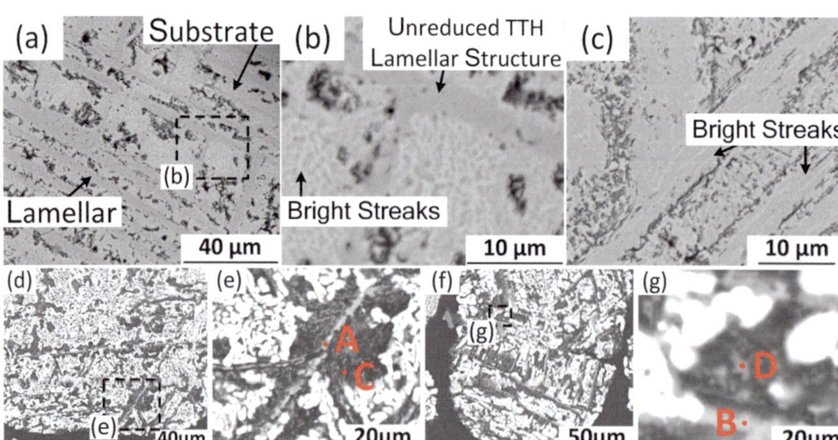

Fig. 8.30 Structural features of inhomogeneous ironsands particles containing TTH lamellar phases at various reduction stages: **a**, **b** at 1 min; **c** at 2.5 min; **d**–**g** up to 60 min

higher concentrations of titanium, magnesium, and aluminum compared to the rod-like structures (A and B). This compositional difference suggests that the point-like phases have undergone a higher degree of reduction than the rod-like ones.

In summary, the microstructures observed during the reduction process can be categorized according to two key principles: (1) the evolution of microstructure throughout the reduction process, and (2) the influence of the initial composition on the final microstructural forms. These classifications are illustrated in Fig. 8.32.

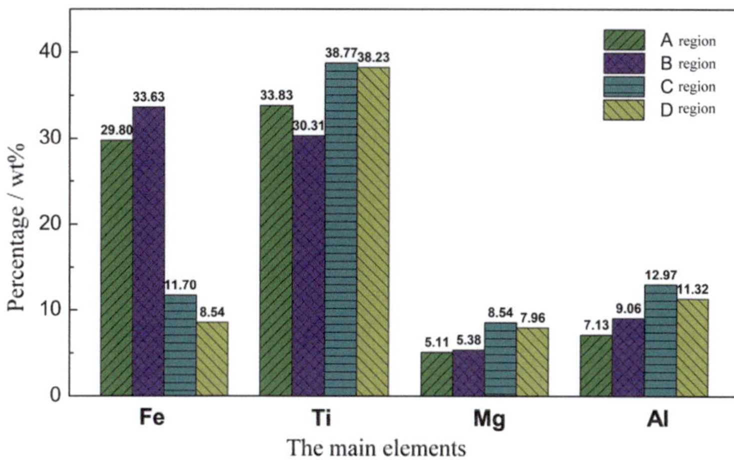

Fig. 8.31 Major elemental contents of the four zones identified in Fig. 6.18

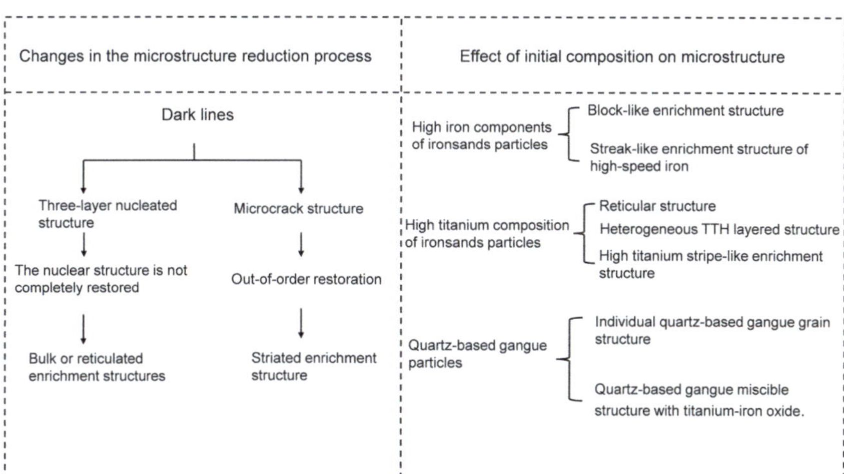

Fig. 8.32 Classification of ironsands microstructures during reduction by two paths: (i) microstructural changes during reduction; (ii) influence of initial composition

The dark-colored structures represent the initial reduction products common to both raw and pre-oxidized ironsands. However, significant differences emerge as the reduction progresses. In raw ironsands, following the initial stages of triple-phase nucleation and partial reduction associated with reactive nucleation, the resulting structures appear as either massive or reticulated elemental enrichment morphologies. The distinction between massive and reticulated forms is largely governed by the relative contents of titanium and iron within the mineral grains at the end stage of reduction.

In contrast, the pre-oxidized ironsands follow a distinctly different reduction pathway. The formation of microcracks during the pre-oxidation treatment introduces a disordered reduction mechanism, which facilitates gas diffusion and results in the development of streak-like (striated) enrichment structures of titanium and iron in the final microstructure.

Furthermore, the initial compositional characteristics of the ironsands particles play a decisive role in determining the type of microstructure formed at the reduction endpoint. For instance:

(1) Particles with high iron content predominantly form blocky enrichment structures or high-iron striated enrichment structures.
(2) Particles rich in titanium are more likely to develop reticulated structures, heterogeneous TTH lamellar structures, or high-titanium striated enrichment structures.
(3) Quartz-based veinstone particles typically exist either as discrete quartz-rich grains or as interwoven structures composed of quartz-based veinstones and titanium-iron oxides.

8.6.5 Comparison of Raw and Pre-oxidized Ironsands in Pore Properties

As demonstrated in Sects. 8.4–8.6, pre-oxidized ironsands exhibit superior reduction kinetics compared to raw ironsands. Microscopic observations during the reduction process suggest that pre-oxidation induces the formation of extensive crack-like pores within the particles, enabling internal diffusion of reducing gases and facilitating reduction reactions throughout the particle. This enhancement significantly accelerates the overall reduction rate. However, both kinetic modeling and scanning electron microscopy (SEM) offer only indirect evidence of this phenomenon.

To provide a more direct and quantitative understanding of the structural differences between raw and pre-oxidized ironsands, this section employs nitrogen adsorption analysis to determine their specific surface area, pore size distribution, and pore volume. These results reveal essential microstructural differences and clarify the effect of pre-oxidation treatment on the pore architecture of ironsands.

The physical gas adsorption method is widely used to investigate the pore characteristics of solid materials. By measuring the volume of gas adsorbed onto a solid surface, this method enables direct characterization of pore size, volume, and distribution. It is highly reproducible under varying equipment and operator conditions. Commonly used adsorbates include nitrogen, argon, and carbon dioxide, with the choice depending on the target material and measurement objectives. Nitrogen, which operates effectively across a broad partial pressure range at − 196 °C, allows characterization of micropores, mesopores, and macropores (0.5–200 nm), making it suitable for this study. Argon provides more accurate results for micropores but is less effective for meso- and macropores, while CO_2 is typically used for carbon-based materials.

For this analysis, nitrogen was selected due to its suitability for characterizing larger pore structures. The Barrett–Joyner–Halenda (BJH) method—developed by Barrett, Joyner, and Halenda in 1951—is used to analyze the mesoporous and macroporous distributions. Based on the Kelvin equation with modifications for multilayer adsorption, it remains one of the most commonly applied models for evaluating pore size distributions in this range.

Fig. 8.33a, b display the nitrogen adsorption–desorption isotherms for raw and pre-oxidized ironsands, respectively. According to IUPAC classification, both materials exhibit type V isotherms with H3-type hysteresis loops, typically associated with non-rigid aggregates of plate-like particles and slit-shaped pores [37]. These curves are indicative of multilayer adsorption involving capillary condensation and evaporation processes in randomly distributed cleft-like and lamellar mesoporous structures.

Although both materials display similar isotherm types, there is a notable difference in the desorption behavior. In raw ironsands (Fig. 8.33a), the forced closure of the hysteresis loop occurs more abruptly near $P/P_0 = 0.42$. In contrast, pre-oxidized ironsands (Fig. 8.33b) exhibit a gentler closure, suggesting a more open and interconnected mesoporous network, such as funnel-shaped or through-pores. Conversely, the mesoporous structures in raw ironsands appear more occluded and restricted [38, 39].

To quantify the specific surface area, the Brunauer–Emmett–Teller (BET) method—formulated in the 1940s—is employed. It calculates the volume of gas

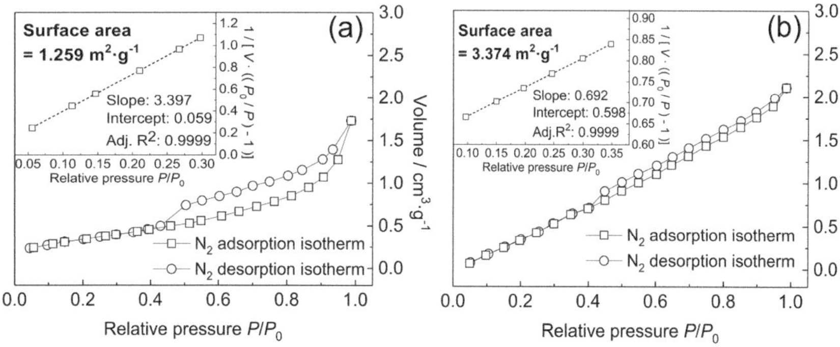

Fig. 8.33 Nitrogen adsorption–desorption isotherms at − 196 °C and BET-specific surface area analysis for: **a** raw ironsands; **b** pre-oxidized ironsands

required for monolayer adsorption and subsequently determines the material's specific surface area. The calculation process is presented in Eqs. 8.8–8.18 and 8.19.

$$\frac{1}{V(P_0/P - 1)} = \frac{1}{CV_m} + \frac{C - 1}{CV_m} \times \frac{P}{P_0} \tag{8.18}$$

$$S = \frac{\mathrm{V_m} \cdot \mathrm{N} \cdot \mathrm{A_m}}{22400} \times 10^{-18} \tag{8.19}$$

In the formula:

V	Volume of nitrogen adsorbed on the sample surface per unit mass of sample condition, ml g^{-1};
V_m	Monolayer saturation adsorption volume of nitrogen on the sample surface per unit mass, ml g^{-1};
P_0	Saturated vapor pressure of nitrogen at liquid nitrogen temperature (−196 °C), Pa;
P	Partial pressure of nitrogen, Pa;
C	BET constant, related to the adsorption energy of the material;
S	Specific surface area of the material, m^2 g^{-1};
N	Avergadro's constant, 6.02×10^{23} mol^{-1};
A_m	Cross-sectional area of nitrogen molecules, 0.162 nm^2;
22400	Molar volume of gas under standard conditions, ml mol^{-1}.

The fitted and final results of the specific surface area analysis are presented in the inset of Fig. 8.33a, b. The specific surface area of the pre-oxidized ironsands is 3.374 m^2 g^{-1}—approximately three times that of the raw ironsands (1.259 m^2 g^{-1}). This significant increase in surface area corresponds to a substantially larger active reduction area and therefore a notably enhanced reduction rate following the pre-oxidation treatment.

In addition, the pore size distributions of raw and pre-oxidized ironsands, calculated using the BJH method, are shown in Fig. 8.34. These data are derived from the nitrogen adsorption isotherms, with the adsorption branch being used instead of the desorption branch to avoid artifacts caused by the tensile strength effect. As shown in Fig. 8.34, pre-oxidized ironsands exhibit increased pore volume across the entire mesoporous range (4–50 nm), as well as higher cumulative pore volume (see inset of Fig. 8.34). This enhancement in pore volume further contributes to improved mass transport conditions and alters the rate-controlling step of the reduction process, thereby optimizing the reduction kinetics overall.

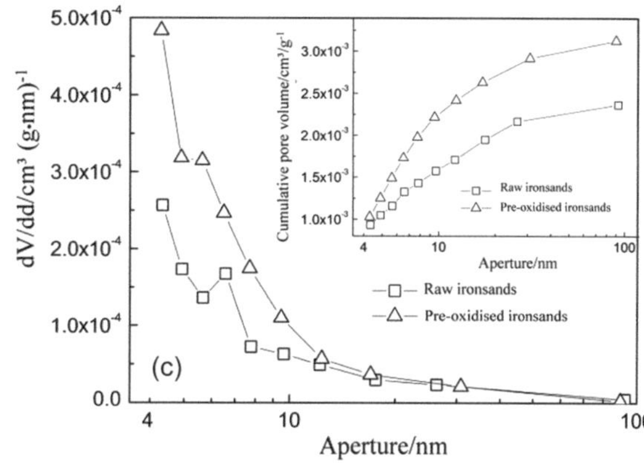

Fig. 8.34 Pore size distribution trends for raw and pre-oxidized ironsands based on BJH analysis; inset shows cumulative pore volume versus pore diameter

8.7 Summary

When both the reduction rate and economic considerations are taken into account, a hydrogen concentration of 40 vol% and a reaction temperature of 900 °C are identified as the optimal reduction conditions. Increasing the reaction temperature enhances molecular thermal motion, thereby accelerating the reaction rate and expanding the temperature-controlled chemical reaction regime. In contrast, increasing hydrogen concentration enhances the reaction rate but does not alter the chemical reaction control regime. Thus, the degree of reduction at the reaction turning point is determined solely by temperature, while the time to reach the turning point is influenced by both temperature and hydrogen concentration.

The Fe_2O_3 solid solution phase in TTH is rapidly reduced to Fe_3O_4. The subsequent reduction of Fe_3O_4—whether originating from the solid solution in TTM or as a product from TTH—is also facile, with FeO and metallic Fe appearing at 2.5 min and FeO disappearing by 15 min. The initially present $FeTiO_3$ is fully reduced to Fe and TiO_2 within 5 min. However, newly formed $FeTiO_3$, generated between 5–10 min via the reduction of Fe_2TiO_4 in TTM and the residual solid solution $FeTiO_3$ in TTH, accumulates until $FeTi_2O_5$ appears around 25 min, indicating further reduction of the newly generated $FeTiO_3$. Ultimately, by 50 min, $FeTiO_3$, $FeTi_2O_5$, and Fe_2TiO_5 are fully reduced and disappear. As a primary reduction product of $FeTiO_3$, TiO_2 appears within the first 5 min, but further reduction to lower-valence titanium oxides (Ti_3O_4 and Ti_2O_3) occurs only at 40–50 min. This delay is attributed to the presence of iron oxides, which inhibit the reduction potential for titanium oxides, preventing the immediate formation of reduced titanium species.

Three types of ironsand particles with different reduction microstructures were investigated. In non-homogeneous particles, the lamellar structure exhibits inferior reducibility compared to the basal region. At the point when reduced iron is already abundant in the basal area, it appears only sporadically in the lamellae. Furthermore, the preferential site of lamellar reduction—either center-first or edge-first—depends on the spatial distribution of chalcopyrite-related elements. In contrast, homogeneous particles undergo outward-to-inward reduction with well-defined boundaries between reduced and unreduced regions. Reduction preferentially initiates along interphase streak-like dark regions due to uneven distribution of chalcopyrite elements and their inhibitory effect on reduction.

Pre-oxidation treatment significantly improves the reduction kinetics of ironsands. Specifically, pre-oxidized ironsands exhibit higher conversion rates and higher apparent reaction rate constants than raw ironsands at the same reaction time. From kinetic modeling, the apparent activation energy (E) and pre-exponential factor (A) for pre-oxidized ironsands are lower: $E = 67.20\ kJ{\cdot}mol^{-1}$ and $A = 127.61\ min^{-1}$, compared to $E = 80.94\ kJ\ mol^{-1}$ and $A = 346.29\ min^{-1}$ for raw ironsands. This demonstrates that pre-oxidation lowers the energy barrier for gas-phase reduction. Additionally, iso-conversional analysis reveals that the apparent activation energy for raw ironsands increases from 50 to 105 kJ mol^{-1} as the conversion rate increases from 0.2 to 0.6, indicating a transition from reaction product nucleation and growth control (early stage), to chemical reaction control (intermediate), and finally to gas diffusion control (late stage). In contrast, pre-oxidized ironsands consistently follow a first-order chemical reaction control mechanism throughout the reduction process, maintaining an apparent activation energy of approximately 67 kJ mol^{-1}. The enlarged pore structure and formation of microcracks suppress the early-stage nucleation-growth control and prevent late-stage gas diffusion control, thereby stabilizing kinetic behavior.

For both raw and pre-oxidized ironsands, dark granular structures emerge early in the reduction process, rapidly progressing from the particle surface to the core before the appearance of metallic iron. Metallic iron phases then develop along the paths defined by these dark granular structures. This suggests that the reducing atmosphere penetrates the interior of nucleating particles at the early stages, rather than acting solely at the outer interface between metallic iron and the unreduced core. An incompletely reduced reaction nucleus model is introduced to characterize intermediate structures. The initial composition and phase of the ironsands significantly influence the reduction microstructure. In pre-oxidized ironsands, finer, randomly distributed incompletely reduced nuclei form due to the expanded pore network and microcracks, promoting disordered reduction behavior, shifting the kinetic rate-limiting step, and improving kinetic conditions.

Elemental migration and enrichment during reduction lead to different end structures. Raw ironsands typically develop alternating Fe-rich and Ti–rich massive regions. In high-Ti ironsands, a reticulated titanium oxide phase forms, in which metallic iron appears either uniformly or in agglomerated forms within the network. Quartz-based veinstones, due to the difficulty in forming TiO_2–SiO_2 crystalline compounds, remain as discrete particles or appear in interwoven structures with titanium–iron oxides. In non-homogeneous particles, lamellar TTH phases reduce more slowly than basal TTM phases.

Finally, compared with raw ironsands, pre-oxidized ironsands exhibit significantly larger specific surface area and increased pore volume in the mesopore range (4–50 nm), confirming that pre-oxidation effectively enlarges internal pore structures.

This not only improves the reduction kinetics but also fundamentally alters the rate-controlling mechanism during the reduction process.

References

1. Geng C, Sun TC, Yang HF, et al. Effect of Na_2SO_4 on the embedding direct reduction of beach titanomagnetite and the separation of titanium and iron by magnetic separation. ISIJ Int. 2015;55(12):2543–9.
2. Gonzalez LM, Forssberg KSE. Utilization of a vanadium-containing titanomagnetite: possibilities of a beneficiation-based approach. T I Min Metall C. 2001;110:C63–72.
3. Wang Z, Pinson D, Chew S, et al. Behavior of New Zealand ironsand during iron ore sintering. Metall Mater Trans B-Proc Metall Mater Proc Sci. 2016;47(1):330–43.
4. Yu ZW, Li GH, Liu C, et al. Comparison of reduction disintegration characteristics of TiO_2-rich burdens prepared with sintering process and composite agglomeration process. Int J Min Met Mater. 2016;23(4):389–98.
5. Yu ZW, Li GH, Jiang T, et al. Effect of basicity on titanomagnetite concentrate sintering. ISIJ Int. 2015;55(4):907–9.
6. Ren S, Zhang JL, Wu LS, et al. Effect of TiO_2 on equilibrium phase sinter at oxygen partial pressure of 5x10(-3)atm. Ironmak Steelmak. 2014;41(2):132–7.
7. Bold T, Kuziak R. Phase transformations during oxidizing roasting of titanomagnetite pellets. Vanadium Recovery Process 1990:167–78.
8. Sun Y, Zheng HY, Dong Y, et al. Melting and separation behavior of slag and metal phases in metallized pellets obtained from the direct-reduction process of vanadium-bearing titanomagnetite. Int J Miner Process. 2015;142:119–24.
9. Zhou LH, Zeng FH. Reduction mechanisms of vanadium-titanomagnetite-non-coking coal mixed pellet. Ironmak Steelmak. 2011;38(1):59–64.
10. Park E, Ostrovski O. Reduction of titania-ferrous ore by carbon monoxide. ISIJ Int. 2003;43(9):1316–25.
11. Wang YM, Yuan ZF. Reductive kinetics of the reaction between a natural ilmenite and carbon. Int J Miner Process. 2006;81(3):133–40.
12. Kucukkaragoz CS, Eric RH. Solid state reduction of a natural ilmenite. Miner Eng. 2006;19(3):334–7.
13. Wang YM, Yuan ZF, Guo ZC, et al. Reduction mechanism of natural ilmenite with graphite. Trans Nonferrous Metals Soc China. 2008;18(4):962–8.
14. Paunova R. Thermodynamic study of the reduction of titanium magnetite concentrate with solid carbon. Metall Mater Trans B-Proc Metall Mater Proc Sci. 2002;33(4):633–8.
15. Schwebel GL, Filippou D, Hudon G, et al. Experimental comparison of two different ilmenites in fluidized bed and fixed bed chemical-looping combustion. Appl Energy. 2014;113:1902–8.
16. Longbottom RJ, Ostrovski O, Zhang JQ, et al. Stability of cementite formed from hematite and titanomagnetite ore. Metall Mater Trans B-Proc Metall Mater Proc Sci. 2007;38(2):175–84.
17. Park E, Ostrovski O. Reduction of titania-ferrous ore by hydrogen. ISIJ Int. 2004;44(6):999–1005.
18. Wang ZY, Zhang JL, Xing XD, et al. Phase transitions and atomic-scale migration during the preoxidation of a titania/ferrous oxide solution. JOM. 2016;68(2):656–67.
19. Bernasowski M. Theoretical study of the hydrogen influence on iron oxides reduction at the blast furnace process. Steel Res Int. 2014;85(4):670–8.
20. Dang J, Hu X, Zhang G, et al. Kinetics of reduction of titano-magnetite powder by H_2. High Temp Mater Process (London). 2013;32(3):229–36.
21. Bonalde A, Henriquez A, Manrique M. Kinetic analysis of the iron oxide reduction using hydrogen-carbon monoxide mixtures as reducing agent. ISIJ Int. 2005;45(9):1255–60.

22. Parisi DR, Laborde MA. Modeling of counter current moving bed gas-solid reactor used in direct reduction of iron ore. Chem Eng J. 2004;104(1–3):35–43.
23. Kim WH, Lee S, Kim SM, et al. The retardation kinetics of magnetite reduction using H_2 and H_2–H_2O mixtures. Int J Hydrogen Energy. 2013;38(10):4194–200.
24. Kuila SK, Chatterjee R, Ghosh D. Kinetics of hydrogen reduction of magnetite ore fines. Int J Hydrogen Energy. 2016;41(22):9256–66.
25. Rajput P, Bhoi B, Sahoo S, et al. Preliminary investigation into direct reduction of iron in low temperature hydrogen plasma. Ironmak Steelmak. 2013;40(1):61–8.
26. Vyazovkin S, Burnham AK, Criado JM, et al. ICTAC Kinetics Committee recommendations for performing kinetic computations on thermal analysis data. Thermochim Acta. 2011;520(1–2):1–19.
27. Piotrowski K, Mondal K, Wiltowski T, et al. Topochemical approach of kinetics of the reduction of hematite to wustite. Chem Eng J. 2007;131(1–3):73–82.
28. Vyazovkin S, Wight CA. Model-free and model-fitting approaches to kinetic analysis of isothermal and nonisothermal data. Thermochim Acta. 1999;340:53–68.
29. Galwey AK, Brown ME. Application of the Arrhenius equation to solid state kinetics: can this be justified? Thermochim Acta. 2002;386(1):91–8.
30. Brown ME, Maciejewski M, Vyazovkin S, et al. Computational aspects of kinetic analysis Part A: The ICTAC kinetics project-data, methods and results. Thermochim Acta. 2000;355(1–2):125–43.
31. Wright J. Iron-titanium oxides in some New Zealand ironsands. NZ J Geol Geophys. 1964;7(3):424–44.
32. Gow AJ. Petrographic studies of ironsands and associated sediments near Hawera, South Taranaki. NZ J Geol Geophys. 1967;10(3):675–96.
33. Cruz-Sanchez E, Alvarez-Castro JF, Ramirez-Picado JA, et al. Study of titanomagnetite sands from Costa Rica. J Alloys Compd. 2004;369(1–2):265–8.
34. Longbottom RJ, Ostrovski O, Park E. Formation of cementite from titanomagnetite ore. ISIJ Int. 2006;46(5):641–6.
35. Sun HY, Dong XJ, She XF, et al. Solid state reduction of titanomagnetite concentrate by graphite. ISIJ Int. 2013;53(4):564–9.
36. Wang ZY, Zhang JL, Ma JF, et al. Gaseous reduction of titania-ferrous solution ore by H-2-Ar mixture. ISIJ Int. 2017;57(3):443–52.
37. Leofanti G, Padovan M, Tozzola G, et al. Surface area and pore texture of catalysts. Catal Today. 1998;41(1–3):207–19.
38. Groen JC, Peffer LAA, Perez-Ramirez J. Pore size determination in modified micro- and mesoporous materials. Pitfalls and limitations in gas adsorption data analysis. Microporous Mesoporous Mater 2003;60(1–3):1–17.
39. Kruk M, Jaroniec M. Gas adsorption characterization of ordered organic-inorganic nanocomposite materials. Chem Mater. 2001;13(10):3169–83.

Chapter 9
High-Temperature Melt Separation of Pre-reduced Ironsands

Following pre-reduction, high-temperature smelting is required to achieve the separation of molten iron and titanium-bearing slag. This chapter investigates the effects of reduction degree on slag composition, viscosity, and microstructure during melting. Experimental results are presented for the production of hot metal and high-titanium slag, with detailed analysis of their physical and chemical characteristics.

9.1 Influence of Pre-reduction of Ironsands on the Properties of Melt Separation Slag

The pre-reduction degree (PRD) of ironsands during the initial stage significantly influences the chemical composition of the slag in the subsequent smelting and separation process, thereby affecting both the slag properties and the overall operation of the melting–splitting procedure. Given that the original ironsands contain approximately 11 wt% TiO_2, the resulting slag after smelting inevitably has a high titanium content. In the context of producing titanium–iron alloys from titanium-rich slag, it is essential to ensure a sufficiently high TiO_2 grade in the final slag. Consequently, the use of fluxing agents such as CaO must be avoided during the slagging process to prevent dilution of TiO_2, leading to the formation of high-titanium slag containing approximately 40–50 wt% TiO_2 during the high-temperature smelting of pre-reduced ironsands.

Numerous studies [1–4] have investigated the behavior of ironmaking slag containing TiO_2. When the TiO_2 content is below 30 wt%, its addition disrupts the silicate network structure, exhibiting a depolymerizing effect that leads to a gradual decrease in slag viscosity. However, as the TiO_2 content increases beyond 30 wt%—particularly in the range of 30–50 wt% [5, 6]—the thermal stability of the slag deteriorates. As the temperature drops from 1600 to 1400 °C, the viscosity of high-TiO_2 slag increases rapidly and surpasses that of slag with lower TiO_2 content.

Z. Wang et al., *Smelting and Utilization Technology of Ironsands*,
https://doi.org/10.1007/978-981-95-2868-4_9

This effect is further exacerbated in the presence of carbon and nitrogen, which facilitate the formation of TiC and Ti(C,N) phases—high-melting-point particulates that disperse within the molten slag and significantly elevate its viscosity [6, 7]. Therefore, it can be hypothesized that slag containing approximately 40 wt% TiO_2 may exhibit substantially reduced operability due to excessive viscosity and phase stability issues.

To optimize slag–iron separation and improve the reaction kinetics during smelting, this chapter investigates the influence of pre-reduction degree as a key variable. PRD reflects the extent of ironsand reduction prior to its entry into the smelting furnace and directly determines the composition of the separation slag—particularly the concentrations of FeO, TiO_2, and SiO_2. As a result, the physical and chemical properties of the slag also vary with PRD. In this study, the melting behavior, viscosity, and structural characteristics of slag under different pre-reduction degrees were systematically evaluated using in-situ high-temperature observation, rotational torque measurements, and Raman spectroscopy. These investigations provide deeper insight into the smelting behavior of ultra-high-titanium slag and support the optimized selection of the appropriate pre-reduction degree for ironsands.

9.1.1 Influence of Pre-reduction Degree on Slag Composition

During the pre-reduction process of ironsands, the iron oxide phases can undergo partial or complete reduction, depending on technical parameters such as reduction time, temperature, and atmosphere. As a result, the pre-reduction degree (PRD) is a controllable variable that significantly influences the slag properties during the subsequent melting stage. In this study, five preset pre-reduction degrees were considered. By assuming that a portion of the iron oxides is reduced during pre-reduction—corresponding to the designated PRD—and that the resulting metallic iron enters the liquid metal phase during melting and separation, the remaining slag components were normalized. The calculated slag compositions for each PRD scenario are summarized in Table 9.1.

The chemical reagents used in this section included FeC_2O_4, TiO_2, SiO_2, MgO, Al_2O_3, CaO, and MnO, all with purities of $\geq$ 99.50 wt%. These powdered reagents were first dried at 150 °C for 10 h to ensure complete moisture removal. They were then weighed according to the proportions in Table 9.1 and thoroughly mixed using a planetary ball mill for one hour. In preliminary trials, the mixed slag samples were heated to 1550 °C and held for 3 h in an argon atmosphere (1 L min^{-1}) to ensure complete homogenization. After quenching the molten slag in water, X-ray fluorescence (XRF) analysis confirmed that the resulting compositions closely matched the target values in Table 9.1.

To investigate the influence of PRD on the properties of the melt-separation slag, three types of characterization experiments were conducted as described below:

Table 9.1 Initial composition of ironsands and melt slag composition at different pre-reduction degrees/wt%

Component	FeO	TiO_2	SiO_2	MgO	Al_2O_3	CaO	MnO	TFe	The rest
Initial composition of ironsands	29.60	11.41	4.13	3.74	3.38	0.60	0.50	55.63	0.06
100% PRD	0.00	48.02	17.38	15.74	14.23	2.53	2.10	–	–
97% PRD	10.46	43.00	15.56	14.09	12.74	2.26	1.88	–	–
95% PRD	16.29	40.20	14.55	13.18	11.91	2.11	1.76	–	–
90% PRD	28.02	34.57	12.51	11.33	10.24	1.82	1.51	–	–
70% PRD	53.87	22.15	8.02	7.26	6.56	1.17	0.97	–	–

Slag Melting

The melting characteristics of the slag were evaluated by in-situ observation of height changes in cylindrical samples during non-isothermal heating. The experimental setup is illustrated in Fig. 9.1 and consists of a combined heating and imaging system. Five sets of slag samples of equal mass were pressed into cylindrical pellets (1 cm in height) under identical pressure and with the addition of an alcohol-based binder. These were placed on a molybdenum plate and heated at a rate of 5 °C min^{-1} under argon. As the temperature increased and the slag softened and melted, the height of the sample decreased. Infrared imaging was used to monitor this change, and the temperatures at which the sample height decreased to 75, 50, and 25% of its original value were defined as the softening temperature, melting temperature, and flow temperature, respectively [8].

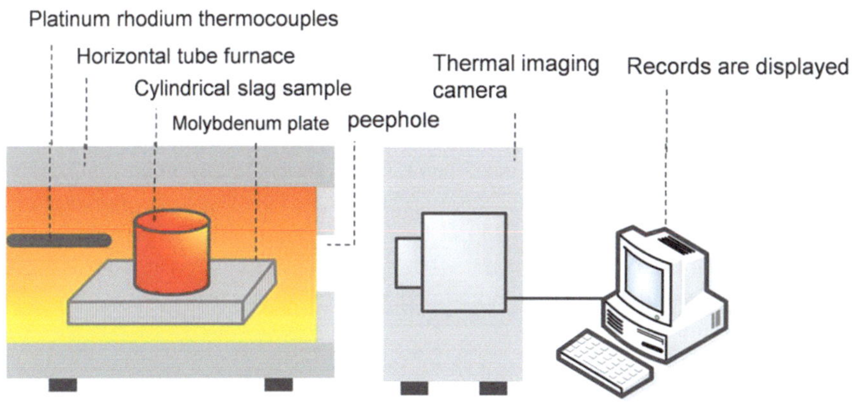

Fig. 9.1 Experimental setup for measuring the melting characteristics of slag

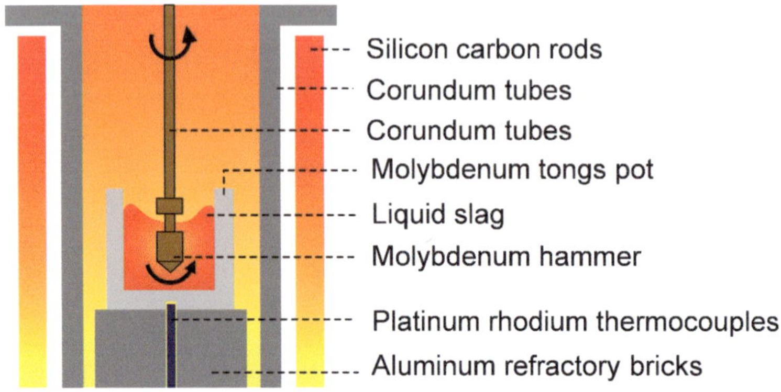

Fig. 9.2 Experimental apparatus for slag viscosity characterization

Slag Viscosity

Due to its high repeatability, the rotational torque method was adopted for slag viscosity measurement [9], with the apparatus shown in Fig. 9.2. Prior to the experiment, the device was calibrated using a standard fluid (vegetable oil) to establish the torque–viscosity relationship. Slag samples were placed in a molybdenum crucible and heated from room temperature to 1550 °C under an argon atmosphere (1 L min^{-1}) at a rate of 5 °C min^{-1}. After holding at 1550 °C for 3 h to ensure thermal and compositional equilibrium, a molybdenum spindle was immersed into the molten slag and rotated at 200 rpm. Viscosity measurements were taken at 5 °C cooling intervals, with a stabilization time of 10 min before each measurement. No significant discrepancies were observed between viscosity values obtained during heating and cooling; therefore, only the cooling curve data were used for subsequent analysis.

Slag Structure

Five groups of slag samples were heated to 1550 °C and held for 3 h under argon, followed by rapid quenching in ice water to form amorphous slag. X-ray diffraction (XRD) analysis confirmed the absence of crystalline phases. Previous studies [10–12] have demonstrated that the structural characteristics of amorphous slag at room temperature are representative of the molten state, especially when analyzed by Raman spectroscopy. Therefore, the structural features of the slag were characterized using Raman spectroscopy (HORIBA) with a spectral range of 100–4000 cm^{-1} and a 532 nm confocal laser excitation source. The spectra were recorded at room temperature and fitted with Gaussian peaks to deconvolute the contributions of different slag structural units.

9.1.2 Melting Characteristics of Melt Separation Slag

Using an infrared thermal imaging system, the morphological evolution of cylindrical slag samples during the melting process is shown in Fig. 9.3. By comparing the contact angle (i.e., the angle between the molten slag and the molybdenum plate) under identical residual height conditions, it is evident that slags derived from ironsands with higher degrees of pre-reduction exhibit greater surface tension and reduced flowability, indicating more difficult melting behavior.

To further assess and compare the melting behaviors of the different slag groups, the trends in softening, melting, and flow temperatures for the five pre-reduction conditions are presented in Fig. 9.4.

As shown in Figs. 9.3 and 9.4a, both the melting temperature and the flow temperature increase progressively with increasing pre-reduction degree. This increase becomes particularly pronounced when the pre-reduction degree rises from 90 to 100%. However, the trend for softening temperature is notably different: at 95

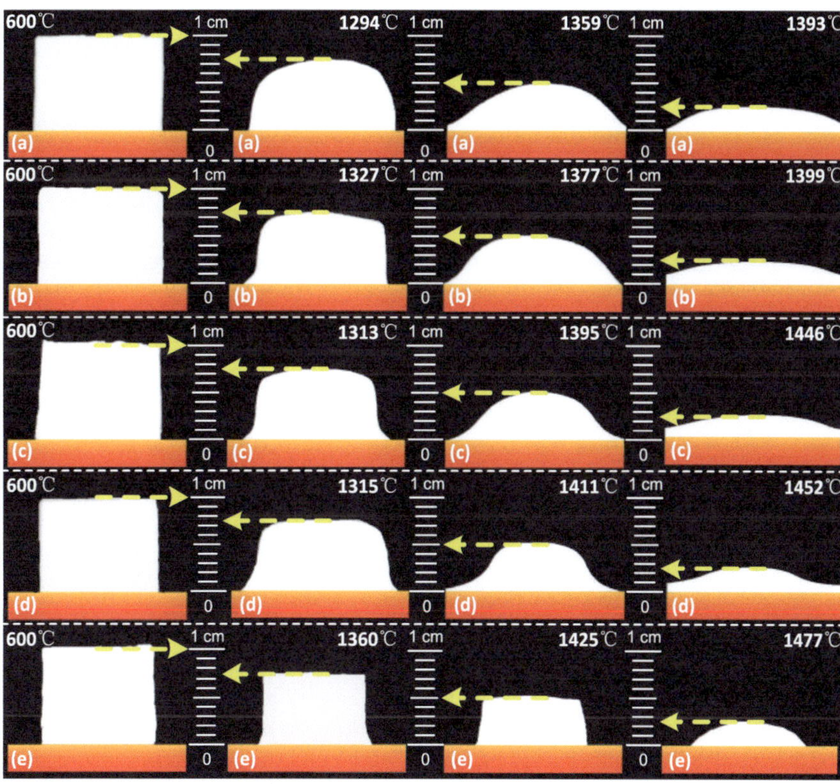

Fig. 9.3 Temperature profiles corresponding to the reduction in height of cylindrical slag samples to 75, 50, and 25% of their original height during heating and melting. **a–e** correspond to slag samples with 70%, 90%, 95%, 97%, and 100% pre-reduction, respectively

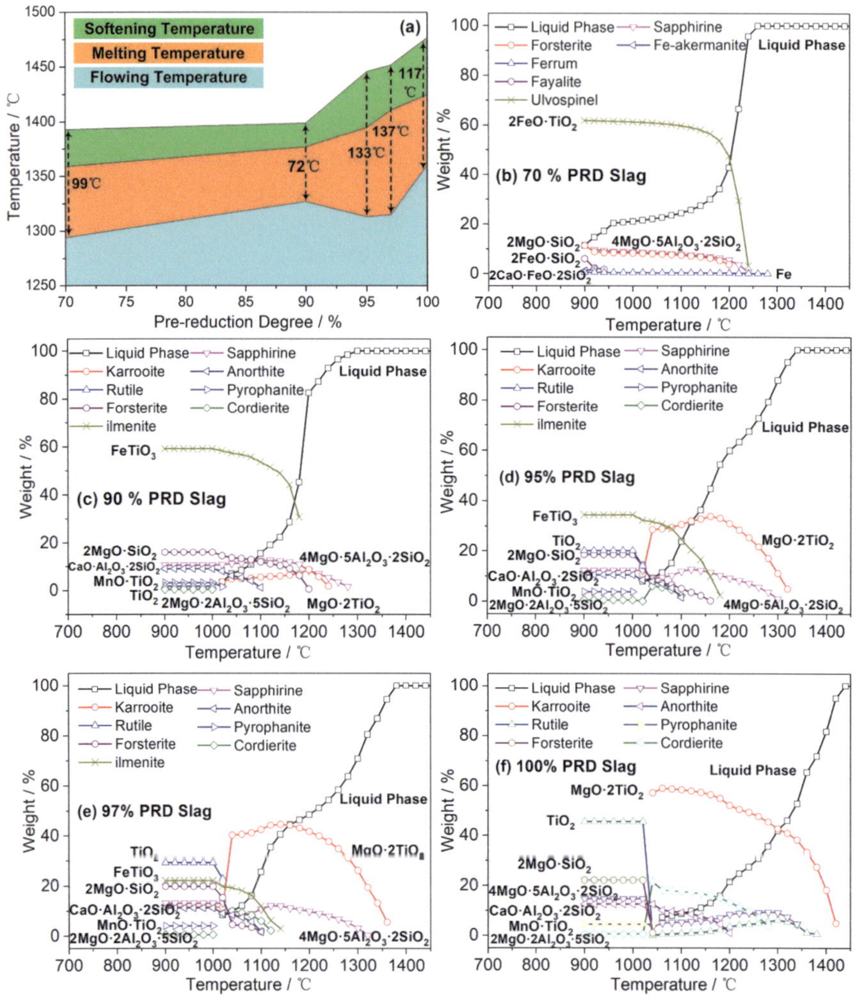

Fig. 9.4 **a** Trends in softening temperature, melting temperature, flow temperature, and melting interval for five groups of melt-fractionated slag samples with varying degrees of pre-reduction; **b–f** Equilibrium phase contents of the five slag compositions at different temperatures, predicted using FactSage thermodynamic analysis software

and 97% pre-reduction, the softening temperature decreases compared to other compositions.

To investigate the mechanisms underlying these temperature trends, the thermodynamic equilibrium behavior of the five slag compositions during heating was analyzed using FactSage 7.0 thermodynamic modeling software. The results are shown in Figs. 9.4b–f.

With increasing pre-reduction degree from 70 to 100%, the slag composition evolves toward higher TiO_2 and lower FeO content. At higher pre-reduction levels

(95, 97, and 100%), the formation of the refractory $MgO–2TiO_2$ phase persists even at elevated temperatures. Due to its poor meltability, this phase significantly elevates the melting and flow temperatures.

In the lower temperature range—corresponding to the softening stage—slags with 70, 90, and 100% pre-reduction contain appreciable amounts of titanium–iron spinel ($2FeO \cdot TiO_2$), ilmenite ($FeTiO_3$), and rutile (TiO_2). These crystalline phases inhibit the early formation of liquid slag, resulting in higher softening temperatures. In contrast, for slags with 95 and 97% pre-reduction, the relative absence of such solid phases in this temperature range allows for earlier onset of melting, thus explaining the observed drop in softening temperature.

Importantly, molten slag in the intermediate state—between softening and complete liquefaction—can significantly reduce slag permeability. A prolonged or excessive molten phase is more prone to trapping gases, thereby increasing the likelihood of slag foaming, which adversely affects smelting operations. Therefore, it is desirable to minimize the temperature interval between softening and flow points.

Among the five compositions, the slag corresponding to 90% pre-reduction shows the narrowest melting temperature range—only 72 °C—due to its more favorable phase assemblage. In contrast, the other slag groups exhibit melting temperature ranges exceeding 100 °C. Hence, from the perspective of melting characteristics and process stability, a 90% pre-reduction degree represents the most suitable condition for optimizing slag fluidity and minimizing operational issues during the melting stage.

9.1.3 Viscosity Properties of Melt Separation Slag

Figure 9.5 illustrates the viscosity trends of molten slags prepared from ironsands with different pre-reduction degrees (PRD), measured from 1550 °C down to the temperature at which viscosity exceeds 2.0 Pa s, at which point the test was terminated. For the slag derived from 100% pre-reduced ironsands, viscosity began to rise immediately upon cooling. In contrast, slags from the other pre-reduction groups exhibited little to no viscosity change during the initial cooling phase, and their viscosities remained similar during this stage.

However, as the temperature continued to decline beyond a certain point, the viscosity of slags with 70–97% PRD began to increase sharply. This behavior aligns with previous findings [2, 9, 13–15], which indicate that slag viscosity is influenced by the degree of polymerization and structural complexity of the silicate network. As temperature decreases, thermal motion weakens, promoting silicate network polymerization and resulting in a rapid increase in viscosity—consistent with the Arrhenius-type viscosity relationship ($\eta = A \cdot \exp(E\eta/R \cdot T)$).

From a process engineering perspective, a slag viscosity of 1.0 Pa s or lower ensures sufficient fluidity for effective slag-metal separation, thereby improving reaction kinetics and enhancing the recovery of valuable elements. Accordingly, the temperature at which each slag reaches 1.0 Pa s was extracted and is plotted in

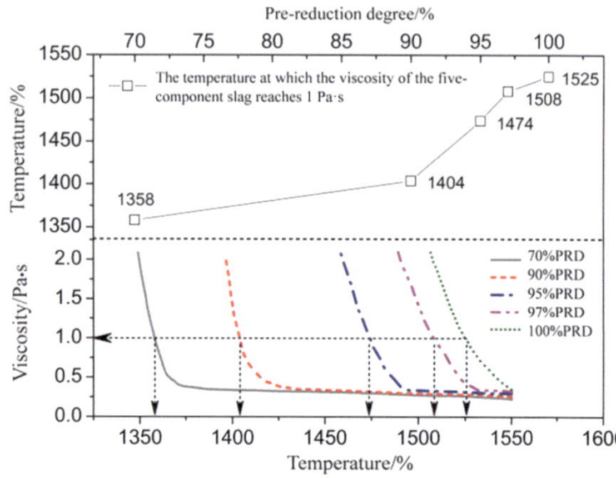

Fig. 9.5 Variation in slag viscosity with temperature at different pre-reduction levels; horizontal line indicates a viscosity of 1 Pa s

Fig. 9.5. Notably, increasing the PRD from 90 to 100% raises this critical temperature sharply, from 1404 °C to 1525 °C, while increasing PRD from 70 to 90% results in only modest changes.

The figure also demonstrates that at a given temperature, slag viscosity decreases with lower PRD. In particular, the viscosity reduction between 95 and 90% PRD is the most pronounced—even greater than that observed when PRD decreases from 90 to 70%. This underscores 90% PRD as the optimal condition for achieving lower viscosity and improved slag fluidity.

The significant influence of ironsands' pre-reduction degree on melt-separated slag composition—particularly the type and concentration of cations (Mn^+) and free oxygen ions (O^{2-})—further explains these trends. In analyzing the effects of cations on slag properties, two critical factors are typically considered: the ionic valence (Z) and the ionic radius (r), as both directly influence the strength of ionic bonding.

Longer bond lengths and lower cation valence result in weaker ionic bonds, which make it easier for oxides (e.g., alkali, alkaline earth, or transition metal oxides) to dissociate and release O^{2-}. These free oxygen ions can depolymerize the silicate network by reacting with bridging oxygen (O^0) to form non-bridging oxygen (O^-). Meanwhile, to maintain charge neutrality, cations coordinate with these non-bridging oxygens and become integrated into the silicate structure. This cation–non-bridging oxygen linkage significantly affects slag viscosity.

As shown in previous studies [9, 16], shorter cation–non-bridging oxygen bonds result in lower shear viscosity, thus reducing the overall viscosity of the molten slag. To account for both bond length and ionic valence, the effective ionic field strength parameter Z/r^2 is introduced (see Table 9.2 [9]). The table shows that Fe^{2+} has a Z/r^2 value similar to that of Ca^{2+} and Mg^{2+}, indicating that FeO—like CaO and MgO has a strong capacity to disrupt the silicate network. Consequently, lowering the PRD increases FeO content in the slag, enhancing its ability to depolymerize the silicate framework and thereby effectively reducing slag viscosity.

Table 9.2 Ionic radii and effective ionic radii

Ionic	Ionic radius/nm	$Z{\cdot}r^{-2}$/nm^{-2}
Ca^{2+}	0.100	200.0
Mg^{2+}	0.072	385.8
Fe^{2+}	0.061	537.5
Al^{3+}	0.053	1068.0
Ti^{4+}	0.061	1075.0
Si^{4+}	0.040	2500.0

9.1.4 Structural Properties of Melt Separation Slag

Raman spectroscopy provides valuable insights into the silicate network and other structural units within slag. Different Raman bands correspond to specific structural motifs, allowing qualitative interpretation of slag polymerization and cation substitution behavior. The vibrational modes of the Raman peaks listed in Table 9.3 are based on relevant literature and closely align with the slag systems investigated in this study. While slight deviations in band positions are expected due to compositional differences between the referenced systems and the present study, the vibrational assignments in Table 9.3 remain the most representative for interpreting the observed spectra.

As indicated in Table 9.3, two possible titanium-containing structural units are identified:

(1) Ti^{4+} substituting for Si^{4+} within the silicate tetrahedral framework, forming Ti–O–Si linkages;
(2) Ti^{4+} forming independent tetrahedral units, resulting in Ti–O–Ti structures. Both configurations typically produce Raman bands in the 790–850 cm^{-1} range.

Table 9.3 Raman spectral characteristics of different slag structural units

Peak band /cm^{-1}	Structural unit	Vibration mode	NBO/Si, Q^n
790–850	Ti–O–Si or Ti–O–Ti	Symmetric elastic vibration [10, 14, 15, 17–21]	–
850–880	$(SiO_4)^{4-}$	Symmetric elastic vibration [10, 15, 18, 21]	4, Q^0
900–930	$(Si_2O_7)^{6-}$	Symmetric elastic vibration [10, 15, 18, 21]	3, Q^1
950–986	Si–O–M (metal)	Asymmetric elastic vibration [22, 23]	
950–990	$(Si_2O_6)^{4-}$	Symmetric elastic vibration [10, 15, 18, 19, 21]	2, Q^2
1050–1100	$(Si_2O_5)^{2-}$	Symmetric elastic vibration [10, 15, 18, 19, 21]	1, Q^3

In addition to Ti-containing structures, the silicate tetrahedral network is characterized by symmetric stretching modes corresponding to varying degrees of polymerization, expressed as NBO/Si (the number of non-bridging oxygen atoms per tetrahedrally coordinated silicon). These structural motifs include:

(1) $(SiO_4)^{4-}$: monomeric units with NBO/Si = 4;
(2) $(Si_2O_7)^{6-}$: dimer structures with NBO/Si = 3;
(3) $(Si_2O_6)^{4-}$: chain structures with NBO/Si = 2;
(4) $(Si_2O_5)^{2-}$: layered structures with NBO/Si = 1.

Alternatively, silicate polymerization can be described using the Q^n notation, where n denotes the number of bridging oxygen atoms per tetrahedrally coordinated silicon. In molten slags, alkaline and alkaline earth metal oxides tend to dissociate into metal cations (M^{2+}, e.g., Fe^{2+}, Mg^{2+}, Ca^{2+}) and free oxygen ions (O^{2-}). These free oxygen ions react with bridging oxygen (BO) atoms in the silicate network, converting them into non-bridging oxygens (NBO), thereby depolymerizing the network structure. Simultaneously, the released cations migrate into the silicate framework to bond with NBOs in order to maintain charge neutrality.

This cation–oxygen interaction results in the formation of Si–O–M (M=Fe, Mg, Ca) linkages, which are typically observed in Raman spectra in the 950–986 cm^{-1} range, as summarized in Table 9.3.

Figure 9.6a presents the Raman spectra of slag samples corresponding to five different pre-reduction degrees (PRD). The spectral bands are primarily concentrated in the range of 850–1050 cm^{-1}. As the PRD increases, no new peaks emerge, nor do any existing peaks disappear; instead, only minor variations in peak intensity are observed. This indicates that the relative contents of different structural units fluctuate with the degree of pre-reduction, but the fundamental slag structure remains unchanged.

Due to the amorphous nature of the rapidly water-quenched glassy slag, the Raman spectra do not exhibit sharp, well-defined peaks typical of crystalline phases. Instead, the spectra show broad, overlapping bands that require peak deconvolution for structural interpretation. In this study, a Gaussian fitting method was employed to separate and quantify individual vibrational components, following the approach described by Mysen [15, 20, 21]. This method assumes that each Raman band conforms to a Gaussian distribution and incrementally adds new components based on discernible shoulders or inflections in the experimental spectrum.

The Raman spectra were fitted based on both this Gaussian method and the vibrational mode assignments from the literature listed in Table 9.3. The resulting deconvoluted spectra for each PRD are shown in Fig. 9.6b–f, and detailed fitting parameters are provided in Table 9.4.

The Raman peaks identified in Table 9.4 can be categorized into three main types:

(1) Si–O–M structures (M = Ti, Fe, Mg, Ca), representing metal–oxygen–silicon linkages;
(2) Silicon tetrahedral units, classified by the Q^n species (Q^0–Q^3);
(3) Titanium-based units, particularly Ti–O–Ti and Ti–O–Si linkages.

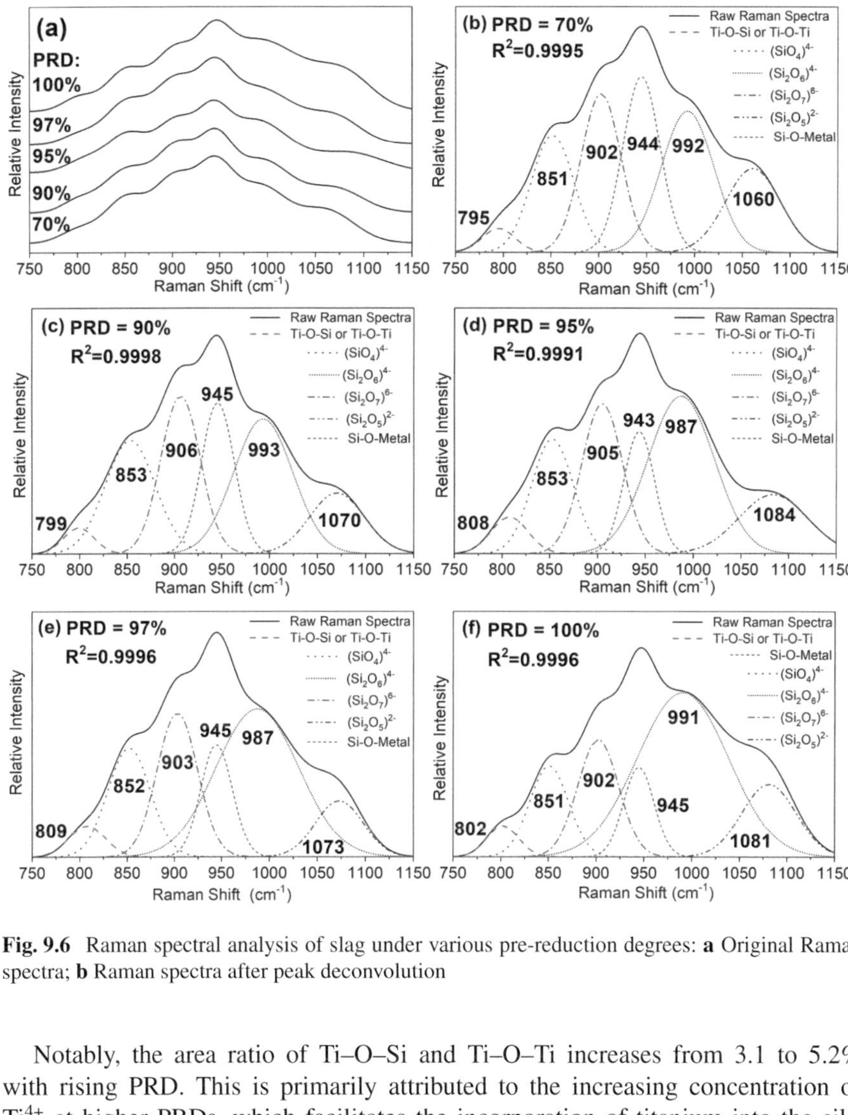

Fig. 9.6 Raman spectral analysis of slag under various pre-reduction degrees: **a** Original Raman spectra; **b** Raman spectra after peak deconvolution

Notably, the area ratio of Ti–O–Si and Ti–O–Ti increases from 3.1 to 5.2% with rising PRD. This is primarily attributed to the increasing concentration of Ti^{4+} at higher PRDs, which facilitates the incorporation of titanium into the silicate network either through substitution or by forming simple polymeric structures, such as monomeric $(TiO_4)^{4-}$ or chain-like $(Ti_2O_6)^{4-}$ [17].

In contrast, the content of Si–O–M units exhibits a more pronounced decreasing trend with increasing PRD. This is largely due to the progressive reduction in FeO, a key network modifier. As FeO content decreases, its ability to depolymerize the silicate network diminishes, resulting in a lower concentration of non-bridging oxygen (NBO) and, consequently, a reduced formation of Si–O–M linkages. This depolymerization mechanism is illustrated in Eq. 9.1, which represents the breakdown of silicate dimer units into monomeric forms via metal oxide modification.

Table 9.4 Raman spectrum peak parameters

Pre-oxidation	Structural unit	Band of a spectrum (cm^{-1})	Height	Half-height and width	Area ratio (%)	$(Q^3 + Q^2)/(Q^1 + Q^0)$
70	Ti–O–Si or Ti–O–Ti	795	19.6	40.6	3.1	0.99
	Q^0	851	95.3	52.5	17.4	
	Q^1	902	129.2	49.0	20.7	
	Si–O–M	944	143.2	45.7	21.1	
	Q^2	992	115.1	64.2	23.7	
	Q^3	1060	68.4	68.3	14.0	
90	Ti–O–Si or Ti–O–Ti	799	21.2	38.2	3.8	1.01
	Q^0	853	92.2	62.5	18.1	
	Q^1	906	128.7	48.7	21.2	
	Si–O–M	945	122.9	40.8	17.2	
	Q^2	993	109.7	72.6	27.3	
	Q^3	1070	49.5	67.0	12.4	
95	Ti–O–Si or Ti–O–Ti	808	25.4	45.5	4.5	1.34
	Q^0	853	79.3	51.2	15.7	
	Q^1	905	103.9	50.1	20.1	
	Si–O–M	943	84.4	37.8	11.8	
	Q^2	987	109.2	80.6	35.3	
	Q^3	1084	41.1	85.3	12.6	
97	Ti–O–Si or Ti–O–Ti	809	25.9	46.1	4.8	1.63
	Q^0	852	88.6	53.3	13.8	
	Q^1	903	117.0	50.4	18.6	
	Si–O–M	945	91.3	39.5	10.0	
	Q^2	987	120.5	101.7	40.9	
	Q^3	1073	45.8	58.3	11.9	
100	Ti–O–Si or Ti–O–Ti	802	28.6	39.5	5.2	2.34
	Q^0	851	83.4	48.2	10.8	
	Q^1	902	107.3	50.6	14.9	
	Si–O–M	945	82.1	37.9	8.9	
	Q^2	991	150.5	111.7	47.1	
	Q^3	1081	66.6	67.4	13.1	

$$M^{2+} + O^{2-} + [Si_2O_7]^{6-} = [SiO_4]^{4-} - M^{2+} - [SiO_4]^{4-} \quad (9.1)$$

It is particularly noteworthy that the area fraction of Si–O–M structures decreases gradually from 21.1 to 17.2% as PRD increases from 70 to 90%, but declines sharply to 8.9% at 100% PRD. This sharp drop correlates well with the viscosity behavior discussed in Sect. 9.1.3, where a significant viscosity increase occurs as PRD exceeds 90%. This confirms that a marked increase in slag viscosity is associated with reduced depolymerization of the silicate network, due to lower FeO content.

Furthermore, the Raman spectra indicate the simultaneous presence of various silicate structural units—$(Si_2O_5)^{2-}$, $(Si_2O_6)^{4-}$, $(Si_2O_7)^{6-}$, and $(SiO_4)^{4-}$—corresponding to Q^3, Q^2, Q^1, and Q^0 species, respectively. These polymeric units can interconvert depending on the balance between network modifiers and network formers, with higher degrees of polymerization indicated by higher values of *n* in the Q^n notation.

To quantitatively evaluate the polymerization degree of the silicate network, the ratio $\psi = (Q^3 + Q^2)/(Q^1 + Q^0)$ was introduced, and the results are listed in Table 9.4. With increasing PRD, Q^1 and Q^0 contents decrease, while Q^3 and Q^2 contents increase, resulting in a corresponding rise in ψ. However, the growth rate of ψ varies: from $\psi = 0.99$ at 70% PRD to $\psi = 1.01$ at 90% PRD represents a modest change, whereas the increase from $\psi = 1.01$ at 90% to $\psi = 2.34$ at 100% is much more pronounced.

This suggests that although the content of highly polymerized structures ($Q^3 + Q^2$) increases steadily between 70 and 90% PRD, the simultaneous presence of lower-order species ($Q^1 + Q^0$) maintains a balance that prevents a significant rise in viscosity or ψ. However, at PRD > 90%, the rapid depletion of simple structural units and accumulation of complex network structures leads to a sharp increase in both ψ and slag viscosity.

9.2 Experiment of Melt Separation of Pre-reduced Ironsands

After undergoing direct reduction, ironsands are introduced into a high-temperature furnace—typically an electric arc furnace—in the form of pre-reduced ironsands for slag-iron melt separation. As demonstrated in Sect. 9.1, the fused primary slag derived from ironsands with a 90% pre-reduction degree exhibits superior smelting properties. Accordingly, ironsands pre-reduced to 90% were selected as the experimental feedstock. To investigate the effect of flux addition, three comparative groups were established, each with different proportions of co-solvent CaF_2. To ensure uniform experimental conditions (melting temperature and holding time), all melt separation experiments were conducted using a graphite porous crucible, the dimensions and morphology of which are shown in Fig. 9.7.

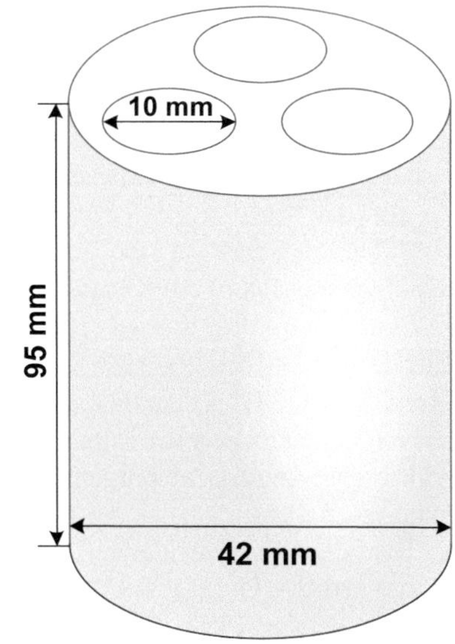

Fig. 9.7 Morphology and dimensional parameters of the porous graphite crucible used in slag melting experiments

The experiments were carried out using the same heating apparatus employed during the pre-oxidation process. Under an argon atmosphere, the crucible containing the sample was heated to the designated temperature and held for a specified time as outlined in Table 9.5. Upon completion, the crucible was promptly removed and quenched in water, ensuring rapid cooling to preserve the phase distribution and chemical composition of slag and metal as in the hot state. After drying, the crucible was opened, and the slag-metal composite was extracted. The macroscopic morphologies of the slag-metal separation products for the three fluxing conditions are displayed in Fig. 9.8.

When 4 wt% CaF_2 was added (Fig. 9.8a), no clear slag-metal separation occurred, and bright metallic phases remained dispersed throughout the slag matrix. At 8 wt% CaF_2 addition (Fig. 9.8b), initial phase separation was observed: the metallic phase largely settled at the bottom, while slag accumulated at the top. However, residual slag in the metal and metal inclusions in the slag indicated incomplete separation. With 12 wt% CaF_2 addition (Fig. 9.8c), the slag and metal phases achieved complete separation. The metal phase exhibited high luster and uniformity, and slag inclusions in the metal were negligible.

Table 9.5 Experimental conditions for melting and separation of three groups of pre-reduced ironsands

Groups	Temperature/°C	Holding time/h	Carbon content/wt%	CaF_2/wt%
a	1550	3	4	4
b				8
c				12

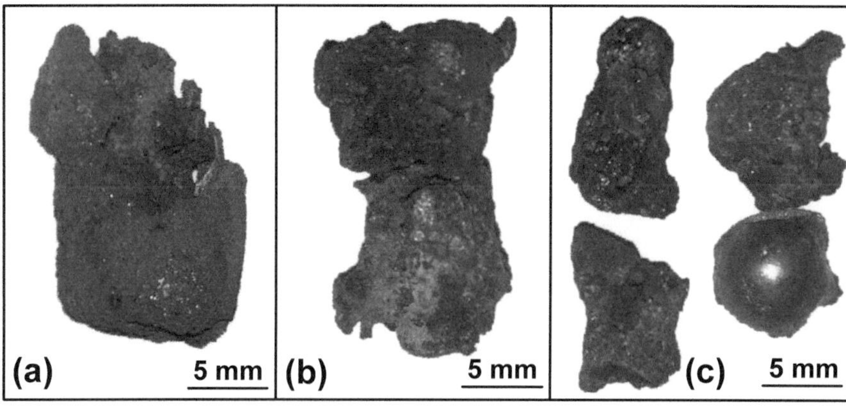

Fig. 9.8 Macroscopic morphology of three groups of pre-reduced ironsands following melt-fractionation experiments

In summary, effective slag-metal separation at 1550 °C was achieved when 90%-pre-reduced ironsands were fluxed with 12 wt% CaF_2. The resulting metal and slag from Group C were ground into powder samples for chemical composition analysis using neutralization titration. The analytical results are summarized in Table 9.6.

As shown in Table 9.6, the reduction of FeO and SiO_2 leads to a further enrichment of TiO_2 in the slag, with its content approaching 44 wt%. The combined content of other gangue components—MgO, Al_2O_3, SiO_2, and CaO—reaches approximately 45 wt%. Owing to the carbon-deficient conditions employed during the melting process, a small amount of FeO remains in the slag. However, since iron oxides are required in the subsequent titanium–iron alloy smelting process, this residual FeO has minimal impact on downstream operations.

Additionally, volatilization of CaF_2 during high-temperature melting results in a residual CaF_2 content of about 7.5 wt% in the slag. Regarding the metallic phase, the use of acidic slag (i.e., without CaO) promotes a high content of elemental silicon. Moreover, the ultra-high TiO_2 concentration in the slag contributes to a relatively high titanium content in the metal phase. Both Si and Ti contents exceed typical levels found in conventional blast furnace hot metal, but this deviation has negligible impact on the subsequent converter steelmaking process. The carbon content in the metal is approximately 3 wt%, indicating that the carbon is not yet saturated.

Table 9.6 Chemical composition of slag and metal phases from pre-reduced ironsands melt with 12 wt% flux/wt%

Slag phase	FeO	TiO_2	MgO	Al_2O_3	SiO_2	CaO	CaF_2
	3.28	43.95	16.57	14.97	10.99	2.66	7.58
Metal phase	Fe		C		Si		Ti
	94.74		3.16		0.97		1.13

9.3 Summary

During the gas-based pre-reduction process, the pre-reduction degree directly influences the chemical and phase compositions of the pre-reduced ironsands, and consequently affects the composition of the primary slag formed during subsequent high-temperature smelting. At higher pre-reduction levels, the resulting slag exhibits higher surface tension and reduced meltability and fluidity. However, the softening behavior displays an irregular trend, primarily due to the varying melting characteristics of individual phases with different compositions.

At a 90% pre-reduction degree, the separation slag demonstrates the narrowest temperature range between softening and flow temperatures. Furthermore, the increase in slag viscosity at 1 Pa s becomes more pronounced when the pre-reduction degree rises from 90 to 95%. Therefore, taking into account both the melting behavior and viscosity evolution of the separation slag, a 90% pre-reduction degree is considered the most suitable.

As the pre-reduction degree increases, the concentration of Ti^{4+} in the separation slag gradually rises. These titanium ions either partially incorporate into the silicate network or form simple polymeric structures. Meanwhile, FeO—like CaO and MgO—acts as an effective network modifier. Its decreasing content with increasing pre-reduction degree leads to a reduced depolymerization effect on the silicate network and a corresponding decline in the formation of Si–O–M (metal–oxygen–silicon) species. The calculated polymerization index (ψ) reveals that network polymerization increases slowly as the pre-reduction degree rises from 70 to 90%, but shows a marked increase beyond 90%, a trend that aligns well with the observed rise in slag viscosity.

References

1. Shankar A, Gornerup M, Lahiri AK, et al. Experimental investigation of the viscosities in CaO-SiO_2-MgO-Al_2O_3 and CaO-SiO_2-MgO-Al_2O_3-TiO_2 slags. Metall Mater Trans B-Proc Metall Mater Proc Sci. 2007;38(6):911–5.
2. Park H, Park JY, Kim GH, et al. Effect of TiO_2 on the viscosity and slag structure in blast furnace type slags. Steel Res Int. 2012;83(2):150–6.
3. Migas P. High-temperatures rheometric analysis of selected heterogeneous slag systems. Arch Metall Mater. 2015;60(2):697–704.
4. Liao JL, Li J, Wang XD, et al. Influence of TiO_2 and basicity on viscosity of Ti bearing slag. Ironmaking Steelmaking. 2012;39(2):133–9.
5. Dong XJ, Sun HY, She XF, et al. Viscosity and viscosity estimation model of fully liquid slags in TiO_2-Al_2O_3-CaO-SiO_2 and TiO_2-Al_2O_3-CaO-SiO_2-MgO systems with high TiO_2 concentration and low mass ratio of CaO to SiO_2. Ironmaking Steelmaking. 2014;41(2):99–106.
6. He SP, Huang QY, Zhang GX, et al. Solidification properties of CaO-SiO_2-TiO_2 based mold fluxes. J Iron Steel Res Int. 2011;18(7):15.
7. Li Y, Li YQ, Fruehan RJ. Formation of titanium carbonitride from hot metal. ISIJ Int. 2001;41(12):1417–22.
8. Pan C, Lv X, Bai C, et al. Melting features and viscosity of SiO_2-CaO-MgO-Al_2O_3-FeO nickel slag in laterite metallurgy. J Min Metall B. 2013;49(1):9–12.

9. Sohn I, Min DJ. A review of the relationship between viscosity and the structure of calcium-silicate-based slags in ironmaking. Steel Res Int. 2012;83(7):611–30.
10. Li JL, Shu QF, Chou KC. Structural study of glassy CaO-SiO_2-CaF_2-TiO_2 slags by Raman spectroscopy and MAS-NMR. ISIJ Int. 2014;54(4):721–7.
11. Taylor M, Brown Jr G E, Fenn PM. Structure of mineral glasses—III. $NaAlSi_3O_8$ supercooled liquid at 805 C and the effects of thermal history. Geochim Cosmochim Acta. 1980;44(1):109–17.
12. Riebling E. Structural similarities between a glass and its melt. J Am Ceram Soc. 1968;51(3):143–9.
13. Huang WJ, Zhao YH, Yu S, et al. Viscosity property and structure analysis of FeO-SiO_2-V_2O_3-TiO_2-Cr_2O_3 slags. ISIJ Int. 2016;56(4):594–601.
14. Zheng K, Zhang ZT, Liu LL, et al. Investigation of the viscosity and structural properties of CaO-SiO_2-TiO_2 slags. Metall Mater Trans B-Proc Metall Mater Proc Sci. 2014;45(4):1389–97.
15. Mysen BO, Virgo D, Scarfe CM. Relations between the anionic structure and viscosity of silicate melts—a Raman spectroscopic study. Am Mineral. 1980;65(7–8):690–710.
16. Yasui I, Hasegawa H, Imaoka M. X-ray diffraction study of the structure of silicate glasses. Pt. 1: alkali metasilicate glasses. Phys Chem Glasses. 1983;24:65–71.
17. Zheng K, Liao JL, Wang XD, et al. Raman spectroscopic study of the structural properties of CaO-MgO-SiO_2-TiO_2 slags. J Non-Cryst Solids. 2013;376:209–15.
18. McMillan P. A Raman spectroscopic study of glasses in the system CaO–MgO–SiO_2. Am Mineral. 1984;69(7–8):645–59.
19. McMillan P. Structural studies of silicate glasses and melts—applications and limitations of Raman spectroscopy. Am Mineral. 1984;69(7–8):622–44.
20. Mysen BO, Virgo D, Seifert FA. The structure of silicate melts: implications for chemical and physical properties of natural magma. Rev Geophys. 1982;20(3):353–83.
21. Mysen BO, Finger LW, Virgo D, et al. Curve-fitting of Raman spectra of silicate glasses. Am Mineral. 1982;67(7–8):686–95.
22. Mohassab Y, Sohn HY. Analysis of slag chemistry by FTIR-RAS and raman spectroscopy: effect of water vapor content in H_2–H_2O-CO-CO_2 mixtures relevant to a novel green ironmaking technology. Steel Res Int. 2015;86(7):740–52.
23. Ueda S, Koyo H, Ikeda T, et al. Infrared emission spectra of CaF_2-CaO-SiO_2 melt. ISIJ Int. 2000;40(8):739–43.

Chapter 10
Preparation of Titanium–Iron Alloy by Aluminum-Thermal Reduction Using Titanium-Containing Slag After Melt Separation

Utilizing the high-titanium slag obtained from prior experiments, this chapter explores the preparation of titanium–iron alloys via aluminothermic reduction. Topics include the calculation and design of material ratios, the influence of reducing agents, heating additives, and fluxes, as well as the experimental verification of process feasibility for titanium resource recovery from slag.

10.1 Experimental Scheme for the Preparation of Titanium–Iron Alloy

Currently, the primary titanium-bearing mineral resources include ilmenite, titanomagnetite, rutile, and chalcocite, among others. Among these, ilmenite serves as the predominant raw material for titanium product manufacturing, accounting for approximately 85–90% of total usage. The remaining sources consist mainly of rutile, high-titanium slag, and similar materials.

To comprehensively evaluate the grade characteristics and utilization potential of titanium-bearing materials, researchers have proposed two key assessment indicators:

(1) Commensurate value;
(2) Mass ratio (f) of titanium oxides to iron oxides.

These parameters, defined by Eqs. 10.1 and 10.2, are commonly used to evaluate the suitability of titanium-containing feedstocks. In this section, traditional titanium-bearing raw materials from four different sources are selected for comparison with two alternative materials—ironsands and fused titanium slag derived from ironsands. The results of this comparative analysis are presented in Table 10.1.

Z. Wang et al., *Smelting and Utilization Technology of Ironsands*,
https://doi.org/10.1007/978-981-95-2868-4_10

Table 10.1 Comparison of compositional indicators of titanium-containing materials

Raw materials	Canada Quebec	Panzhihua steel company	Chengde	Beihai, Guangxi province	This book ironsands	This book titanium slag
TiO_2/wt%	36	47	–	59	12	44
Commensurate/wt%	95.4	87.0	92.0	93.6	87.6	47.2
f	0.89	1.54	1.33	2.35	0.21	12.38

The commensurate value (wt%) is calculated as:

$$\text{Commensurate(wt\%)} = \text{Titanium oxide mass percent(wt\%)} + \text{Mass percent iron oxides(wt\%)} \quad (10.1)$$

$$f = \frac{\text{Mass fraction of titanium oxide(wt\%)}}{\text{Mass fraction of iron oxide(wt\%)}} \quad (10.2)$$

As shown in Table 10.1, although the total content of titanium and iron oxides in the ironsands used in this study is comparable to that of the four traditional titanium-bearing raw materials, the significantly higher iron grade in ironsands results in a relatively low TiO_2 content, approximately 12 wt%. This is substantially lower than the TiO_2 grades typically found in conventional feedstocks for titanium product smelting, which range from 35 to 60 wt%. Consequently, the f-value (TiO_2/FeO_x ratio) of ironsands is also relatively low. Even after the majority of iron oxides are reduced (with a reduction degree of ~ 90%), the resulting TiO_2 grade in the titanium slag remains only around 44 wt%, comparable to the unreduced state of other titanium-containing materials.

Therefore, the titanium slag produced from ironsands in this study is not suitable for the production of high-grade rutile-type titanium products (TiO_2 > 70–80 wt%), which is typically the aim when using high-quality feedstocks. As a result, this work proposes an alternative utilization pathway: producing low- to medium-grade titanium–iron alloys from ironsands-derived titanium slag. This is achieved by introducing Fe_2O_3 and Al into the slag, using aluminum as the reducing agent, and utilizing the exothermic aluminothermic reaction between Fe_2O_3 and Al to provide the necessary reaction heat.

This study is divided into three main stages:

Theoretical Calculations Using FactSage phase diagram analysis, material and heat balance calculations, and national standard comparisons, the aluminothermic reduction of a simplified ternary slag system (TiO_2–Fe_2O_3–CaO) with aluminum was modeled. Relevant empirical data were used to estimate the required raw material compositions and the expected final slag composition, based on the target phase composition of the titanium–iron alloy.

Preliminary Experiments Based on the optimized composition and dosage schemes obtained from the theoretical stage, aluminothermic reduction experiments were carried out using the ternary slag system (TiO_2–Fe_2O_3–CaO) and Al as the reductant. The resulting metal and slag phases were analyzed chemically to refine the empirical parameters used in the theoretical calculations.

Experimental Verification Using Fused Titanium Slag In this step, the corrected empirical data were applied to the real composition of fused titanium slag obtained from pre-reduced ironsands. The alloy phase composition of a qualified titanium–iron alloy was set as the target, and calculations were performed to determine the required amounts of Al (reducing agent) and Fe_2O_3 (heat-generating agent). Using these theoretical values, aluminothermic reduction experiments were conducted, and the final alloy and slag compositions were chemically analyzed to confirm the feasibility of producing titanium–iron alloys directly from ironsands-derived fused titanium slag.

10.2 Preparation of Titanium–Iron Alloy from TiO_2–Fe_2O_3–CaO Ternary System Slag

The chemical reaction for the production of metallic titanium using aluminum as a reducing agent is shown in Eq. 10.3, while the reaction employing iron oxide as a heat-generating agent in the aluminothermic reaction is represented by Eq. 10.4.

$$3TiO_2 + 4Al = 3Ti + 2Al_2O_3 \tag{10.3}$$

$$Fe_2O_3 + 2Al = 2Fe + Al_2O_3 \tag{10.4}$$

China's national standards for titanium–iron alloy grades and their chemical compositions are summarized in Table 10.2. For this section, where the simplified ternary slag system TiO_2–Fe_2O_3–CaO is used as the raw material, the alloy design is based on the FeTi40-A and FeTi40-B grade specifications. After eliminating trace impurity elements from the standard and normalizing the major components, the targeted alloy composition was set at Ti:Al:Fe = 48 wt%: 10 wt%: 42 wt%.

Table 10.2 Grades and chemical composition of titanium–iron alloys in China/wt%

Brand	Ti	Al	Si	P	S	C	Cu	Mn
FeTi30-A	25–35	8.0	1.5	0.05	0.03	0.10	0.40	2.5
FeTi30-B	25–35	8.5	5.0	0.06	0.04	0.15	0.40	2.5
FeTi40-A	35–45	9.0	3.0	0.03	0.03	0.10	0.40	2.5
FeTi40-B	35–45	9.5	4.0	0.04	0.04	0.15	0.40	2.5

However, given the use of high-purity chemical reagents in laboratory-scale tests and the abundant heat released by the aluminothermic reaction (Eq. 10.4), it is assumed that part of the metallic iron in the product may be substituted by titanium. Therefore, an enhanced alloy composition was proposed, with a titanium content exceeding the upper limit of the national standard, to provide a performance margin for the experimental process. This designed alloy composition is detailed in Table 10.3.

At the same temperature, different alloy compositions may exhibit different phase structures. To ensure favorable kinetic conditions for the slag–metal reaction and to promote effective mass transfer and phase separation, it is crucial that the metal phase remains in a fully liquid state during the smelting process. This guarantees smooth alloy formation and separation from the slag.

To assess this, thermodynamic equilibrium phase diagrams of the Ti–Fe–Al ternary alloy system were calculated using FactSage 7.0 thermochemical software at temperatures of 1200, 1300, 1400, and 1550 °C. The results, presented in Fig. 10.1, indicate that the designed alloy composition (Table 10.3) remains in a completely liquid phase at temperatures of 1200 °C and above, ensuring appropriate physical conditions for efficient alloy smelting.

Dosage of aluminum metal as the reducing agent

Based on the designed composition of the Ti–Fe–Al ternary alloy (Table 10.3) and the stoichiometric relationships provided by Eqs. 10.3 and 10.4, the required quantity of aluminum metal as the reducing agent can be calculated as follows:

Amount of aluminum required to reduce titanium oxide: $5\,\text{g}/\left(48\,\text{g}\cdot\text{mol}^{-1}\times 3\right)\times\left(27\,\text{g}\cdot\text{mol}^{-1}\times 4\right)=6\,\text{g}$.
Amount of aluminum required to reduce iron oxide: $1\,\text{g}/\left(56\,\text{g}\cdot\text{mol}^{-1}\times 2\right)\times\left(27\,\text{g}\cdot\text{mol}^{-1}\times 2\right)=0.48\,\text{g}$.
Aluminum content included in the alloy product: 1 g.
Total required aluminum dosage: 6 g + 0.48 g + 1 g = 7.48 g.

Design of TiO_2, CaO dosages, and smelting slag composition for titanium–iron alloy production

In the smelting of titanium–iron alloys, the addition of CaO as a flux is essential for two main reasons:

Firstly, if TiO_2–Fe_2O_3 is used as the raw material for titanium–iron alloy smelting, the aluminothermic reduction reaction produces a refractory TiO_2–Al_2O_3 binary

Table 10.3 Designed composition of Ti–Al–Fe alloys

Ingredient	Ti	Al	Fe	Total
Absolute quality/g	8	1	1	10
Percentage/wt%	80	10	10	100

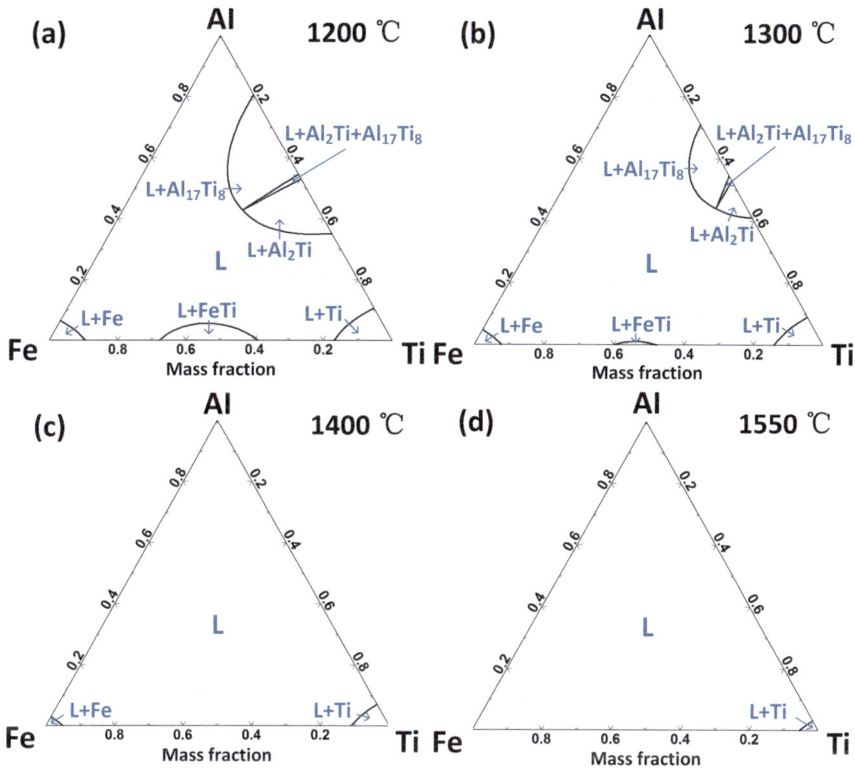

Fig. 10.1 Phase diagram of Ti–Fe–Al ternary alloy **a** 1200 °C; **b** 1300 °C; **c** 1400 °C; **d** 1550 °C

slag, which significantly impedes slag fluidity and hinders the smelting process. According to the phase diagram generated using FactSage thermodynamic software (see Fig. 10.2), the TiO_2–Al_2O_3 system belongs to a class of refractory slags, exhibiting high liquidus temperatures across all compositional ratios. Even at a TiO_2 molar fraction of 0.84—where the system reaches its lowest eutectic point—the melting temperature remains as high as 1706 °C, which exceeds the practical operating temperatures of typical smelting processes. Therefore, the TiO_2–Al_2O_3 binary slag system alone cannot meet the requirements for a low-melting-point slag composition necessary for efficient titanium–iron alloy production. To address this, CaO is introduced as a fluxing agent to effectively reduce the liquidus temperature of the slag system, thereby enhancing melt fluidity and promoting smoother slag–metal separation.

Secondly, during the reduction of TiO_2 by Al, TiO may form as an intermediate product. As a strongly basic oxide, TiO readily reacts with Al_2O_3 to form a stable compound, which impedes the progression of the reduction reaction, as illustrated in Eq. 10.5. To overcome this issue, it is necessary to introduce a more basic fluxing agent, calcium oxide, which can preferentially react with Al_2O_3 to form calcium

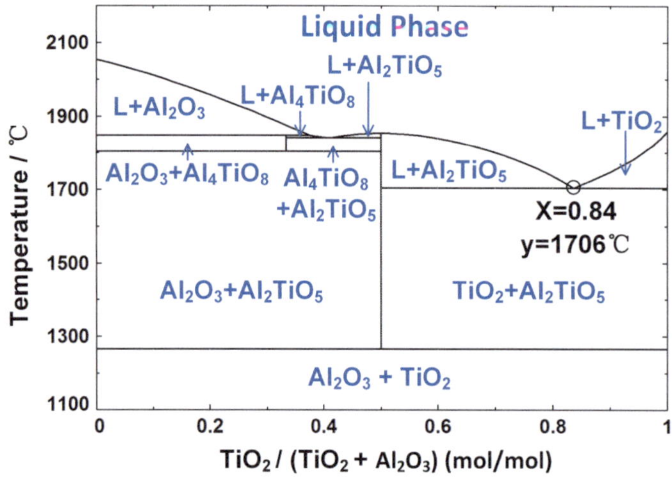

Fig. 10.2 Binary phase diagram of TiO_2–Al_2O_3 between 1100 and 2200 °C

aluminate, thereby liberating TiO and facilitating the continuation of the reduction process. The corresponding reaction is shown in Eq. 10.6.

$$2TiO_2 + \frac{4}{3}Al = 2TiO + \frac{2}{3}Al_2O_3 \tag{10.5}$$

$$TiO_2 + \frac{4}{3}Al + \frac{2}{3}CaO = Ti + \frac{2}{3}(CaO \cdot Al_2O_3) \tag{10.6}$$

Figure 10.3 presents the phase diagrams of the ternary slag system formed after the addition of CaO at temperatures of 1500, 1550, 1600, and 1650 °C, where the grey-shaded area represents the pure liquid-phase region. At 1500 °C, the liquid-phase region consists of two small, isolated zones located in the CaO-rich and TiO_2-rich areas. As the temperature increases to 1550 °C, the liquid-phase region gradually expands. At 1600 °C, the two previously separate regions merge into a continuous liquid-phase zone, which further enlarges at 1650 °C.

Based on the expected alloy composition and the calculated amount of aluminum metal used as the reductant, the final contents of Al_2O_3 and TiO_2 in the slag can be determined. By introducing an appropriate amount of CaO, the resulting ternary slag composition falls within the pure liquid-phase region at 1550 °C—the target smelting temperature for titanium–iron alloy used in subsequent experiments. This provides a theoretical basis for determining the actual dosage of the fluxing agent CaO.

The alumina in the slag primarily originates from the oxidation of metallic aluminum during the reduction of iron and titanium oxides, producing Al_2O_3 via the reaction: $2Al \rightarrow Al_2O_3$. The total mass of alumina formed can be calculated as follows:

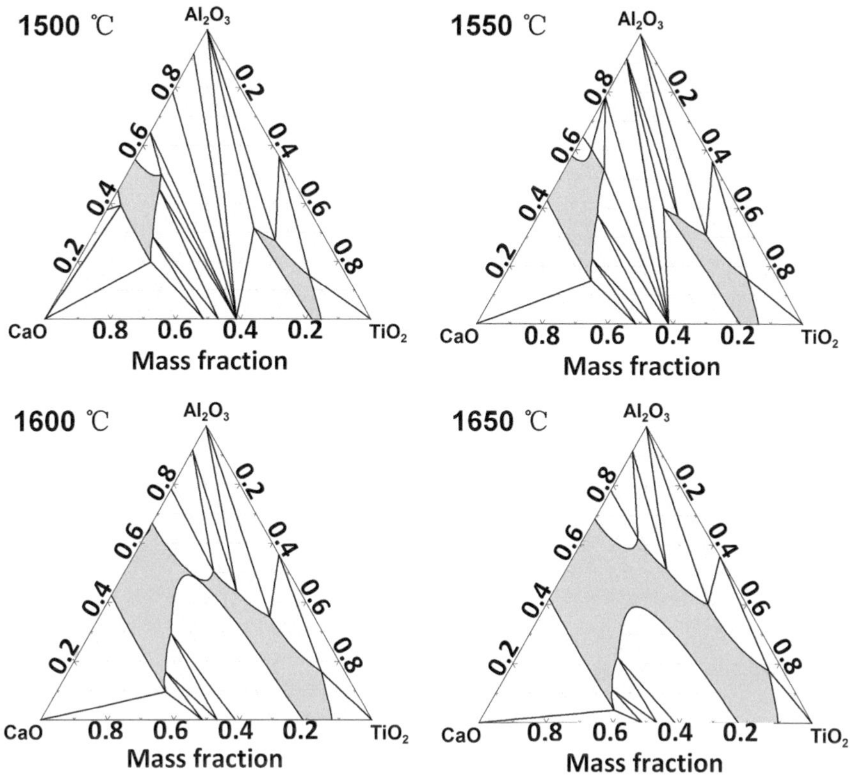

Fig. 10.3 Changes in the liquid-phase region of the TiO_2–Al_2O_3–CaO ternary slag at 1500 °C, 1550 °C, 1600 °C, and 1650 °C (gray-shaded areas indicate liquid-phase regions)

$$\begin{aligned}&(6\,\text{g}(\text{Aluminum metal needed to reduce TiO}_2)\\&\quad+0.48\,\text{g}\,(\text{Aluminum metal needed to reduce Fe}_2\text{O}_3))/54\,\text{g}\cdot\text{mol}^{-1}\\&\quad\times 102\,\text{g}\cdot\text{mol}^{-1}=12.24\,\text{g}\end{aligned}$$

According to Table 10.3, the titanium content in the designed alloy composition is 8 g. Based on literature data, the reduction reaction described in Eq. 10.6 has a conversion rate of approximately 77%. Therefore:

The required mass of TiO_2 in the feedstock is calculated as: 8 g/48 g·mol^{-1} × 80 g·mol^{-1}/77% = 17.32 g.
The mass of TiO_2 remaining in the slag is: 8 g/48 g·mol^{-1} × 80 g·mol^{-1}/77% × 23% = 3.98 g.

Having determined the absolute masses of TiO_2 and Al_2O_3 in the slag system, and referring to the Al_2O_3–TiO_2–CaO ternary phase diagram at 1550 °C (Fig. 10.3), the addition of 15 g of CaO positions the slag composition within the complete liquid-phase region in the 1500–1650 °C range. This condition is favorable for both the thorough progression of the slag-metal reduction reaction and efficient slag-metal separation. The corresponding mass percentage composition of the three oxides is summarized in Table 10.4.

Iron oxide dosage

The addition of iron oxide serves two primary purposes:

(1) To provide the iron component in the titanium–iron alloy through the reduction reaction between metallic aluminum and iron oxide;
(2) To generate substantial heat via the aluminothermic reaction, thereby rapidly increasing the local reaction temperature. This promotes the reduction of titanium dioxide and facilitates effective separation between the metallic and slag phases.

The chemical reaction between iron oxide and metallic aluminum is represented by:

$$Fe_2O_3 + Al = 2Fe + Al_2O_3 \tag{10.7}$$

Given that the designed iron content in the alloy (Table 10.3) is 1 g, and based on literature indicating a reduction conversion rate of approximately 99% for Eq. (10.7), the required dosage of Fe_2O_3 can be calculated as follows:

$$1\,g/112\,g \cdot mol^{-1} \times 160\,g \cdot mol^{-1}/99\% = 1.43\,g$$

In summary, the optimized mass ratio of the TiO_2-Fe_2O_3-CaO ternary slag system for smelting titanium–iron alloy is determined, and the corresponding composition is presented in Table 10.5.

Calculation of heat balance for the aluminothermic reduction reaction

The enthalpy changes for the reduction of titanium dioxide and iron oxide using aluminum as the reductant at 1550 °C were obtained using the thermodynamic software FactSage 7.0, as shown in Eqs. 10.8 and 10.9.

Table 10.4 Design slag composition

Ingredient	TiO_2	Al_2O_3	CaO	Total
Mass/g	3.98	12.24	15.00	31.22
Content/wt%	12.75	39.20	48.05	100

Table 10.5 Raw material composition for titanium–iron alloy smelting using ternary system slag

Ingredient	TiO_2	Fe_2O_3	CaO	Al
Mass/g	17.32	1.43	15.00	7.48
Percentage/wt%	40.57	6.77	35.14	17.52

$$\frac{3}{4}TiO_2 + Al = \frac{3}{4}Ti + \frac{1}{2}Al_2O_3$$
$$\Delta H_{1550\,^\circ C} = -138456.7\,J \cdot mol^{-1} \quad (10.8)$$

$$\frac{1}{2}Fe_2O_3 + Al = Fe + \frac{1}{2}Al_2O_3$$
$$\Delta H_{1550\,^\circ C} = -42,3855.2\,J \cdot mol^{-1} \quad (10.9)$$

According to their enthalpy changes, both reactions are exothermic, and the heat released is calculated as follows:

$$6\,g/27\,g \cdot mol^{-1} \times 13,8456.7\,J \cdot mol^{-1} = 30,768.2\,J$$
$$0.48\,g/27\,g \cdot mol^{-1} \times 42,3855.2\,J \cdot mol^{-1} = 7535.2\,J$$

From literature data, the specific heat required to raise the temperature by 100 °C is approximately 125.6 J·g^{-1}. Assuming an initial reaction temperature of 1550 °C and an adiabatic system, the theoretical final temperature after the exothermic reaction is estimated as:

$$1550\,^\circ C + (30,768.2\,J + 7535.2\,J)/125.6\,J \cdot g^{-1}/(10\,g + 31.22\,g) \times 100\,^\circ C = 2289.8\,^\circ C$$

Based on the TiO_2–Fe_2O_3–CaO ternary slag composition presented in Table 10.5, aluminum metal was pressed into flakes and mechanically polished to remove the surface alumina layer prior to the experiment. To ensure thorough reduction and facilitate the aluminothermic reaction, the aluminum flakes and slag materials were arranged alternately within a corundum crucible, as illustrated in Fig. 10.4.

The experiment was conducted in a vertical tube resistance furnace under an argon atmosphere. The system was heated to 1550 °C while maintaining an argon flow rate

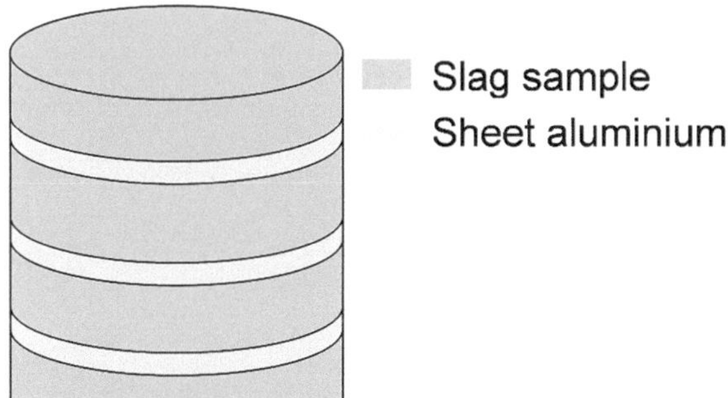

Fig. 10.4 Schematic diagram showing the relative positions of the slag and metallic aluminum sheet inside the corundum crucible

of 0.5 $L \cdot min^{-1}$. After the furnace stabilized for 30 min at the target temperature, the crucible containing the sample was placed on the sample holder and quickly introduced into the constant-temperature zone. Simultaneously, an infrared thermal camera was activated to monitor and record the progression of the aluminothermic reduction process. Figure 10.5 shows the timeline of the reaction: prior to ignition, the exothermic peak during the reaction, and the temperature recovery phase following completion.

As shown in Fig. 10.5, once the sample temperature reached a critical point, the aluminothermic reaction initiated rapidly. The reaction proceeded with significant intensity, causing the temperature to rise sharply within approximately 10 s. Based on thermocouple measurements, the maximum temperature achieved by the sample was approximately 2250 °C, which is in close agreement with the theoretical adiabatic temperature of 2289 °C. Approximately 7 min after the reaction began, the sample temperature gradually decreased and stabilized at the furnace's set temperature of 1550 °C.

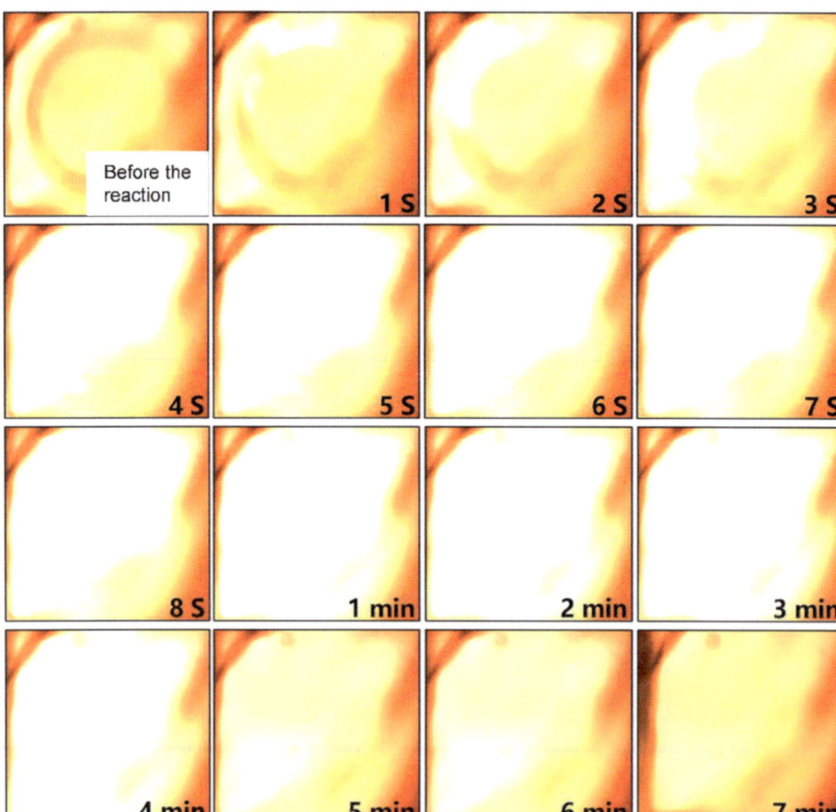

Fig. 10.5 In situ infrared observation of the aluminothermic reduction reaction

At 10 min after the onset of the reaction, the sample was removed from the furnace and rapidly cooled to room temperature under an argon atmosphere to facilitate the separation of the reacted slag phase from the metallic phase. The macroscopic morphology of the resulting product is shown in Fig. 10.6. Following separation, the metallic phase was subjected to microstructural observation and compositional analysis using SEM and EDS, with the results presented in Fig. 10.7.

As evident from Fig. 10.6, the interface between the metallic and slag phases is distinctly defined, with no observable dispersion of the metallic phase into the slag. This indicates that, under the given experimental conditions—specifically, the heat released from the aluminothermic reaction and the controlled reaction time—the viscosity between the slag and alloy phases remained sufficiently low to allow for efficient and thorough phase separation.

The backscattered electron (BSE) imaging function of the SEM produces images with varying shades of gray that reflect differences in the atomic number and composition of the sample. As shown in Fig. 10.7, the distinct grayscale contrast indicates the presence of multiple phases with differing chemical compositions. Within the metallic phase, three distinct compositional regions can be identified: a light gray phase (Region No. 1), a bright white phase (Region No. 2), and a darker phase (Region No. 3).

To analyze these phases, EDS was used to measure the elemental distribution of Ti, Fe, Al, and O. For each phase, five representative areas were selected for surface composition analysis. The average, maximum, and minimum elemental contents for each region are presented in the bar charts of Fig. 10.7.

From a comparative analysis of Regions No. 1 to No. 3, it can be concluded that Region No. 1 (light gray) contains relatively high amounts of both oxygen and titanium, along with moderate levels of aluminum and iron. This suggests that this phase contains residual metal oxides, indicating incomplete reduction. In contrast, Regions No. 2 (bright white) and No. 3 (dark) exhibit very low oxygen content, confirming that they primarily consist of metallic alloy phases. Region No. 2 is enriched in iron and contains a slightly higher aluminum concentration than the other

Fig. 10.6 Macroscopic morphology of the slag-iron phase after aluminothermic reduction

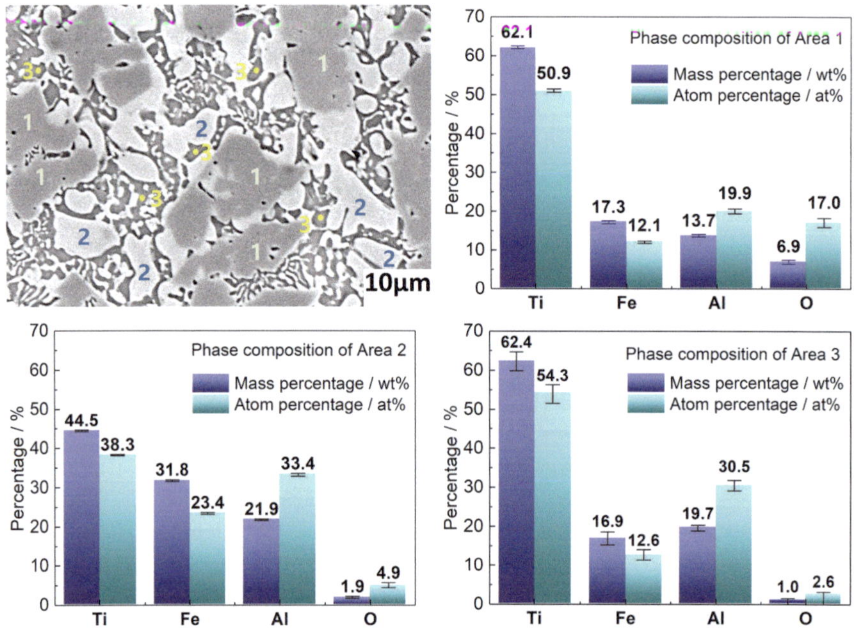

Fig. 10.7 Electron microscopy images of metallic phases and corresponding EDS analysis of phase compositions

regions. Region No. 3, on the other hand, has a higher titanium content, suggesting it is a more titanium-rich titanium–iron-aluminum alloy phase.

The area fractions of these microscopic phases were approximated using image analysis software, with the results shown in Fig. 10.8. Given the assumption that the area ratios correspond closely to volume ratios, the approximate average composition of the metallic phase was calculated based on the elemental distributions and area proportions. The results are summarized in Table 10.6.

Based on this analysis, the titanium–iron alloy obtained through aluminothermic reduction of the TiO_2–Fe_2O_3–CaO ternary slag contains 54.6 wt% titanium—lower than the designed alloy content of 80.0 wt%—and 18.5 wt% aluminum, which exceeds the target value of 10.0 wt%. The iron content was measured at 23.4 wt%, also higher than the expected 10.0 wt%.

In the actual experiment, the total mass of the initial raw material mixture, based on the composition in Table 10.5, was 10 g. The mass of the metallic phase obtained after the reduction reaction was 2.01 g. Using this information, the reduction rate was calculated based on the initial titanium and iron content in the raw materials and the final alloy composition. The calculation process and resulting reduction rates are presented in Eqs. 10.10 and 10.11.

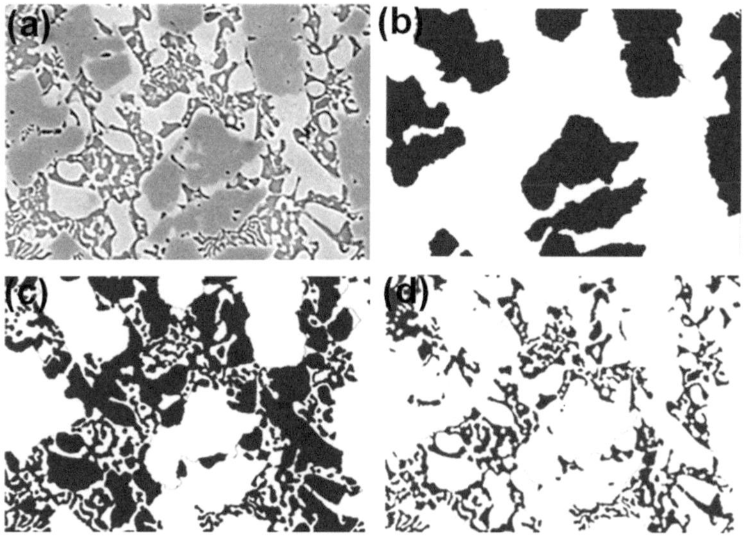

Fig. 10.8 Physical appearance of the three alloy phases identified and extracted using image processing techniques. **a** monolithic alloy; **b** No. 1 grayish phase; **c** No. 2 whiteish phase; **d** No. 3 interphase phase

Table 10.6 Area ratio and composition of the three alloy phases

Physical phase	Ti/wt%	Fe/wt%	Al/wt%	O/wt%	Area ratio/%
No. 1	62.1	17.3	13.7	6.9	35.1
No. 2	44.5	31.8	21.9	1.9	42.9
No. 3	62.4	16.9	19.7	1.0	22.0
Comprehensive	54.6	23.4	18.5	3.5	100.0

$$\begin{aligned}&\text{Reduction rate of Ti}\\&=\frac{\text{Mass of metal phase after reaction}\times\text{Mass fraction of Ti in the metallic phase after reaction}}{\text{Mass of Ti dioxide in raw material before reaction}\times\dfrac{\text{Relative atomic mass of Ti}}{\text{Relative molecular mass of Ti dioxide}}}\\&=\frac{2.01\,\text{g}\times 54.60\,\text{wt\%}}{10\,\text{g}\times 40.57\,\text{wt\%}\times\frac{48}{80}}=45.31\%\end{aligned}\tag{10.10}$$

$$\text{Reduction rate of Fe} = \frac{\text{Mass of metal phase after reaction} \times \text{Mass fraction of Fe in the metal phase after reaction}}{\text{Mass of Fe}_2\text{O}_3 \text{ in raw material before reaction} \times \frac{\text{Relative atomic mass of Fe} \times 2}{\text{Relative molecular mass of Fe}_2\text{O}_3}} = \frac{2.01\,\text{g} \times 23.40\,\text{wt\%}}{10\,\text{g} \times 6.77\,\text{wt\%} \times \frac{56 \times 2}{160}} = 99.74\% \tag{10.11}$$

This reduction rate will serve as an important parameter for subsequent studies, which aim to optimize the smelting of titanium–iron alloys using real fused titanium slag derived from ilmenite concentrate as the raw material, with Fe_2O_3 as a thermal booster and aluminum metal as the reducing agent. These values will inform the initial composition design for future aluminothermic reduction experiments.

10.3 Pre-experiments on the Preparation of Titanium–Iron Alloy

As a preliminary study for the production of titanium–iron alloy via smelting of high-titanium slag derived from the melt separation of ironsands, the previous section (Sect. 7.3.2) explored the feasibility of the aluminothermic reduction method. In that experiment, a synthetic ternary slag system composed of TiO_2–Fe_2O_3–CaO was used, and titanium–iron alloy was successfully produced. The experiment not only confirmed the viability of the method but also provided key data regarding the reduction rates of titanium and iron under the experimental conditions.

In this section, actual high-titanium slag obtained from the melt separation of pre-reduced ironsands is used as the experimental raw material. The objective is to prepare a titanium–iron alloy that meets national standards, using metallic aluminum as the reducing agent and ferric oxide (Fe_2O_3) as the thermal booster. The required dosages of the reducing and heat-generating agents are determined through thermodynamic calculations, based on the titanium and iron reduction rates obtained in Sect. 7.3.2.

Referencing the data in Table 9.6, the titanium slag used here was derived from a process involving gas-based pre-reduction, melt separation, and staged deep reduction. Its composition is provided in Table 10.7.

Table 10.7 Chemical composition of high-titanium slag obtained after melt separation of pre-reduced ironsands/wt%

FeO	TiO_2	MgO	Al_2O_3	SiO_2	CaO	CaF_2
3.28	43.95	16.57	14.97	10.99	2.66	7.58

Using 10 g of fused high-titanium slag as the calculation basis, the necessary quantities of aluminum reductant, ferric oxide, and other ingredients were determined. The chemical reactions of titanium dioxide and ferric oxide with aluminum are described in Eqs. 10.8 and 10.9, respectively. The following thermodynamic calculations were conducted to determine the mass balances during smelting, including both slag and alloy phase formation.

Mass of titanium metal generated, consumption of titanium dioxide, consumption of aluminum metal as a reducing agent, mass of alumina generated

$$\begin{aligned}&\text{Titanium metal generation quality} = \text{high titanium slag titanium}\\&\quad\text{element mass} \times \text{titanium element reduction rate} = 10\,\text{g} \times 43.95\,\text{wt\%}\\&\quad\times \left(48\,\text{g}\cdot\text{mol}^{-1}/80\,\text{g}\cdot\text{mol}^{-1}\right) \times 45.31\,\text{wt\%} = 1.19\,\text{g}\end{aligned}$$

$$\begin{aligned}&\text{Titanium dioxide consumption} = \text{titanium metal mass} \times \text{conversion rate}\\&\quad = 1.19\,\text{g} \times \left(80\,\text{g}\cdot\text{mol}^{-1}/48\,\text{g}\cdot\text{mol}^{-1}\right) = 1.98\,\text{g}\end{aligned}$$

$$\begin{aligned}&\text{Aluminum metal consumption} = \text{mass of titanium metal}\\&\quad\times \text{titanium - aluminum substitution ratio in the chemical reaction equation}\\&\quad = 1.19\,\text{g}/\left(3/4 \times 48\,\text{g}\cdot\text{mol}^{-1}\right) \times 27\,\text{g}\cdot\text{mol}^{-1} = 0.89\,\text{g}\end{aligned}$$

$$\begin{aligned}&\text{Mass of alumina generated} = \text{mass of aluminum metal consumed}\\&\quad\times \text{conversion rate} = 0.89\,\text{g} \times \left(102\,\text{g}\cdot\text{mol}^{-1}/2/27\,\text{g}\cdot\text{mol}^{-1}\right) = 1.68\,\text{g}\end{aligned}$$

Mass of metallic iron produced, consumption of ferric oxide, consumption of metallic aluminum as a reducing agent, mass of alumina produced

$$\begin{aligned}&\text{Mass of iron metal produced} = \text{mass of titanium metal produced}\\&\quad/2(\text{design calculation based on pre-experimental results},\\&\quad\text{titanium} - \text{iron ratio } 2:1) = 1.19\,\text{g}/2 = 0.60\,\text{g}\end{aligned}$$

$$\begin{aligned}&\text{Ferrous oxide consumption} = (\text{mass of metallic iron generated}\\&\quad-\text{mass of iron generated from ferrous oxide in high titanium slag})\\&\quad\times \text{conversion rate} = \left(0.60\,\text{g}{-}0.33\,\text{g}/72\,\text{g}\cdot\text{mol}^{-1} \times 56\,\text{g}\cdot\text{mol}^{-1}\right)\\&\quad\times \left(160\,\text{g}\cdot\text{mol}^{-1}/112\,\text{g}\cdot\text{mol}^{-1}\right) = 0.49\,\text{g}\end{aligned}$$

$$\text{Aluminum Metal Consumption} = \text{Iron Oxide Consumption} \times \text{Iron Oxide to Aluminum Replacement Ratio in Chemical Reaction Equation} = 0.33\,\text{g}/\left(3 \times 72\,\text{g} \cdot \text{mol}^{-1}\right) \times \left(2 \times 27\,\text{g} \cdot \text{mol}^{-1}\right) + 0.49\,\text{g}/160\,\text{g} \cdot \text{mol}^{-1} \times \left(2 \times 27\,\text{g} \cdot \text{mol}^{-1}\right) = 0.25\,\text{g}$$

$$\text{Mass of alumina generated} = \text{mass of aluminum metal consumed} \times \text{conversion rate} = 0.25\,\text{g} \times \left(102\,\text{g} \cdot \text{mol}^{-1}/2/27\,\text{g} \cdot \text{mol}^{-1}\right) = 0.47\,\text{g}$$

Mass of silicon monomers produced, silicon dioxide consumed, aluminum metal consumed as a reducing agent, mass of alumina produced

$$\text{Monolithic silicon generation mass} = \text{high titanium slag in the mass of silicon element} \times \text{silicon element reduction rate} = 10\,\text{g} \times 10.99\,\text{wt\%} \times \left(28\,\text{g} \cdot \text{mol}^{-1}/60\,\text{g} \cdot \text{mol}^{-1}\right) \times 90\% = 0.46\,\text{g}$$

$$\text{Silicon dioxide consumption} = \text{mass of singlet silicon} \times \text{conversion rate} = 0.46\,\text{g} \times \left(60\,\text{g} \cdot \text{mol}^{-1}/28\,\text{g} \cdot \text{mol}^{-1}\right) = 0.99\,\text{g}$$

$$\text{Aluminum metal consumption} = \text{mass of singlet silicon} \times \text{ratio of silicon to aluminum substitution in the chemical reaction equation} = 0.46\,\text{g} \times \left(4 \times 27\,\text{g} \cdot \text{mol}^{-1}/3/28\,\text{g} \cdot \text{mol}^{-1}\right) = 0.59\,\text{g}$$

$$\text{Mass of alumina generated} = \text{mass of aluminum metal consumed} \times \text{conversion rate} = 0.59\,\text{g} \times \left(102\,\text{g} \cdot \text{mol}^{-1}/2/27\,\text{g} \cdot \text{mol}^{-1}\right) = 1.11\,\text{g}$$

Based on the above calculations, the full set of raw material ratios and resulting slag compositions are summarized in Tables 10.8 and 10.9.

Table 10.8 Composition ratios of primary ingredients

Dosage	High titanium slag	Al	Fe_2O_3
Mass/g	10.00	1.73	0.49
Percentage/wt%	81.8	14.2	4.0

Table 10.9 Predicted slag composition after aluminothermic reduction reactions

Ingredient	TiO_2	MgO	Al_2O_3	SiO_2	CaO	CaF_2
Mass/g	2.42	1.66	4.76	0.11	0.27	0.76
Percentage/wt%	24.2	16.6	47.8	1.1	2.7	7.6

Calcium oxide dosage and excess aluminum metal dosage

Using the equilibrium module in the FactSage 7.0 thermodynamic calculation software, the phase compositions at thermodynamic equilibrium were obtained under the condition of 1550 °C, based on the slag composition and contents provided in Table 10.9. The resulting equilibrium phase data are summarized in Table 10.10.

To ensure adequate slag fluidity, it is essential that the slag remains in the liquid phase at the experimental temperature, thereby preventing the formation of solid phases. As indicated in the phase diagram in Fig. 10.3, increasing the CaO content in the slag enhances its liquid-phase proportion. Therefore, when smelting titanium–iron alloys using high-titanium slag as the raw material, the addition of CaO remains necessary to achieve full slag liquefaction and maintain its fluidity.

Using the FactSage 7.0 thermodynamic software and taking the slag composition in Table 10.9 as the basis for calculations, the CaO content was gradually increased until a fully liquid slag phase was obtained at 1550 °C. At this condition, the required CaO mass was determined to be 1.43 g, indicating that an additional 1.16 g of CaO must be added to the original high-titanium slag.

Moreover, to ensure complete reduction of titanium, it is necessary to maintain a minimum concentration of metallic aluminum (approximately 10 wt%) in the metal phase during the smelting process. Therefore, an additional 0.3 g of aluminum reductant was added in excess.

In summary, the final smelting feed composition, along with the predicted alloy and slag compositions, are presented in Table 10.11.

The raw materials were weighed and prepared according to the composition ratios listed in Table 10.11 and placed into corundum crucibles following the spatial arrangement illustrated in Fig. 10.4. After the tubular vertical resistance furnace was preheated to 1550 °C under an argon atmosphere, the crucibles containing the raw materials were inserted into the isothermal zone of the furnace and held at this temperature for 30 min. During this period, an aluminothermic reduction reaction occurred,

Table 10.10 Phase compositions at 1550 °C under the compositional conditions of Table 10.9

Physical phase	Mass/g	Percentage/wt%	Note
Liquid phase	8.45	84.7	CaO% = 6.5; SiO_2% = 1.3; MgO% = 15.6; Al_2O_3% = 44.7; TiO_2% = 28.7; CaF_2% = 3.3
$CaO·2MgO·8Al_2O_3$	1.15	11.5	Solid phase
MgF_2	0.38	3.8	Solid phase

Table 10.11 Design of raw material ratios, alloy phases, and slag phase compositions for titanium ferroalloy smelting

Proportion of raw materials	High titanium slag	Al	Fe_2O_3	CaO
Mass/g	10.00	2.03	0.49	1.16
Percentage/wt%	73.1	14.8	3.6	8.5

Alloys	Ti	Al	Fe	Si
Mass/g	1.19	0.30	0.60	0.46
Percentage/wt%	46.7	11.8	23.5	18.0

Slag	TiO_2	MgO	Al_2O_3	SiO_2	CaO	CaF_2
Mass/g	2.42	1.66	4.76	0.11	1.43	0.76
Percentage/wt%	21.7	14.9	42.7	1.0	12.8	6.8

releasing a substantial amount of heat. Metallic aluminum reduced the iron and titanium oxides, after which the system gradually returned to the set furnace temperature. The metallic and slag phases separated and aggregated distinctly. Upon completion of the reaction, the crucibles and samples were removed and cooled to room temperature under argon atmosphere. The solidified products were subsequently analyzed using SEM and EDS, and the results are presented in Fig. 10.9.

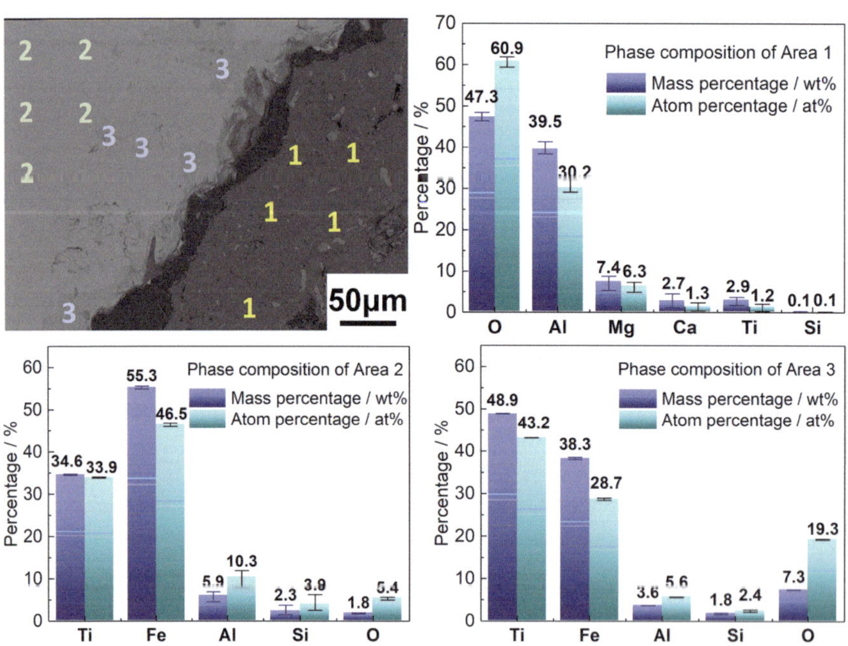

Fig. 10.9 SEM morphology and EDS analysis of the titanium–iron alloy and slag phases

In Fig. 10.9, two homogeneous phases within the alloy (Phases 2 and 3) and one slag phase (Phase 1) were characterized. Their elemental compositions are shown in the corresponding bar graph. Phase 2 was identified as the predominant constituent of the alloy. When compared with the Chinese national titanium–iron alloy standards in Table 10.2, the composition of Phase 2 aligns with the FeTi30-B grade and is approaching the FeTi40-A grade. These results confirm the feasibility of producing titanium–iron alloy from high-titanium slag derived from ironsands via aluminothermic reduction. However, several challenges were observed during the experiment:

Presence of Residual Oxides in the Alloy Phase

Phase 3 in Fig. 10.9, although high in titanium content (48.9 wt%), also shows a relatively high oxygen content (7.3 wt%), suggesting incomplete reduction and the presence of metal oxides. Given the comparative reducibility of Ti, Fe, and Si oxides—where titanium oxides are notably more difficult to reduce—it is likely that the remaining oxide phase is primarily composed of titanium oxides. This indicates that some titanium was not fully reduced and remained in the alloy as oxides.

Incomplete Aggregation of the Alloy Phase

Post-reaction sampling revealed that the alloy phase did not form a large, continuous metallic mass as anticipated (see Fig. 10.4), but instead was dispersed within the slag as fine particles. As shown in Fig. 10.10, under low magnification, the metallic particles were embedded within the slag, with particle sizes ranging from 10 μm to 1 mm. This dispersion may be attributed to the relatively high viscosity and surface tension of the slag, which hindered the coalescence of the metal phase and thus impeded efficient metal-slag separation.

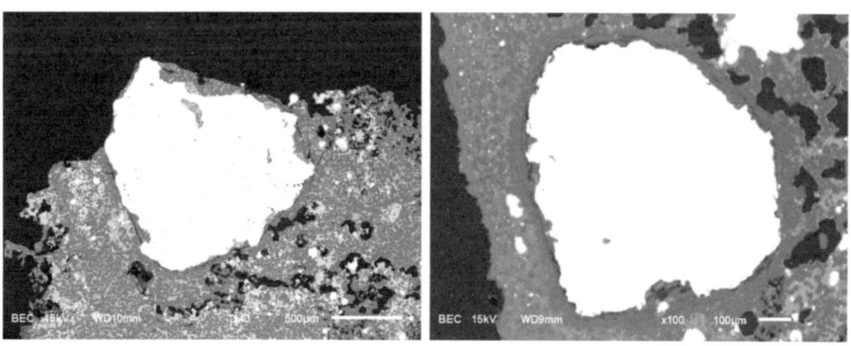

Fig. 10.10 Alloy metal phases exhibiting point-like distribution characteristics

10.4 Factors of Aluminum-Thermal Reduction

10.4.1 Dosage of Reducing Agent

The Factsage thermodynamic software was employed to simulate the effect of aluminum (Al) dosage on the aluminothermic reduction reaction at 1550 °C, using 100 g of slag with varying amounts of Al addition.

As shown in Fig. 10.11, increasing the Al dosage from 60 to 120 wt% led to a significant rise in both alloy yield and Ti metal reduction rate. Specifically, the alloy yield increased from 27.71 to 94.23 wt%, while the Ti reduction rate rose from 18.95 to 94.70%. This trend indicates that higher Al input promotes the forward reactions of TiO_2, SiO_2 and Fe_2O_3 with Al, thereby enhancing the overall reduction and alloy formation.

Figure 10.12 illustrates the effect of Al dosage on intermediate product formation and system temperature. With increasing Al dosage, the content of TiO in the liquid phase gradually declined until it completely disappeared. Conversely, the formation of Ti_5Si_3 in the solid phase initially increased with rising Al content but later decreased and eventually vanished. This behavior is attributed to the exothermic nature of the aluminothermic reactions: as more Al is added, greater heat is released, elevating the system temperature. The temperature rise suppresses the intermediate reduction product TiO, accelerating the direct reduction of TiO_2 to Ti metal. The increase in available Ti also initially favors the formation of Ti_5Si_3 (from Ti and Si), which peaks when Al dosage reaches ~ 70 wt%. Beyond this point, as the temperature continues to rise, Ti_5Si_3 melts or reacts further, leading to its gradual disappearance.

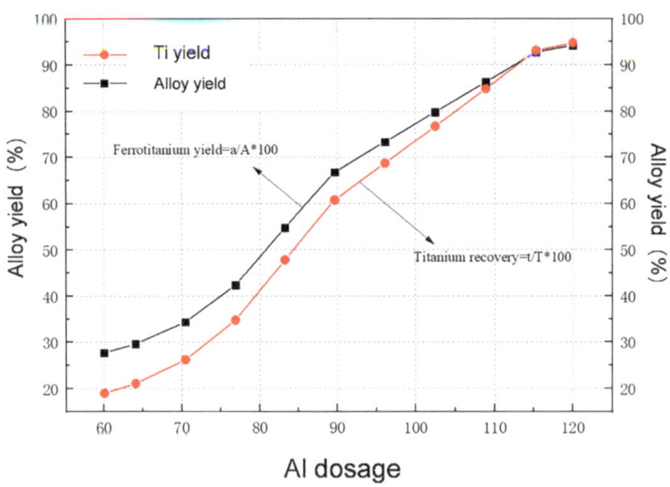

Fig. 10.11 Effect of aluminum addition on alloy yield and titanium recovery

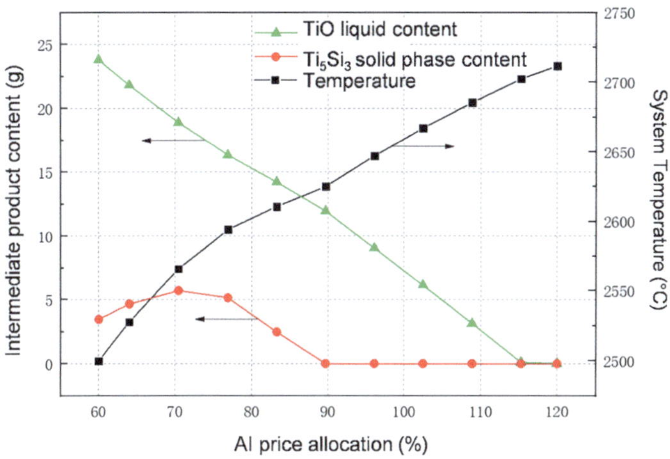

Fig. 10.12 Effect of aluminum addition on system temperature and intermediate reaction products

The effect of Al dosage on the elemental composition of the alloy is illustrated in Fig. 10.13. As the Al dosage increased from 60 wt% to 120 wt%, the Ti content in the alloy rose from 46.98 wt% to a peak of 60.52 wt%, before slightly declining to 59.33 wt%. Meanwhile, the Si content decreased from 26.58 wt% to 16.43 wt%, and the Fe content declined from 20.68 wt% to 5.24 wt%. In contrast, the aluminum content increased steadily from 5.71 wt% to 18.96 wt%.

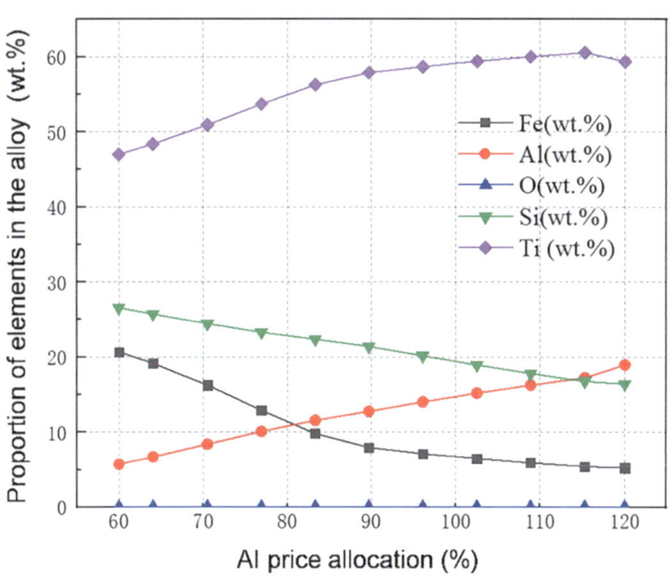

Fig. 10.13 Influence of aluminum addition on the elemental composition of the alloy

These trends are primarily attributed to the sequential reduction behavior of the metal oxides. Thermodynamically, Fe_2O_3, SiO_2, and TiO_2 are reduced by Al in the order of Fe → Si → Ti. As a result, with increasing Al dosage, the earlier-reduced Fe and Si become less dominant in the alloy composition, while Ti and Al contents increase. The slight decline in Ti content when the Al dosage exceeds 110 wt% is due to the complete disappearance of TiO in the liquid phase and a reduction of TiO_2 content in the slag to approximately 2.5 wt%. At this stage, the excess Al enters the alloy phase more readily than Ti, thereby diluting the Ti content in the final alloy.

10.4.2 Dosage of Heat Generating Agent

Factsage software was used to simulate the effect of Fe_2O_3 dosage on the aluminothermic reduction reaction at 1550 °C, using 100 g of slag and 23 g of Al, while varying the amount of Fe_2O_3 added.

The influence of Fe_2O_3 dosage on alloy yield and Ti reduction efficiency is illustrated in Fig. 10.14. As the Fe_2O_3 content increased from 5 to 20 g, the alloy yield rose from 52.50 wt% to 63.84 wt%, while the Ti reduction rate declined from 33.26% to 14.37%. This is primarily attributed to the preferential and enhanced reaction between Fe_2O_3 and Al, which increases Fe incorporation into the alloy and consequently raises alloy yield. However, this reaction consumes a greater portion of the available Al, thereby reducing the amount available for the reduction of TiO_2, resulting in a decreased Ti reduction rate.

The effect of Fe_2O_3 dosage on system temperature and intermediate products is shown in Fig. 10.15. As Fe_2O_3 dosage increases, the exothermic reaction between Fe_2O_3 and Al progresses positively, raising the system temperature. Simultaneously,

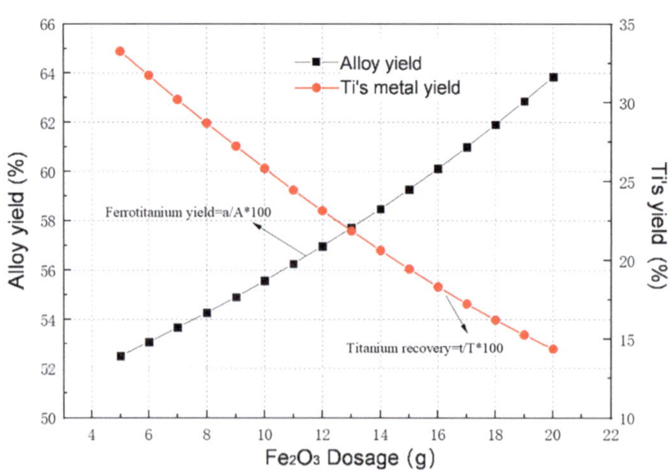

Fig. 10.14 Effect of Fe_2O_3 addition on alloy yield and titanium recovery

less Al is available for the reduction of TiO_2, which leads to increased formation of intermediate TiO species due to the higher reaction temperature and insufficient reducing agent.

The elemental composition of the resulting alloy as a function of Fe_2O_3 dosage is presented in Fig. 10.16. With Fe_2O_3 increasing from 5 to 20 g:

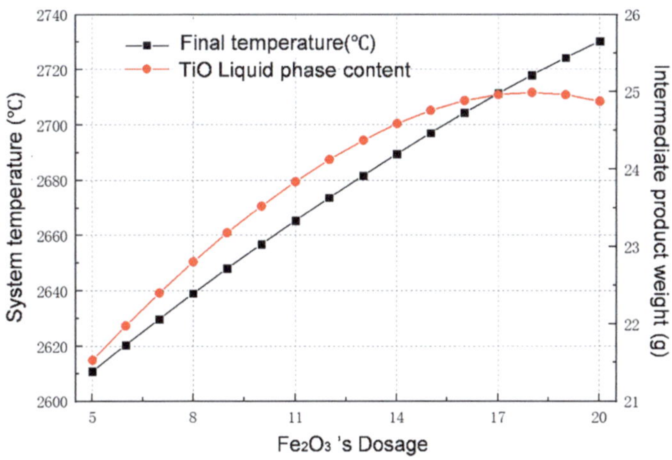

Fig. 10.15 Effect of Fe_2O_3 addition on system temperature and intermediate products

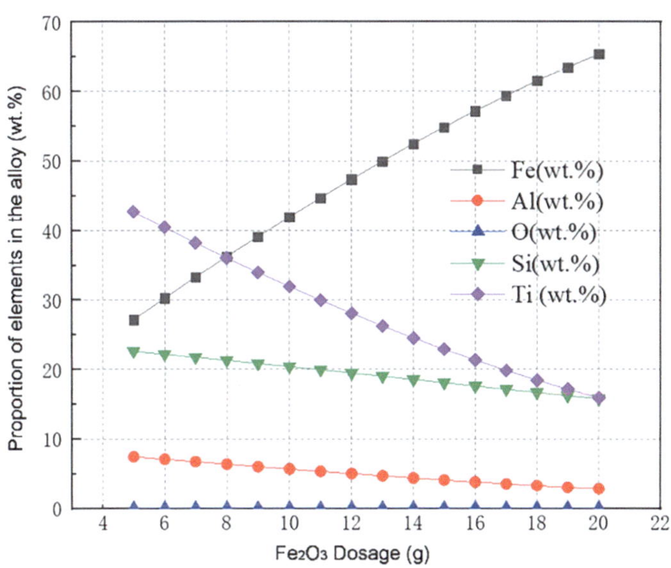

Fig. 10.16 Influence of Fe_2O_3 addition on the elemental composition of the alloy

(1) Ti content decreased from 42.70 to 15.93 wt%,
(2) Si content declined from 22.63 to 15.76 wt%,
(3) Fe content increased significantly from 27.17 to 65.39 wt%,
(4) Al content dropped from 7.44 to 2.83 wt%.

These trends further confirm that Fe_2O_3 is preferentially reduced, dominating alloy composition at the expense of Ti and Si reduction.

10.4.3 Dosage of Flux Agent

Further simulations were performed to examine the influence of CaO addition at 1550 °C using a base of 100 g slag, 23 g Al, and 10 g Fe_2O_3.

The relationship between the CaO/Al_2O_3 weight ratio and both alloy recovery and Ti reduction rate is shown in Fig. 10.17. As the CaO/Al_2O_3 ratio increases from 0.5 to 1.3, both the alloy yield and Ti reduction rate increase. However, when the ratio is raised from 1.3 to 1.5, both parameters decline. This behavior is explained by a decrease in slag viscosity at moderate CaO levels, which promotes more efficient aluminothermic reduction and facilitates alloy separation. Conversely, excessive CaO promotes the formation of $Ca_2Ti_2O_5$, a stable compound that consumes TiO_2 and thereby hinders Ti reduction, leading to reduced alloy formation.

Figure 10.18 shows the effect of CaO/Al_2O_3 ratio on intermediate products and system temperature. Increasing the CaO/Al_2O_3 ratio reduces TiO intermediates and lowers the system temperature overall. A small temperature peak occurs at a ratio

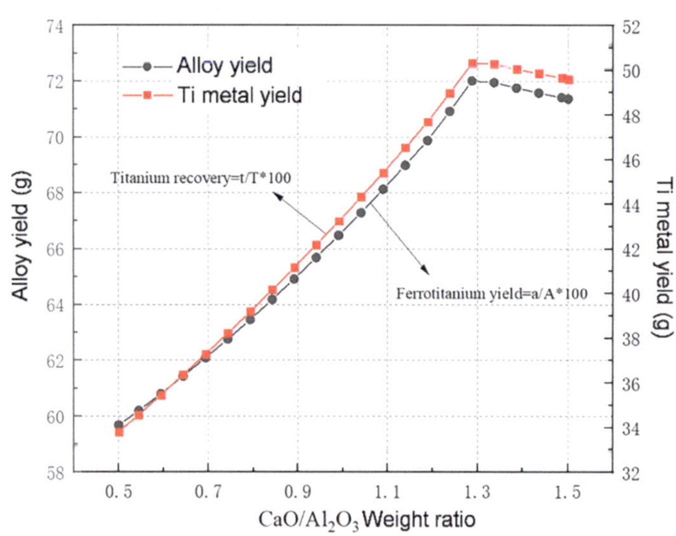

Fig. 10.17 Effect of CaO addition on alloy yield and titanium recovery

of 1.2–1.3, followed by a decline. Notably, the formation of $Ca_2Ti_2O_5$ begins at a ratio of 1.23, where CaO reacts exothermically with TiO_2. The quantity of $Ca_2Ti_2O_5$ continues to grow as CaO content increases.

The impact of CaO/Al_2O_3 ratio on alloy composition is shown in Fig. 10.19:

(1) Ti content increases from 37.63 to 45.55 wt%, then stabilizes,
(2) Si content decreases from 15.77 to 13.05 wt%,

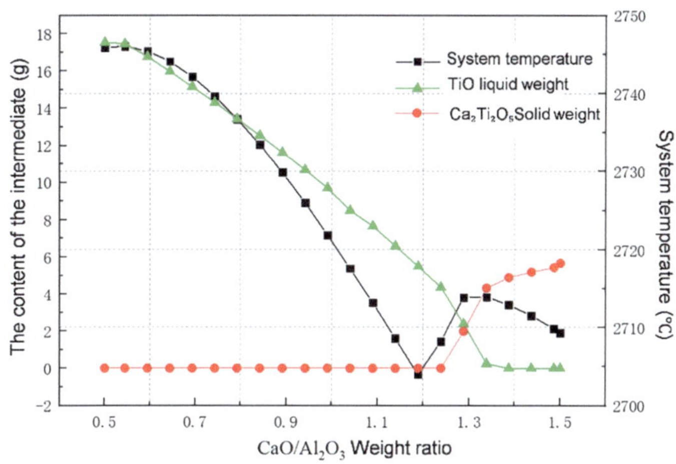

Fig. 10.18 Effect of CaO addition on system temperature and intermediate products

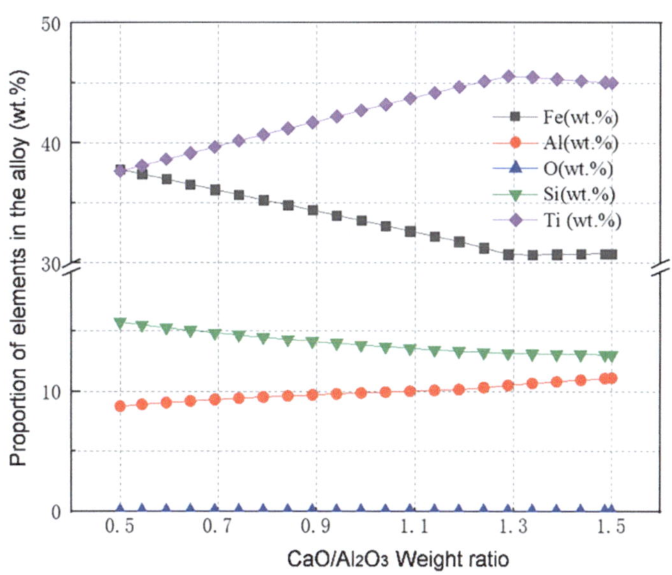

Fig. 10.19 Influence of CaO addition on the elemental composition of the alloy

(3) Fe content decreases from 37.78 to 30.79 wt%,
(4) Al content increases from 8.99 to 11.12 wt%.

The increase in Ti content results from reduced slag viscosity and weakened interaction between TiO and Al_2O_3 as CaO—a stronger basic oxide—displaces TiO from stable complexes. This enhances the Ti reduction reaction while reducing the activity of Al_2O_3. The relative increase in Ti and Al thus reduces the proportions of Fe and Si in the alloy.

In summary, the thermodynamic simulations using Factsage software at 1550 °C demonstrate the significant effects of reductant (Al), pyrophoric agent (Fe_2O_3), and flux (CaO) on aluminothermic reduction. Key findings include:

(1) Increasing Al dosage from 60 to 120 wt% raises alloy yield from 27.71 to 94.23 wt%, and Ti reduction from 18.95 to 94.70%.
(2) Increasing Fe_2O_3 from 5 to 20 g increases alloy yield from 52.50 to 63.84 wt%, but reduces Ti reduction from 33.26 to 14.37%.
(3) Increasing the CaO/Al_2O_3 ratio from 0.5 to 1.3 enhances both alloy yield and Ti reduction, but further increase to 1.5 causes both to decline due to the formation of $Ca_2Ti_2O_5$.

These results provide important insights into the optimization of raw material ratios for efficient Ti–Fe alloy production via aluminothermic reduction.

10.5 Preparation of Titanium–Iron Alloys by Aluminum-Thermal Method

Titanium-iron alloy is a ferrous alloy primarily composed of iron and titanium, with a titanium content typically ranging from 20 to 75 wt% [1]. Due to titanium's high chemical reactivity with sulfur, carbon, oxygen, and nitrogen, titanium–iron alloys are widely employed in the steelmaking industry as effective deoxidizers and desulfurizers for removing impurities from molten steel [2]. Additionally, ferro-titanium serves as a Ti alloying additive, enhancing the mechanical strength, corrosion resistance, heat resistance, and other performance characteristics of steel [2, 3].

The extensive application of titanium–iron alloys has made their production a key area of research. Table 10.12 summarizes recent studies on the synthesis of titanium–iron alloys, including starting materials, preparation methods, and the titanium content in the final product. Among these, electrochemical reduction in molten $CaCl_2$-based electrolytes has emerged as a major preparation method. Mohanty et al. [4] and Ma et al. [5] demonstrated the feasibility of directly reducing ilmenite and titanium dioxide slag to titanium–iron alloys via electrochemical techniques. Building on the Fray-Farthing-Chen (FFC) process [6], Panigrahi et al. [1, 7] successfully synthesized high-purity Fe–Ti intermetallic alloys using mixed oxide precursors such as $FeTiO_3/TiO_2$ and Fe_2O_3/TiO_2. Subsequently, Zhou et al. [8] and Li et al. [9] provided detailed investigations into phase transformation behavior and the influence of CaO addition during electrochemical reduction in $CaCl_2$–NaCl molten salt systems.

However, electrochemical reduction methods present inherent limitations, including prolonged electrolysis duration and the formation of porous product structures, which have spurred interest in alternative techniques such as aluminothermic reduction (ATR). For instance, Pourabdoli et al. [10] utilized titanium dioxide slag derived from ilmenite smelting as the raw material and optimized ATR parameters to produce titanium–iron alloys with titanium contents ranging from 45 to 54 wt%. Similarly, Sokolov [3] and Babyuk et al. [11] reported the successful production of titanium–iron alloys with approximately 48 wt% titanium via ATR treatment of ilmenite.

Titanium ranks fourth among metal reserves, following iron, aluminum, and magnesium. Global reserves of titanium oxide are estimated at approximately 650 billion tons, with rutile (TiO_2) and ilmenite ($FeTiO_3$) commonly extracted through electric furnace smelting and hydrometallurgical methods [12]. Previous studies [13–15] have achieved titanium enrichment by employing pre-reduction and high-temperature melt separation of ironsands, resulting in titanium-containing slag with a TiO_2 content of approximately 42.5 wt%. In the present study, this titanium-enriched slag is utilized as the primary titanium source for the production of titanium–iron alloy via aluminothermic reduction (ATR).

In the earlier work, gas-based reduction of ironsands followed by high-temperature separation of molten phases was conducted to yield high-purity iron and titanium-rich slag [13–15]. The compositional details of the resulting direct reduced iron (DRI) and fused titanium slag are presented in Table 10.13. Following mineral beneficiation, gas reduction, and melt separation, the TiO_2 content in the titanium slag was successfully increased to 43 wt%, offering a viable feedstock for downstream titanium alloy production.

The ATR reaction system consisted of pre-reduced and melt-separated titanium-containing slag, a reducing agent (aluminum metal, purity > 99.5 wt%, Si < 0.2 wt%, Zn < 0.04 wt%, Cu < 0.04 wt%), a fluxing agent (CaO, purity > 99.9 wt%), and an exothermic agent (hematite, Fe_2O_3, purity > 99.9 wt%). To remove any residual hydrates or carbonates, CaO and Fe_2O_3 were pre-calcined at 1000 °C for 2 h under an argon atmosphere. The raw materials were then weighed, ground, and mixed in a ball mill for 1 h. The resulting powder mixture (particle sizes: Al: 0.6–1.0 mm; others < 0.15 mm) was loaded into an alumina crucible, which was nested inside a graphite crucible to serve as a protective layer.

A thermocouple positioned beneath the graphite crucible monitored the sample zone temperature. Once the target furnace temperature was reached and held constant for 2 h, the crucible assembly was swiftly inserted into the furnace from the bottom using an alumina platform. The aluminothermic reaction was triggered instantaneously at the reaction temperature, releasing a large amount of heat in a short time, which ensured favorable kinetic conditions for metal reduction. After the reaction onset, the sample was held at the reaction temperature for an additional 2 h to ensure

Table 10.12 Summary of titanium–iron preparation results in different studies

Project	Ingredients, Content, Method
Raw material (ilmenite)	$FeTiO_3$ (Content:99.9 wt%), TiO_2 (Content:99 wt%), $FeTiO_3$:TiO_2 = 2.78:1 or 0.58:1
Extraction methods	Direct electrochemical reduction to remove water for 8–10 h, with the electrolysis process lasting 25–30 h
Titanium-iron alloy composition/wt%	Ti:45–70, Fe:26–51, Ca:0.13–0.24, O:0.19–0.46, C:0.03–0.34
Raw material (ilmenite)	Chemically pure $FeTiO_3$
Extraction methods	Thermal reduction by electrolytic reduction at 4.4 V, 973 K for 24 h
Titanium–iron alloy composition	Fe_2Ti and FeTi
Raw material (ilmenite)	Pure $FeTiO_3$
Extraction methods	Electrolytic reduction heat reduction at 3.2 V, 973 K for 15 h
Titanium-iron alloy composition	FeTi
Raw material (high-titanium slag)/wt%	TiO_2:63, FeO:17, SiO_2:10, Al_2O_3:2.5, MgO:1.2, CaO:3.5, V_2O_5:0.4, Cr_2O_3:0.04, P_2O_5:0.2
Extraction methods	Aluminum thermal reduction reaction in an ESCM furnace
Titanium-iron alloy composition/wt%	Ti:54.5, Fe:22.2, Al:6, Si:11.5, Mn:2, C:0.85, P:0.08, S:0.007
Raw material (titanium concentrate)/wt%	TiO_2:63, FeO:27, SiO_2:2.7, Al_2O_3:4.3, MgO:0.9, Cr_2O_3:1.5, MnO:0.4, CuO:0.2
Extraction method 1	The pilot-scale thermal reduction of aluminum was carried out in two steps. In the first step: ilmenite, lime and anthracite were melted; in the second step: the resulting TiO_2 slag was poured into the reactor together with Al and iron ore
Titanium–iron alloy composition1/wt%	Ti:48.3, Al:13.8, Si:2.57, Mn:0.64, Cr:0.95

(continued)

Table 10.12 (continued)

Project	Ingredients, Content, Method
Extraction method 2	Thermal reduction of aluminum on an industrial scale is carried out in two steps. In the first step: ilmenite, lime and anthracite are melted; in the second step: the resulting TiO_2 slag is poured into the reactor together with Al and iron ore
Titanium–iron alloy composition2/wt%	Ti:31.0, Fe:57.2, Al:4.6, Si:4.9, Mn:0.24, V:0.13, Cu:1.6
Raw material (ilmenite)/wt%	TiO_2:62.85, MgO:1.24 SiO_2:2.25, Al_2O_3:4.08, Cr_2O_3:1.72, MnO:0.71, CuO:0.35
Extraction methods	Addition of the reducing agent aluminum metal after ilmenite is completely melted down
Titanium–iron alloy composition	① Utilization of ilmenite: 11.4 kg, CaO:0.6 kg, Al:3 kg Production of titanium–iron alloys Ti:30.5wt%, Al:4.5wt%, Si:1.4wt%, Cr:1.0wt%, Mn:1.0wt% ② Utilization of ilmenite: 39.9 kg, CaO:2.1 kg, Al:18 kg Production of titanium–iron alloys Ti:48.5 wt%, Al:13.3 wt%, Si:2.4 wt%, Cr:0.86 wt%, Mn:0.76 wt% ③ Utilization of ilmenite:114 kg, CaO:6kg, Al:40 kg Production of titanium–iron alloys Ti:48.5wt%, Al:13.8wt%, Si:2.6wt%, Cr:0.95wt%, Mn:0.64wt%

Table 10.13 Chemical composition of directly reduced iron and titanium dioxide slag after gas-based reduction and melt separation processes/wt%

Ingredient	TFe	FeO	TiO_2	SiO_2	MgO	Al_2O_3	CaO	CaF_2	Margin
Direct iron reduction	64.63	8.31	13.26	4.80	4.35	3.93	0.70	–	0.03
Fused titanium slag	2.14	2.75	42.50	15.73	15.83	14.31	2.54	6.35	0.13

complete separation of the metallic and slag phases. The crucible was then transferred to a cooling chamber and allowed to cool to room temperature under an argon atmosphere.

Throughout the experiment, argon gas (purity 99.5%) was continuously purged into the furnace as a protective atmosphere. Simultaneously, temperature changes in the sample were recorded using an infrared thermal imaging system, schematically illustrated in Fig. 10.20. Following cooling, the samples were ground and polished, and the microstructure and chemical composition of the alloy and slag phases were analyzed using a scanning electron microscope (SEM, FEI Quanta 250) equipped with an energy-dispersive X-ray spectrometer (EDS, EDAX).

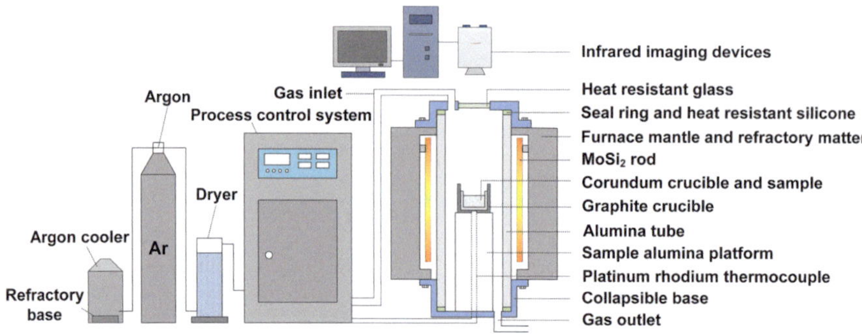

Fig. 10.20 Schematic diagram of the experimental setup for titanium–iron alloy preparation via aluminothermic reduction of titanium slag

Previous studies [11, 16, 17] have demonstrated that thermodynamic data calculated using the FactSage software are reliable for simulating aluminothermic reduction processes. Accordingly, the TiO_2 and SiO_2 components in titanium slag, as well as the Fe_2O_3 additive, can spontaneously react with metallic Al at the experimental temperature, as indicated by the Gibbs free energy changes calculated by FactSage and shown in Fig. 10.21a. Thermodynamically, Fe_2O_3 is the most readily and preferentially reduced by Al, while the reactivities of SiO_2 and TiO_2 toward Al are comparable. The enthalpy changes of the aluminothermic reduction reactions, presented in Fig. 10.21b, confirm that all three reactions are exothermic, with the reduction of Fe_2O_3 being particularly exothermic, releasing approximately 350–370 $kJ{\cdot}mol^{-1}$.

$$\mathrm{Al} + 0.75\mathrm{TiO_2} \Rightarrow 0.75\mathrm{Ti} + 0.5\mathrm{Al_2O_3} \tag{10.12}$$

$$\Delta G_1 = \Delta G_1^{\theta} + \mathrm{R} \cdot T \cdot \ln \frac{a_{\mathrm{Ti}}^{0.75} \cdot a_{\mathrm{Al_2O_3}}^{0.5}}{a_{\mathrm{Al}} \cdot a_{\mathrm{TiO_2}}^{0.75}} \tag{10.13}$$

$$\mathrm{Al} + 0.5\mathrm{Fe_2O_3} \Rightarrow \mathrm{Fe} + 0.5\mathrm{Al_2O_3} \tag{10.14}$$

$$\Delta G_2 = \Delta G_2^{\theta} + \mathrm{R} \cdot T \cdot \ln \frac{a_{\mathrm{Fe}} \cdot a_{\mathrm{Al_2O_3}}^{0.5}}{a_{\mathrm{Al}} \cdot a_{\mathrm{Fe_2O_3}}^{0.5}} \tag{10.15}$$

$$\mathrm{Al} + 0.75\mathrm{SiO_2} \Rightarrow 0.75\mathrm{Si} + 0.5\mathrm{Al_2O_3} \tag{10.16}$$

$$\Delta G_3 = \Delta G_3^{\theta} + \mathrm{R} \cdot T \cdot \ln \frac{a_{\mathrm{Si}}^{0.75} \cdot a_{\mathrm{Al_2O_3}}^{0.5}}{a_{\mathrm{Al}} \cdot a_{\mathrm{SiO_2}}^{0.75}} \tag{10.17}$$

$$a_i = \frac{n_i}{\sum n_i} \cdot \gamma_i \tag{10.18}$$

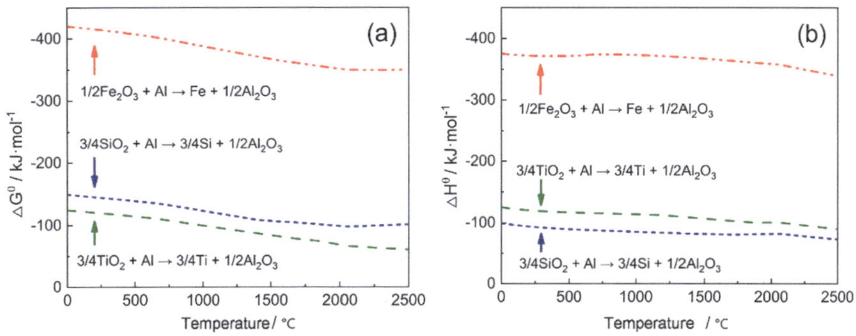

Fig. 10.21 Thermodynamic changes for key reactions: **a** Gibbs free energy and **b** enthalpy

$$\min .G = \sum_{r=1}^{3} n_r \cdot \Delta G_r \tag{10.19}$$

$$s.t. \sum_{i=1}^{j} \sum_{e=1}^{k} m_{ie} \cdot n_i = C \tag{10.20}$$

Equations 10.12 to 10.17 describe the aluminothermic reduction reactions and their corresponding Gibbs free energy changes. As the slag and alloy phases are in the liquid state during the reaction, the activity of each component is calculated using Eq. 10.18, where i denotes a component (e.g., Si, Al, Fe, Ti, Al_2O_3, Fe_2O_3, SiO_2, TiO_2), and n_i, a_i, and γ_i represent the mole number, activity, and activity coefficient, respectively. According to the principle of minimum Gibbs free energy [18], the equilibrium composition of a complex system can be determined by minimizing the total free energy, as expressed in Eq. 10.19, where r corresponds to reactions 10.12, 10.14, and 10.16, and n_r is the number of moles of reaction r. Equation 10.20 serves as a constraint to enforce elemental conservation, where e denotes an element, j and k represent the total number of components and elements, and m_{ie} is the atomic number of element e in component i. These equations allow the exploration of the thermodynamic equilibrium state of the aluminothermic reduction process, including the content of each component, the masses of the slag and alloy phases, and thermodynamic parameters such as the adiabatic temperature.

Figure 10.22 presents the changes in alloy and slag compositions as the amount of added Al increases from 5 to 20 g under constant conditions for 100 g of titanium slag. Figure 10.22a, b indicate that increasing Al addition results in elevated Ti, Si, and Al contents in the alloy phase and a corresponding reduction in TiO_2 and SiO_2 in the slag, suggesting that higher Al content enhances the reduction of Ti and Si oxides. However, when the Al_2O_3 content in the slag exceeds 12 wt%, the increase in Ti content in the alloy begins to plateau, while Al content continues to rise, deviating from the optimal alloy composition. Figure 10.22c shows that at a slag Al_2O_3 content of 12 wt%, the slag/alloy weight ratio reaches approximately 9.8, and the adiabatic temperature of the system approaches 2500 °C—conditions favorable for slag–alloy separation. Thus, a 12 wt% Al_2O_3 slag ratio was selected for validation experiments, considering alloy quality, slag/alloy ratio, and thermal characteristics. The range and optimization scheme for the aluminothermic reduction are summarized in Table 10.14. Based on this scheme, the calculated thermodynamic equilibrium compositions of the alloy and slag phases are provided in Table 10.15. Furthermore, using the data from Table 10.15, Factsage calculations were performed to simulate phase changes during cooling (from 2400 to 100 °C), as shown in Fig. 10.22d. These results indicate that both slag and alloy remain in a liquid state above 1900 °C, which is beneficial for their separation.

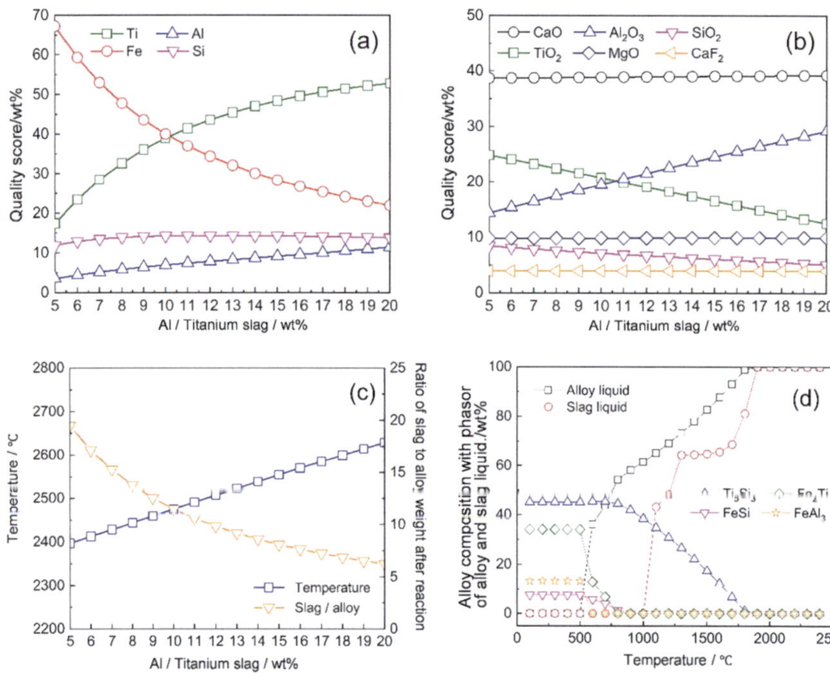

Fig. 10.22 Variations in slag and alloy phase characteristics with aluminum addition under thermodynamic equilibrium conditions: **a** alloy composition and content, **b** slag composition and content, **c** slag/alloy weight ratio and system temperature under adiabatic conditions, and **d** alloy and slag compositions and contents at various temperatures

Table 10.14 Thermodynamic optimization scheme for aluminothermic reduction of titanium slag

Ingredient	Titanium slag	CaO	Fe_2O_3	Al
Calculated thermodynamic range/g	100	60	5	5–20
Optimization solutions/g	100	60	5	12
Percentage/wt%	56	34	3	7

Table 10.15 Alloy and slag phase compositions and contents based on the thermodynamic optimization scheme

Slag	TiO_2/wt%	CaO/wt%	Al_2O_3/wt%	MgO/wt%	SiO_2/wt%	CaF_2/wt%
	19.1	38.9	21.5	9.9	6.7	4.0
Alloy	Ti/wt%	Fe/wt%	Al/wt%	Si/wt%	Alloy slag ratio	Temperature/°C
	43.6	34.3	7.8	14.3	0.10	2507

A schematic of the reaction process is shown in Fig. 10.23. The aluminothermic reduction reaction releases a large amount of heat in a short time, resulting in a rapid and intense reaction. The peak temperature is reached within 10 s and maintained for approximately 4 min, with the total temperature fluctuation lasting around 7 min. After cooling, the reaction heat is sufficient to ensure separation of the alloy and slag phases, and the majority of the alloy aggregates effectively. The alloy was physically separated from the slag, and a product yield of approximately 12.7% (alloy/initial TiO_2 slag = 12.7 g/100 g) was achieved, which is about 3.3% lower than the theoretical value.

The SEM morphology and EDS surface elemental mapping of the alloy sample, rapidly cooled under argon, are shown in Fig. 10.24a. The alloy exhibits a uniform composition with minimal inclusions (3 inclusions within 0.52 mm^2). ICP analysis of major elements reveals discrepancies between the experimental and theoretical compositions (Table 10.15): Ti and Si contents are significantly reduced from 43.6 wt% and 14.3 wt% to 30.5 wt% and 3.2 wt%, respectively, increasing the relative Fe content. The Al content (8.4 wt%) is consistent with theoretical predictions.

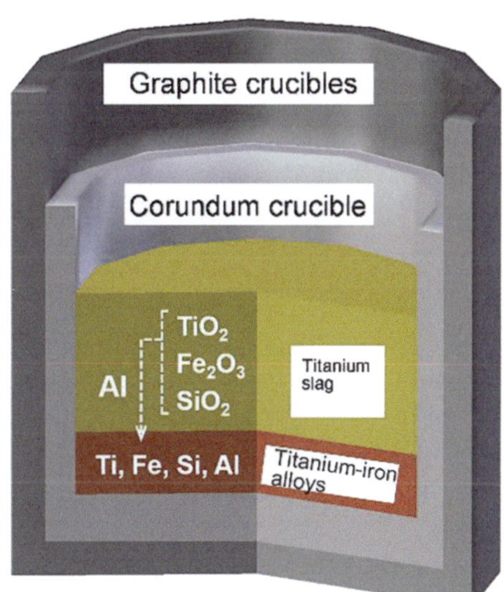

Fig. 10.23 Schematic illustration of the aluminothermic reduction process

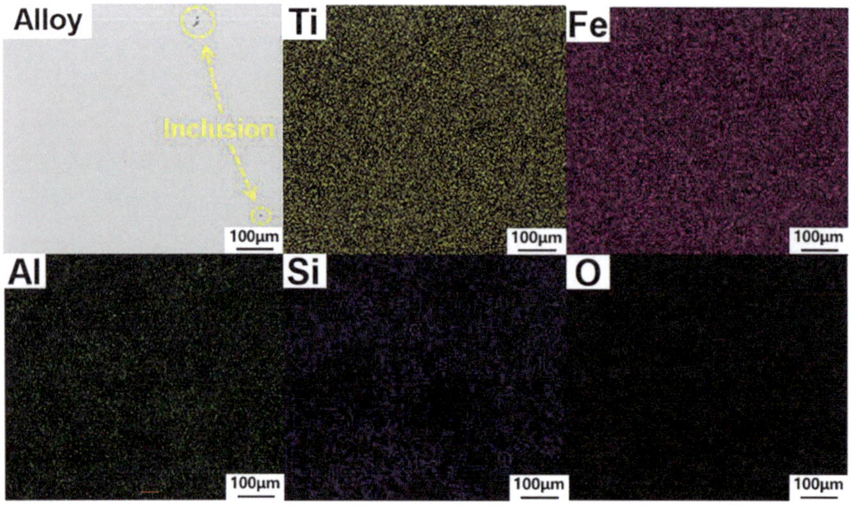

Fig. 10.24 SEM images and EDS scans of alloy phases

The slag was polished, sampled, and examined by SEM and EDS, as shown in Fig. 10.25a. Several alloy particles were observed dispersed within the slag. Image processing (Fig. 10.25b) revealed these particles to be spherical, ranging from 1–50 μm in size. Their compositions, shown in Fig. 10.25c and Table 10.16, reveal significantly higher Si content (20.85 wt%) and lower Ti and Al contents than the main alloy phase. The high Si content is attributed to the reaction between SiO_2 in the slag and the Ti and Al in the entrained alloy droplets, which increases the Si content and reduces the Al content in the dispersed alloy particles.

An enlarged view of the boxed region in Fig. 10.25a, shown in Fig. 10.26, reveals a granular alloy phase composed of two distinct regions: Phase B (Ti–Si-rich) and Phase C (Fe–Si-rich), as identified by EDS and Table 10.16. These phases differ significantly in composition from the main alloy, likely due to the intense reaction mixing liquid alloy into the slag. The high SiO_2 content in the slag also provides

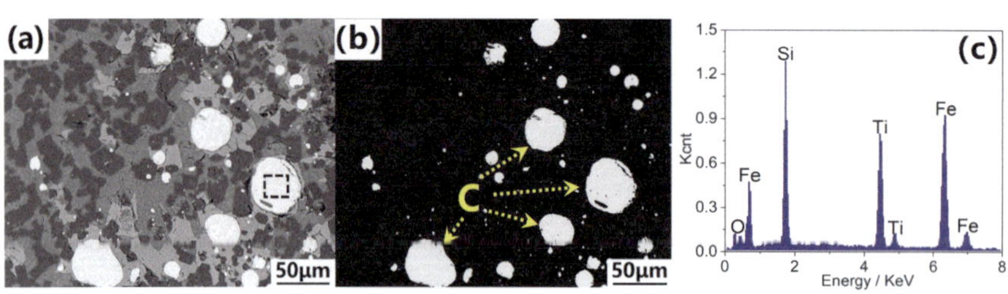

Fig. 10.25 **a** SEM image of slag phase containing alloy particles; **b** alloy phase morphology extracted from (a) using image processing; **c** average alloy particle composition determined via EDS

Table 10.16 Major elements content of alloy particles in slag based on EDS detection method/wt%

Element	Ti	Fe	Si	Al	O	Position
Average alloy composition	23.65	53.36	20.85	0.55	1.59	Figure 10.25c
Ti–Si rich alloy phase	32.23	43.70	22.19	0.39	1.49	Figure 10.26b
Fe–Si rich alloy phase	8.08	74.37	15.74	0.44	1.37	Figure 10.26c

extensive interface contact, facilitating the reduction of SiO_2 by Ti and Al. During cooling, as shown in Fig. 10.22d, Ti_5Si_3 forms at ~ 1800 °C, and Fe_2Si and Fe_2Ti separate below ~ 800 °C. Thus, the observed alloy phases in Fig. 10.26a likely result from differences in precipitation temperature and slag composition.

Figure 10.27a presents the SEM morphology of the slag after aluminothermic reduction, showing three main slag phases (a, b, and c), identified by EDS in Fig. 10.27a–c. XRD analysis (Fig. 10.27b) was used to determine the compositions and elemental contents of these phases (Table 10.17). Phase a, a Ti–rich oxide phase (TiO/Ti_2O_3, > 40 wt% Ti), is primarily distributed around alloy particles and likely forms via reduction of SiO_2 in the Ti slag, consistent with observations in Figs. 10.25 and 10.26. Phase b is mainly composed of MgO and Al_2O_3—MgO originating from the initial slag, and Al_2O_3 from both the reaction product and initial slag. These oxides readily form high-melting spinel ($MgAl_2O_4$), which precipitates and disperses in the slag. Phase c primarily contains Ca and Si, mainly in the form of Ca_2SiO_4. CaO also immobilizes Al_2O_3 and SiO_2 in the form of 2 CaO–Al_2O_3–SiO_2, preventing the formation of $TiAl_2O_4$ and ensuring continuous reduction and improved alloy quality.

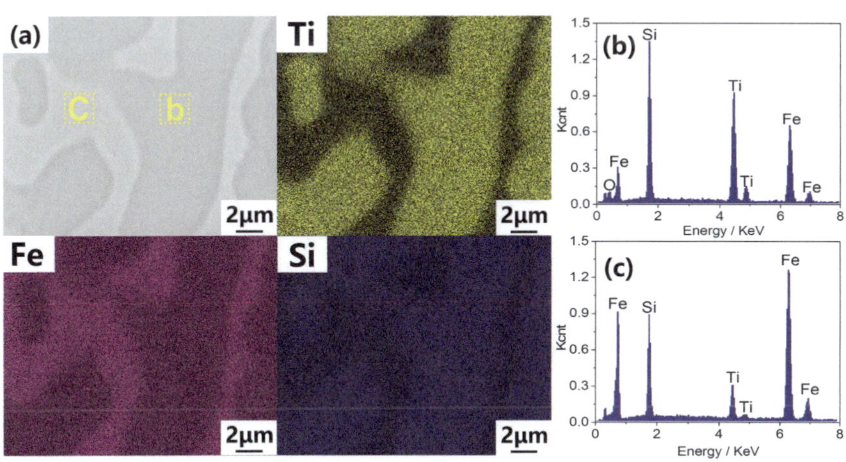

Fig. 10.26 **a** Magnified view of the dashed region in Fig. 10.25a; **b**, **c** EDS results for two distinct alloy phases; (Ti), (Fe), and (Si) indicate EDS surface scans for the corresponding elements

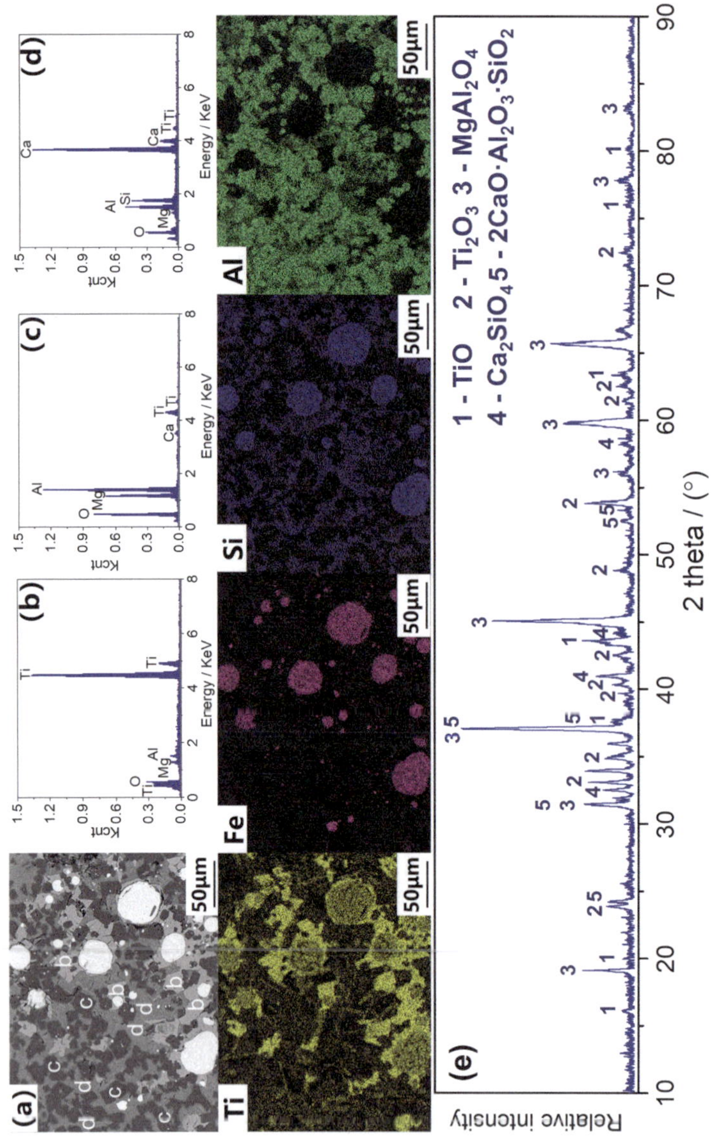

Fig. 10.27 **a** SEM microstructure of the slag following aluminothermic reduction, with a, b, and c representing distinct slag phases; **b–d** are the corresponding EDS results; (Ti), (Fe), (Si), and (Al) show elemental surface scans. **e** XRD pattern of the slag

Table 10.17 Elemental contents of phases a, b, and c with corresponding possible phases in Fig. 10.27

Element	Ti/at.%	Mg/at.%	Al/at.%	Ca/at.%	Si/at.%	Bal./at.%	Main components
Region a	40.70	3.25	2.26	< 0.5	< 0.5	53.79	TiO/Ti_2O_3
Region b	3.50	13.45	26.83	< 0.5	< 0.5	56.22	$MgAl_2O_4$
Region c	0.97	1.19	9.60	23.39	8.72	56.13	Ca_2SiO_4/ $2CaO{\cdot}Al_2O_3{\cdot}SiO_2$

In conclusion, titanium-containing slag obtained from pre-reduction and high-temperature melting of ironsands was used to produce Ti–Fe alloys via aluminothermic reduction. Additions of CaO and Fe_2O_3 enhanced slag fluidity and reaction heat. Increasing the Al-to-slag ratio raised the Ti, Al, and Si contents in the alloy. In situ infrared observation revealed that the exothermic phase lasted nearly 7 min, ultimately yielding an alloy with 30.5 wt% Ti, 8.4 wt% Al, and 3.2 wt% Si. However, some alloy droplets remained trapped in the slag due to intense reaction dynamics, contributing to a yield of 12.7%. These entrained alloy particles exhibited higher Si content (20.9 wt%), leading to the formation of Ti_5Si_3 and Fe_2Si and increasing viscosity. The slag comprises three distinct phases: a Ti–rich oxide phase surrounding the alloy particles; a spinel phase of $MgAl_2O_4$ formed from Al_2O_3 and MgO; and a Ca–Si-rich phase in which CaO immobilizes Al_2O_3 and SiO_2, preventing undesirable compound formation and supporting the overall alloy quality.

10.6 Summary

TiO_2 slag with a grade exceeding 40 wt% was obtained from vanadium-titanium magnetite through gas-based reduction followed by melt separation. When the pre-reduction rate of ironsands reached 90%, and 12 wt% CaF_2 was used as a flux, effective separation of slag and iron was achieved at 1550 °C, yielding titanium slag with a TiO_2 grade of approximately 44 wt%. The feasibility of producing titanium–iron alloys via aluminothermic reduction was verified through preliminary experiments using a ternary slag system (TiO_2–Fe_2O_3–CaO) and metallic aluminum. The optimal Al addition ratio was determined by thermodynamic calculations. In the aluminothermic reduction process, additional CaO and Fe_2O_3 were introduced to enhance slag fluidity, ensure sufficient exothermicity, and achieve satisfactory alloy quality. The reaction proceeded rapidly and intensely, reaching peak temperatures within approximately 10 s and completing within 7 min, resulting in favorable alloy phase aggregation.

Based on the reduction efficiency observed in preliminary experiments, titanium–iron alloy was synthesized using fused titanium slag derived from ironsands as the primary raw material. The resulting alloy contained 30.5 wt% Ti, 8.4 wt% Al, and 3.2 wt% Si. Analysis of the titanium–iron alloy confirmed compliance with the national standard (FeTi30-B). Furthermore, characterization revealed that numerous alloy particles with varying compositions were dispersed throughout the slag, which itself comprised three distinct phases.

References

1. Panigrahi M, Iizuka A, Shibata E, et al. Electrolytic reduction of mixed (Fe, Ti) oxide using molten calcium chloride electrolyte. J Alloys Compd. 2013;550:545–52.
2. Corathers LA, Gambogi J, Kuck PH, et al. Minerals yearbook 2008: ferroalloys. U.S. Department Interior; 2010.
3. Sokolov VM, Babyuk VD, Zhydkov YA, et al. Aluminothermic studies of a liquid partial reduced ilmenite. Miner Eng. 2008;21(2):143–9.
4. Mohanty J. Electrolytic reduction of titania slag in molten calcium chloride bath. JOM. 2012;64(5):582–4.
5. Ma M, Wang DH, Hu XH, et al. A direct electrochemical route from ilmenite to hydrogen-storage ferrotitanium alloys. Chem Eur J. 2006;12(19):5075–81.
6. Chen GZ, Fray DJ, Farthing TW. Direct electrochemical reduction of titanium dioxide to titanium in molten calcium chloride. Nature. 2000;407(6802):361–4.
7. Panigrahi M, Shibata E, Iizuka A, et al. Production of Fe–Ti alloy from mixed ilmenite and titanium dioxide by direct electrochemical reduction in molten calcium chloride. Electrochim Acta. 2013;93:143–51.
8. Zhou ZR, Hua YX, Xu CY, et al. Preparation of ferrotitanium from ilmenite by electrolysis-assisted calciothermic reduction in $CaCl_2$-NaCl molten salt. JOM. 2016;68(2):532–9.
9. Xiong L, Hua Y, Xu C, et al. Effect of CaO addition on preparation of ferrotitanium from ilmenite by electrochemical reduction in $CaCl_2$ NaCl molten salt. J Alloys Compd. 2016;676:383–9.
10. Pourabdoli M, Raygan S, Abdizadeh H, et al. A new process for the production of ferrotitanium from titania slag. Can Metall Q. 2013;46(1):17–23.
11. Babyuk V, Friedrich B, Sokolov V. Investigations of liquid phase aluminothermic reduction of ilmenite. World of Metallurgy. 2007;60:288–94.
12. Sibum H. Titanium and titanium alloys—from raw material to semi-finished products. Adv Eng Mater. 2003;5(6):393–8.
13. Wang ZY, Zhang JL, Ma JF, et al. Gaseous reduction of titania-ferrous solution ore by H-2-Ar mixture. ISIJ Int. 2017;57(3):443–52.
14. Wang ZY, Zhang JL, Jiao KX, et al. Effects of pre-reduction degree of ironsand on slag properties in melting separation process. Steel Res Int. 2018;89(3).
15. Wang ZY, Zhang JL, Jiao KX, et al. Effect of pre-oxidation on the kinetics of reduction of ironsand. J Alloys Compd. 2017;729:874–83.
16. Bale CW, Belisle E, Chartrand P, et al. FactSage thermochemical software and databases, 2010–2016. Calphad-Comp Coupling Phase Diag Thermochem. 2016;54:35–53.
17. Friedrich B, Hammerschmidt J, Stoephasius J-C. Aluminothermic reduction of titanium oxide. Erzmetall. 2003;56(2):82–93.
18. White WB, Johnson SM, Dantzig GB. Chemical equilibrium in complex mixtures. J Chem Phys. 1958;28(5):751–5.

Chapter 11
Industrial Test and Application Cases of Ironsands Using Short Process

This chapter presents two case studies of short-process industrial applications of ironsands: (1) the smelting of ironsands using a rotary kiln–electric furnace route in New Zealand, and (2) the smelting of vanadium-titanium magnetite using a similar process at Panzhihua Iron and Steel in China. These cases provide valuable references for the industrial utilization of ironsands in alternative ironmaking routes.

11.1 Industrial Applications of Ironsands Based on New Zealand Smelting Method

The rotary kiln–electric furnace process employs a rotary kiln for pre-reduction, followed by smelting in an electric furnace under a nitrogen atmosphere. This setup enables rapid melting and reduction of the ore, facilitating efficient smelting of ironsands and the production of high-quality molten iron. New Zealand has adopted this process in accordance with its unique resource and energy conditions, and the system has operated stably for many years.

In New Zealand's rotary kiln–electric furnace steelmaking route, ironsands sourced from the Waikato iron ore deposit serve as the primary raw material. Lignite is used as the reductant, along with selected slagging agents. After drying and preheating in a multi-chamber furnace, the feedstock is introduced into the rotary kiln, where it is reduced to sponge iron. The resulting product is then transferred to a melting furnace for slag-iron separation, and the molten iron is subsequently sent to the steelmaking workshop for vanadium extraction and steelmaking. A distinctive advantage of this process is that it does not require grinding and pelletizing; instead, coarse-grained iron ore concentrate can be directly processed, enabling a shorter and more efficient production route. Additionally, sponge iron from the rotary kiln is directly hot-charged into the electric furnace, avoiding the need for cooling and passivation.

Z. Wang et al., *Smelting and Utilization Technology of Ironsands*,
https://doi.org/10.1007/978-981-95-2868-4_11

The ironsands used in this process are beneficiated using a wet magnetic separation method, followed by secondary separation via spiral and conical separators. The chemical composition of the magnetically concentrated ironsands is presented in Table 11.1. For the reducing agent, low-grade bituminous coal (with volatile matter $\geq$ 30 wt%) or lignite is selected. Partial carbonization of the coal during the reduction process produces semi-coke (charcoal), which serves as an effective reductant. In the rotary kiln, coal is converted into coke particles. The coal composition is listed in Table 11.2, and the chemical composition of the directly reduced iron produced in the rotary kiln is given in Table 11.3.

A process flow diagram of the New Zealand method is shown in Fig. 11.1. Ironsands and lignite are first dried and preheated in a multi-chamber furnace before being introduced into a rotary kiln, which is 75 m in length, 4 m in external diameter, and inclined at 3.5%. The kiln is connected to a 36 $\times$ 35 m rotary cooler via a sealed, inclined transfer chute. Twelve air injection ports are distributed along the kiln length and are linked to a fan equipped with a Venturi inlet valve to regulate combustion air supply.

Feeding methods for the rotary kiln–electric furnace system are categorized into cold-charging and hot-charging processes. In the cold-charging process, sponge iron is cooled from 850 to 25 °C in a rotary cooler. The cooled product is screened into particle sizes of 150 and 25 mm, then crushed, re-mixed, and subjected to magnetic separation. The magnetic fraction (sponge iron) is stored for subsequent electric

Table 11.1 Chemical composition of iron-containing raw material ironsands/wt%

Ingredient	TFe	TiO_2	Al_2O_3	SiO_2	MgO	CaO	V_2O_5	MnO	P	S
Content	54.0	20.8	2.61	3.4	1.81	0.8	0.6	0.6	0.06	0.004

Table 11.2 Industrial and elemental analyses of coal dust used in the New Zealand method

Industrial analysis	Content/wt%	Elemental analysis (dry, ashless)	Content/wt%
Hydration	17.2	C	75.1
Ash	4.2	H	5.3
Volatility	36.1	N	1.1
Fixed carbon	42.6	S	0.3
S	0.23	O	18.2
		Heating value	6897 kcal/kg

Table 11.3 Chemical composition of directly reduced iron after rotary kiln pre-reduction/wt%

TFe	MFe	TiO_2	Al_2O_3	SiO_2	MgO	CaO	V_2O_5	MnO	P	S	C
62.57	57.9	24.44	3.02	3.94	1.9	0.93	0.69	0.69	0.06	0.003	0.19

Note Metallization rate = 88 wt%

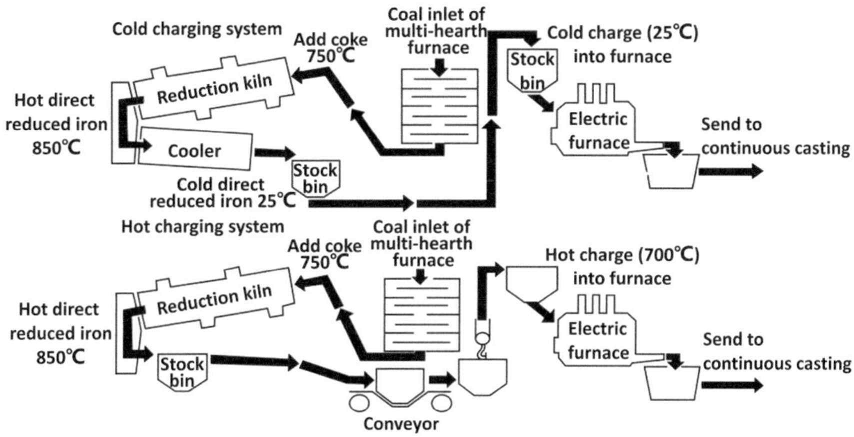

Fig. 11.1 Process flow diagrams for hot-loading and cold-loading operations

furnace smelting. The non-magnetic fraction can be returned to the kiln as recycled carbon or added to the electric furnace charge as a carbon source.

In contrast, the hot-charging process bypasses cooling. Sponge iron is conveyed directly to the electric furnace, where it is smelted into vanadium-bearing ferrous alloy and high-titanium slag. The smelting process requires approximately 200 min and consumes about 650 kWh of electricity. The vanadium-containing alloy is then transported to the steelmaking unit for vanadium extraction and further refining.

Based on the operational experience with the New Zealand method, several critical considerations have been identified:

(1) For titanium-bearing ores, the rotary kiln generally maintains a metallization rate of 85–90 wt%. Achieving further reduction would necessitate higher kiln temperatures, which could cause severe ring formation.
(2) Semi-coke produced via partial carbonization of low-grade bituminous coal or lignite is a highly effective reducing agent.
(3) Due to friction, thermal expansion, and the generation of iron whiskers, pelletized or lump ore tends to promote ring formation. In contrast, the particle size and physical properties of New Zealand ironsands help maintain shape integrity and prevent whisker formation, thereby reducing the risk of ring formation compared to pellets or lump ore.

11.2 Industrial Test Case of Pangang's Rotary Kiln-Electric Furnace Method

Pangang Steel Plant 410 also successfully conducted pilot tests of the rotary kiln–electric furnace process for smelting titanium-bearing iron ore concentrates; however, industrial-scale production was not implemented.

exhibited a linear upward trend. Sinter with unaltered particle size ironsands displayed a typical microstructure of calcium ferrite complexes well bonded with hematite and silicates. However, the addition of 8 wt% fine or intermediate particles led to the emergence of skeletal, rhombic hematite structures, negatively affecting metallurgical performance. While the sinter maintained desirable particle size distribution and drum strength at 8 wt% unmodified particle size, smaller particle sizes led to deterioration in phase composition, particle distribution, and mechanical strength.

(4) The high-temperature performance of sinters under varying ironsands dosages and particle sizes was assessed. Increased TiO_2 content, due to higher ironsands ratios, combined with CaO to form brittle perovskite in highly basic sinters, causing significant low-temperature reduction degradation at 8 wt% unmodified ironsands. Reduced particle size further increased the formation of rhombic skeletal hematite, worsening reduction degradation. Titanium's presence inhibits iron oxide reduction, slightly reducing reducibility at 8 wt% across all particle sizes—though still above quality standards. Smaller particle sizes negatively impacted reducibility due to reduced sinter porosity and decreased gas permeability. An increase in titanium carbonitride precipitates also increased gas penetration resistance in the soft-melt zone, reducing gas permeability. Sinter pyrometallurgical properties met standards at ≤ 5 wt% original particle size; however, increased dosage or finer particles led to varying degrees of property decline.

(5) The grinding characteristics and pelletizing behavior of ironsands were studied. After 35 min of ball milling, 85.63 wt% of particles were < 0.074 mm, with a specific surface area of 2741 cm^2/g—meeting pelletization standards. With 10–100 wt% ironsands, preheating at 950 °C, roasting at 1250 °C, and a 15-min duration for both steps, the raw and fired pellets met industrial strength requirements. As ironsands content increased, Ti and Fe solid solution phases in the pellets also increased, resulting in slow reduction in low-temperature regions and low crystal transformation degrees—favorable for reduction pulverization resistance. However, reduced reducibility was observed, mainly due to titanium inhibiting iron–oxygen migration. Poor reducibility also suppressed iron whisker formation, increasing reductive expansion. When pellets were entirely composed of ironsands, the expansion index was only 10.11%.

(6) During non-isothermal pre-oxidation, the phase transformation sequence is TTM → TTH → PSB. Titanium becomes increasingly concentrated in lamellar regions, while Fe, Al, and O accumulate in homogeneous zones. This not only partially separates titanium from iron and oxygen but also reduces the hindrance of titanium on subsequent reduction. Five sub-reactions identified via DDTG analysis aligned with XRD and thermodynamic results. XRD cell refinement confirmed that phase lattice contraction dominated the structural changes during pre-oxidation.

(7) Hydrogen pre-reduction experiments determined that 40 vol% hydrogen at 900 °C is optimal. At 900 °C, TTH (Fe_2O_3) was reduced to Fe_3O_4 within 2.5 min; Fe_3O_4—including that from both TTM and TTH—was reduced within 15

min. Pre-existing $FeTiO_3$ was fully reduced to Fe and TiO_2 within 5 min, while newly formed $FeTiO_3$ appeared between 5 and 10 min due to intermediate Fe_2TiO_4 reduction. $FeTi_2O_5$ formation at 25 min indicated further $FeTiO_3$ reduction. By 50 min, $FeTiO_3$, $FeTi_2O_5$, and Fe_2TiO_5 had disappeared.

(8) Pre-oxidized ironsands showed significantly higher reduction rates than raw ores under identical conditions. Model-derived kinetic parameters for raw ironsands were $E = 80.94$ kJ·mol^{-1} and $A = 346.29$ min^{-1}; for pre-oxidized ironsands, $E = 67.20$ kJ·mol^{-1} and $A = 127.61$ min^{-1}, indicating a lower energy barrier post–pre-oxidation. Pre-oxidized samples followed a first-order chemical reaction mechanism, while raw ores underwent successive rate-controlling steps: nucleation and growth, first-order chemical reaction, and three-dimensional gas diffusion.

(9) Pre-oxidized ironsands exhibited larger specific surface areas and enhanced pore volumes (mesopore range, 4–50 nm) compared to raw ore. The pre-oxidation process created enlarged pore structures and microcracks, providing more active sites for nucleation and improved gas diffusion channels. As a result, reduction of pre-oxidized ironsands proceeded without rate-limiting gas diffusion or nucleation stages, thereby enhancing reaction kinetics. The cell contraction of primary phases is hypothesized to drive pore development and microstructural evolution.

(10) In both raw and pre-oxidized ironsands, dark structures first emerged at the early reduction stage, developing rapidly from the grain surfaces inward. In raw samples, reduction proceeded radially from central, incompletely reduced nuclei. In contrast, microcracks in pre-oxidized particles led to random, finely dispersed reactive nuclei. Final structures were influenced by the initial phase composition, resulting in alternating Ti rich and Fe-rich zones, TiO_2 networks, and other microstructural features.

(11) Considering melting behavior and slag viscosity, a 90% pre-reduction degree at 1550 °C was optimal for slag-iron separation, yielding titanium slag with ~ 44 wt% TiO_2. Aluminothermic reduction of TiO_2–Fe_2O_3–CaO ternary slag was feasible, with titanium and iron reduction rates of ~ 45% and ~ 100%, respectively. The alloy phase was well-separated and aggregated, with a titanium grade of ~ 55 wt%. Using this titanium slag, aluminothermic reduction produced titanium–iron alloys (FeTi30-B grade) with Ti = 34.6 wt%, Al = 5.9 wt%, and Si = 2.3 wt%. The reaction was highly exothermic, reaching peak temperature within 10 s and completing in 7 min.